T0215327

UNITEXT - La Matematica per il 3+2

Volume 119

The **UNITEXT - La Matematica per il 3+2** series is designed for undergraduate and graduate academic courses, and also includes advanced textbooks at a research level. Originally released in Italian, the series now publishes textbooks in English addressed to students in mathematics worldwide. Some of the most successful books in the series have evolved through several editions, adapting to the evolution of teaching curricula. Submissions must include at least 3 sample chapters, a table of contents, and a preface outlining the aims and scope of the book, how the book fits in with the current literature, and which courses the book is suitable for.

More information about this subseries at http://www.springer.com/series/5418

Rocco Chirivì • Ilaria Del Corso •
Roberto Dvornicich

Selected Exercises
in Algebra

Volume 1

 Springer

Rocco Chirivì
Dipartimento di Matematica e Fisica
University of Salento
Lecce, Italy

Ilaria Del Corso
Dipartimento di Matematica
University of Pisa
Pisa, Italy

Roberto Dvornicich
Dipartimento di Matematica
University of Pisa
Pisa, Italy

Translated by
Alessandra Caraceni
Oxford, UK

ISSN 2038-5714   ISSN 2532-3318   (electronic)
UNITEXT
ISSN 2038-5722   ISSN 2038-5757   (electronic)
La Matematica per il 3+2
ISBN 978-3-030-36155-6   ISBN 978-3-030-36156-3   (eBook)
https://doi.org/10.1007/978-3-030-36156-3

This Springer imprint is published by the registered company Springer Nature Switzerland AG.
The registered company address is: Gewerbestrasse 11, 6330 Cham, Switzerland

"The nice thing about mathematics is doing mathematics."

—Pierre Deligne

To Andrea, who knows what mathematics is

Rocco

To Francesca, with a wish that she will be able to find and nurture her passions

Ilaria

To young people who already love or might come to love mathematics

Roberto

Preface

This is a book of exercises in algebra, accompanied by notes covering the relevant theory to be used as a reference for the solutions proposed. It is the result of several decades of experience of teaching the subject at the University of Pisa. It collects exam questions set over the years, along with their solutions.

The reason for us to write this book is the idea that Mathematics can only be learned by *rediscovering* it, which is most effectively done by working out one's own solutions. Moreover, we believe these exercises possess some features that differentiate them from those of most other books. The first is the lack of "duplicate" questions: there are no exercises that only differ by the value of certain parameters, and whose solutions are hence essentially the same. The second—and perhaps crucial—characteristic is that here no exercise is solved mechanically, by rote. On the contrary, solving each exercise requires a clever idea.

Anyone who has grappled with mathematics knows that this approach has a dramatic impact on both the level of challenge posed by a problem, and also on its intrinsic value. The insight gained by solving an exercise which requires an original idea is infinitely superior to what comes from merely applying methods we have learnt. We also like to think of the ideas necessary for solving our exercises as rather well assorted, a fact we believe is one of the main strengths of the book. In view of all this, one might wish to describe the present text as a collection of problems in algebra rather than exercises.

Because of the peculiar history of the University of Pisa, whose courses are also attended by students of the Scuola Normale Superiore, these exercises might at times be more difficult than the average exam questions at other Italian universities. This, however, is not at odds with the aim of the work, which is to develop problem-solving skills and prepare students for the study of algebra in the best way possible. We therefore hope that the reader will take this very important advice to heart: do *not* give in to temptation and look up the solution to a problem if you have not spent a long time trying to solve it yourself. Understanding a ready-made solution, even an elegant one, does not foster real mathematical growth. As we have already mentioned, in order to learn mathematics one must *do* mathematics. Actively seeking a solution is what expands our understanding and what lets us discover the links between concepts, the web of connections that are fundamental for true learning. We also hope to intrigue students and prompt them to dig deeper into the topics that are only briefly mentioned in each question.

The book's structure mirrors the way that algebra is taught during the first years of a maths degree at the University of Pisa. This structure is the result of historical developments starting from the time when all basic notions in algebra were bundled into a single first-year course. Following the split of Bachelor's and Master's degrees, those contents were divided into two: one part resulted in the first-year "Arithmetic" course, and the other, initially taught under the name "Algebraic Structures", now constitutes the second-year "Algebra 1" course. The two blocks correspond exactly to the two volumes of this book.

The volume dedicated to "Arithmetic" is mostly concerned with elementary tools such as induction, elements of combinatorics, integer numbers and congruences, followed by an introduction to the basic properties of algebraic structures: Abelian groups, rings, polynomials and their roots, field extensions and finite fields. In the second volume, which corresponds to the "Algebraic Structures" and "Algebra 1" courses, we delve deeper into the theory of groups, commutative rings—with a particular emphasis on unique factorisation—and field extensions, as well as introducing the fundamentals of Galois theory.

Each part is accompanied by notes on the theory. Although exhaustive, they do not attempt to make up for an algebra textbook, but rather provide the necessary framework to the exercises. They should, in particular, recall relevant results without providing any proofs. (For further information the reader may instead turn, for example, to Herstein's "Topics in Algebra", Wiley & Sons, or Artin's "Algebra", Pearson.)

This book also includes a series of preliminary exercises. These should be tackled first since their results will often be needed in subsequent problems. We also stress that all solutions provided do not require any theoretical tools other than those recalled in the introductory chapters and the preliminary exercises. Invoking more advanced theorems would simplify certain solutions and occasionally render an exercise trivial, but doing so would be contrary to the very spirit of this work.

Acknowledgements We wish to thank Ciro Ciliberto for the support, Francesca Bonadei at Springer Italia for the precious assistance and all the students who attended our lectures and busied themselves with the exercises during exams.

Updates We warmly invite readers to provide us with feedback and point out mistakes (which, in a book containing the detailed solutions to 250+ exercises, are almost inevitably to be found) by contacting us at the addresses: rocco.chirivi@unisalento.it or ilaria.delcorso@unipi.it or roberto.dvornicich@unipi.it.

Updates and corrections will appear on the web page http://www.dmf.unisalento.it/~chirivi/libroEserciziAlgebra.html.

Lecce, Italy   Rocco Chirivì
Pisa, Italy   Ilaria Del Corso
Pisa, Italy   Roberto Dvornicich
June 2017

Preface to the English Edition

This is the English translation of our book *Esercizi Scelti di Algebra*, Volume 1. We have taken the opportunity of the translation to correct some typos and inaccuracies present in the first edition in Italian.

We would like to express our deep gratitude to our translator Alessandra Caraceni; her impressive work, based on profound mathematical competence, has gone far beyond an impeccable translation.

Lecce, Italy   Rocco Chirivì
Pisa, Italy   Ilaria Del Corso
Pisa, Italy   Roberto Dvornicich
October 2019

Contents

Chapter 1
Theory

1 Fundamentals

1.1 Sets

The concept of a *set* is a primitive notion; we shall neither attempt to define it nor give an axiomatic presentation of set theory. We shall instead adopt the naïve view of a set as a collection of objects, called its elements. The only property one can ascribe to a set X is that it is possible to determine whether or not an object x is an element of X: if it is, we write $x \in X$ and say that x *belongs to X*. If x is not an element of X, we write $x \notin X$ and say that x *does not belong to X*. Also, two sets X and Y are said to be *equal* if they contain the same elements. There is but one set containing no elements, that is, the *empty set*, denoted by $\varnothing$.

A set X is a *subset* of a set Y if every element of X is an element of Y, in which case we write $X \subseteq Y$; conversely, by $X \nsubseteq Y$ we mean that X is *not* a subset of Y. The empty set is a subset of every set X (that is, $\varnothing \subseteq X$), and naturally we also have $X \subseteq X$. The family of all subsets of X is denoted by $\mathcal{P}(X)$ and is called the *power set* of X. The sets $\varnothing$ and X are therefore elements of $\mathcal{P}(X)$.

A subset X of a set Y is often defined by means of some property p; we shall write

$$X \doteq \{y \in Y \mid p(y)\}$$

to mean that X is the set of elements of Y that satisfy property p.

The *union* $X \cup Y$ of two sets X and Y is the set of all elements that belong to X or to Y; that is,

$$x \in X \cup Y \quad \text{if and only if} \quad x \in X \text{ or } x \in Y.$$

© Springer Nature Switzerland AG 2020
R. Chirivì et al., *Selected Exercises in Algebra*, UNITEXT 119,
https://doi.org/10.1007/978-3-030-36156-3_1

Note that the word "or" is used here as an inclusive disjunction: an element of $X \cup Y$ may belong to both X and Y. In this sense the connective "or" is akin to the Latin conjunction *vel*.

The *intersection* $X \cap Y$ is the set of elements belonging to both X and Y:

$$x \in X \cap Y \quad \text{if and only if} \quad x \in X \text{ and } x \in Y.$$

One can take unions and intersections of any number of sets: if $\mathcal{F}$ is a family of sets, we have

$$x \in \bigcup_{X \in \mathcal{F}} X \quad \text{if and only if} \quad \text{there exists a set } X \text{ in } \mathcal{F} \text{ such that } x \in X$$

and, similarly,

$$x \in \bigcap_{X \in \mathcal{F}} X \quad \text{if and only if} \quad \text{for all } X \text{ in } \mathcal{F} \text{ we have } x \in X.$$

Proposition 1.1 *Union and intersection are distributive over each other: given three sets X, Y and Z we have*

$$X \cap (Y \cup Z) = (X \cap Y) \cup (X \cap Z) \quad and \quad X \cup (Y \cap Z) = (X \cup Y) \cap (X \cup Z).$$

Two sets X, Y with no element in common, i.e. such that $X \cap Y = \varnothing$, are called *disjoint*. Each subset X of Y is disjoint from its *complement* $Y \setminus X$, which is the set of elements of Y that do not belong to X. If two sets X and Y are disjoint we shall sometimes write $X \sqcup Y$ for their union, and hence call it a *disjoint* union.

Operations on sets are related in a straightforward manner to logical operations on propositions, as clarified by the following

Proposition 1.2 *Given a set Z and two subsets $X = \{x \in Z \mid p(x)\}$ and $Y = \{y \in Z \mid q(y)\}$ of Z we have*

(i) $X \cup Y = \{z \in Z \mid p(z) \text{ or } q(z)\}$,
(ii) $X \cap Y = \{z \in Z \mid p(z) \text{ and } q(z)\}$,
(iii) $Z \setminus X = \{z \in Z \mid \text{not } p(z)\}$,
(iv) $X \subseteq Y$ if and only if $p(z)$ implies $q(z)$ for all z in Z.

Proposition 1.3 *Given a set Z and two subsets X and Y of Z, De Morgan's laws hold:*

$$Z \setminus (X \cup Y) = (Z \setminus X) \cap (Z \setminus Y) \quad and \quad Z \setminus (X \cap Y) = (Z \setminus X) \cup (Z \setminus Y);$$

that is, taking the complement changes unions into intersections and vice-versa.

The set of all ordered pairs (x, y), where x is an element of X and y an element of Y, is denoted by $X \times Y$ and called the *Cartesian product* of X and Y. The same construction can be performed on any number of sets: $X_1 \times X_2 \times \cdots \times X_n$ is the set of ordered n-tuples $(x_1, x_2, \ldots, x_n)$ with $x_1 \in X_1$, $x_2 \in X_2$, $\ldots$, $x_n \in X_n$. We shall, for the sake of brevity, denote by X^n the Cartesian product of n copies of X.

1.2 Maps

A *map* from a set X to a set Y is some recipe that assigns to each element of X a unique element of Y. More formally, a map f from X to Y is a subset of the Cartesian product $X \times Y$ with the property that for each $x \in X$ there exists a unique $y \in Y$ such that $(x, y) \in f$; the set X is called the *domain* of f and Y is the *codomain* of f. We shall write $f : X \longrightarrow Y$ or $X \xrightarrow{f} Y$ to express that $f \subseteq X \times Y$ is a map from X to Y, and always use the functional notations $f(x) = y$ or $f : x \longmapsto y$ or $x \xmapsto{f} y$ rather than $(x, y) \in f$.

When $f(x) = y$ we shall say interchangeably that y is the *image* of x, or that f *sends* or *maps* x to y, or even that y *is obtained from* x via f. The subset $\mathrm{Im}(f) = \{f(x) \mid x \in X\} \subseteq Y$ of elements of Y reached by f starting from some x in X is called the *image* of f. Notice that, as implied by the definition, two maps f and g are equal if and only if they have same domain and codomain and are such that $f(x) = g(x)$ for all x in the domain.

Given a subset A of X we call $f(A)$ the *image* of A under f, that is, the set of all elements of the form $f(a)$ for some a in A. Conversely, given a subset B of Y, we denote by $f^{-1}(B)$ the set of x in X such that $f(x) \in B$, and call it the *pre-image* of B under f. Taking images and pre-images of sets is in a sense compatible with taking unions and intersections, as detailed by the following

Proposition 1.4 *Taking pre-images commutes with unions and intersections*

$$f^{-1}(A \cup B) = f^{-1}(A) \cup f^{-1}(B), \quad f^{-1}(A \cap B) = f^{-1}(A) \cap f^{-1}(B)$$

while taking images only commutes with unions

$$f(A \cup B) = f(A) \cup f(B).$$

Moreover,

$$f(A \cap B) \subseteq f(A) \cap f(B);$$

note that in this case we only have an inclusion rather than an equality: the set $f(A \cap B)$ can be a proper subset of $f(A) \cap f(B)$.

Given two maps $X \xrightarrow{f} Y$ and $Y \xrightarrow{g} Z$, the *composite* $g \circ f$ of f and g is the map from X to Z defined by the rule

$$X \ni x \xmapsto{g \circ f} g(f(x)) \in Z.$$

Proposition 1.5 *The composition of maps is* associative: *given maps f, g and h such that the composites $g \circ f$ and $h \circ g$ are well defined, we have $h \circ (g \circ f) = (h \circ g) \circ f$.*

A map which sends any two distinct elements of X to distinct elements of Y is called *injective* or *one-to-one* . In order to stress the injectivity of a map from X to Y, we may write $X \hookrightarrow Y$. A map such that every element y of Y is the image of some x in X is said to be *surjective* or *onto*; equivalently, $f : X \to Y$ is surjective when $\text{Im}(f) = Y$. A surjective map is denoted by $X \twoheadrightarrow Y$. When a map f is both injective and surjective we shall say that f is *bijective*, and occasionally express this by writing $f : X \underset{\sim}{\to} Y$.

Proposition 1.6 *The composite of injective maps is injective and the composite of surjective maps is surjective. In particular, the composite of bijective maps is bijective.*

When X is a subset of a set Y we may define the *inclusion* map $X \ni x \xmapsto{i_X} x \in Y$, which is clearly injective. In particular, the inclusion of X into X itself is called the *identity map* and is denoted by Id_X, or simply Id when not ambiguous; the identity is a bijective map.

Given a subset X of a set Y and a map f from Y to a set Z, the *restriction* of f to X is the map $X \ni x \xmapsto{f_{|X}} f(x) \in Z$; clearly, we have $f_{|X} = f \circ i_X$

Given a map f from a set X to a set Y, an *inverse* of f is a map $Y \xrightarrow{g} X$ such that $g \circ f = \text{Id}_X$ and $f \circ g = \text{Id}_Y$. A map for which an inverse exists is called *invertible*. Not all maps admit an inverse, and in fact

Proposition 1.7 *A map is invertible if and only if it is bijective. Moreover, if a map is invertible then its inverse is unique.*

For an invertible map f we shall denote by f^{-1} the unique inverse of f.

Given a map $X \xrightarrow{f} X$ we denote by X^f the set of *fixed points* of f, that is, $X^f = \{x \in X \mid f(x) = x\}$; we shall also use the notation $\text{Fix}(f)$.

A bijective map from a set X to itself is called a *permutation*. The set of permutations $\mathsf{S}(X)$ of X will assume an important role in our study of algebra. In general, the set of all maps $X \longrightarrow Y$ is denoted by Y^X.

A diagramme of sets and maps is said to be *commutative* if composing maps along any two paths of arrows starting and ending at common places yields the same map. For example, the following diagramme

$$
\begin{array}{ccc}
X & \xrightarrow{\ f\ } & Y \\
{\scriptstyle g}\downarrow & & \downarrow{\scriptstyle h} \\
A & \xrightarrow{\ i\ } & B
\end{array}
$$

is commutative if and only if $(h \circ f)x = (i \circ g)x$ for all x in X.

1.3 Relations

Let X be a set and R a subset of the Cartesian product $X \times X$. The set R corresponds to the *relation* $\sim_R$, or simply $\sim$ when no ambiguity arises, defined on X by setting $x \sim_R y$ if and only if $(x, y) \in R$. Of particular importance are equivalence relations. A relation $\sim$ is an *equivalence relation* if it satisfies the following properties:

(i) reflexivity: $x \sim x$ for all $x \in X$,
(ii) symmetry: if $x \sim y$ then $y \sim x$,
(iii) transitivity: if $x \sim y$ and $y \sim z$ then $x \sim z$.

Notice that equality itself is an equivalence relation; as a matter of fact, it might help to think of general equivalence relations as "weaker" versions of equality. Given an element $x \in X$ and an equivalence relation $\sim$ on X, the *equivalence class* of x is the set $C\ell(x)$ of $y \in X$ such that $x \sim y$. Any two distinct equivalence classes are disjoint and the union of all equivalence classes of elements of X is the set X itself.

Let us now introduce further terminology closely linked to relations. A *partition* of a set X is a family $\mathcal{P}$ of nonempty subsets of X satisfying the following properties:

(i) any two distinct sets in $\mathcal{P}$ are disjoint,
(ii) the union of all sets in $\mathcal{P}$ is X.

There is a perfect correspondence between equivalence relations and partitions in the sense of the following

Theorem 1.8 *Given an equivalence relation $\sim$ on a set X, the family of equivalence classes under $\sim$ is a partition of X. Conversely, given a partition $\mathcal{P}$ of a set X, the relation $\sim$ defined by*

$$x \sim y \quad \textit{if and only if} \quad \textit{there exists } C \in \mathcal{P} \textit{ such that } x, y \in C$$

is an equivalence relation on X whose equivalence classes are the sets of the partition $\mathcal{P}$.

The family of the equivalence classes under a relation $\sim$ on X is called the *quotient set* of X by $\sim$ and is denoted by $X/\!\sim$. Moreover, the map

$$X \ni x \xmapsto{\pi} \mathcal{C}\ell(x) \in X/\!\sim$$

which sends any element x to its equivalence class is called the *quotient map*.

A map $X \xrightarrow{f} Y$ is said to be *compatible* with the equivalence relation $\sim$ on X if $f(x) = f(y)$ for all $x, y \in X$ such that $x \sim y$. If f is compatible with $\sim$, then there exists a unique map $\overline{f}$ such that $f = \overline{f} \circ \pi$; in other words, $\overline{f}$ makes the following diagramme commutative:

The same property may be expressed by saying that f *passes to the quotient*. We shall sometimes define a map $\overline{f}$ directly as $\mathcal{C}\ell(x) \longmapsto f(x)$; when doing so, one needs to check that $\overline{f}$ is *well defined*, that is, that f is compatible with $\sim$.

When two relations $\sim$ and $\sim'$ on a set X are such that $x \sim y$ implies $x \sim' y$, the partition $\mathcal{P}$ induced by $\sim$ is *finer* than the partition $\mathcal{P}'$ induced by $\sim'$: in other words, for each class $C \in \mathcal{P}$ there exists a class $C' \in \mathcal{P}'$ such that $C \subseteq C'$. The association $C \longmapsto C'$ thus defined is a surjective map ϵ that renders the following diagramme commutative:

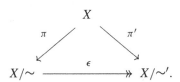

A *set of representatives* $\mathcal{R}$ for an equivalence relation $\sim$ on X is a subset of X such that restricting the projection map to $\mathcal{R}$ yields a bijection between $\mathcal{R}$ and $X/\!\sim$. That is, $\mathcal{R}$ has been assembled by selecting, for each equivalence class, a unique representative in X.

Other important relations we shall frequently use are *order relations*, or simply *orders*. An *order relation* on a set X is a relation $\leq$ which is

(i) reflexive: $x \leq x$ for all $x \in X$,
(ii) anti-symmetric : if $x \leq y$ and $y \leq x$ then $x = y$,
(iii) transitive if $x \leq y$ and $y \leq z$ then $x \leq z$.

Notice we do not require all elements of X to be comparable to each other: it may indeed be that neither $x \leq y$ nor $y \leq x$ for some x, y in X. If for all x, $y \in X$ we do have $x \leq y$ or $y \leq x$ then the relation is a *total order*. Sometimes, when we need to emphasise that a certain order relation might not be a total order, we shall call it a *partial order* .

A *strict order* is a relation $<$ on X which is

(i) irreflexive: there is no $x \in X$ such that $x < x$, and
(ii) transitive: if $x < y$ and $y < z$ then $x < z$.

Each order relation corresponds to a strict order and vice-versa. Indeed, given an order $\leq$, by setting $x < y$ whenever $x \leq y$ and $x \neq y$, we create a strict order $<$; conversely, given a strict order $<$ we may set $x \leq y$ whenever $x < y$ or $x = y$ to recover an order relation. This correspondence between $\leq$ and $<$ will be tacitly exploited throughout the book.

1.4   The Principle of Induction

We shall denote by $\mathbb{N}$ the set of natural numbers $\{0, 1, 2, \ldots\}$; we will not give an axiomatic presentation of natural numbers but rather take their basic properties as a given. Let us only mention that one possible formalisation is the one due to Giuseppe Peano, and briefly remind the reader of its fundamental fifth axiom, the so-called induction principle:

Axiom 1.9 (Induction Principle) *Let $p(n)$ be a property depending on a natural number n; suppose that $p(0)$ holds and that for all m in $\mathbb{N}$ we have that $p(m)$ implies $p(m + 1)$. Then $p(n)$ holds for all n.*

When using this principle to prove a statement, verifying that $p(0)$ holds is often called the *base* (of the induction), while proving that $p(m)$ implies $p(m + 1)$ constitutes the *inductive step*. The principle of induction may be stated under various equivalent forms. For example,

Proposition 1.10 (Induction Principle—Second Form) *Let $p(n)$ be a property depending on a natural number n; suppose that $p(0)$ holds and that for all natural numbers m we have that $p(m + 1)$ is implied by the conjunction of $p(0)$, $p(1)$, ..., $p(m - 1)$, $p(m)$. Then $p(n)$ holds for all n.*

In this second version we may assume that $p(0)$, $p(1)$, ..., $p(m)$ all hold in order to prove $p(m + 1)$ in the inductive step.

Another equivalent formulation which can prove useful is the so-called *well-ordering principle*

Proposition 1.11 (Well-Ordering Principle) *Every nonempty subset A of $\mathbb{N}$ contains a least element, i.e. an element $a \in A$ such that $a \leq b$ for all $b \in A$.*

We shall call a *sequence* in a set X a map $\mathbb{N} \longrightarrow X$; we will use the notation $(a_n)_n$ rather than $n \longmapsto a_n$ when dealing with sequences.

We now wish to define a sequence $(a_n)_n$ in X by choosing its initial value $a_0 = x \in X$ and, for each n, by somehow constructing the term a_{n+1} from the preceding terms $a_0, a_1, \ldots, a_{n-1}, a_n$. This may be done by considering maps $f_n : X^n \longrightarrow X$ with $n \geq 1$ and setting $a_{n+1} = f_{n+1}(a_0, a_1, \ldots, a_{n-1}, a_n)$ for each n. When a sequence $(a_n)_n$ is constructed in this way, we shall say it is *recursive* or defined *by recursion*. The fact that the process of recursion is well-founded constitutes yet another equivalent form of the principle of induction:

Proposition 1.12 (Principle of Recursion) *Let X be a set and x an element of X; suppose we are given, for each $n \in \mathbb{N}$, a map $f_n : X^n \longrightarrow X$. Then there is a unique sequence $(a_n)_n$ for which $a_0 = x$ and $a_{n+1} = f_{n+1}(a_0, a_1, \ldots, a_{n-1}, a_n)$ for all $n \in \mathbb{N}$.*

An example of a recursive definition is given by the sequence $(F_n)_n$ of *Fibonacci numbers*: these are defined so that $F_0 = 0$, $F_1 = 1$ and, for all $n \geq 1$, $F_{n+1} = F_n + F_{n-1}$. The first few terms of the sequence are

$$0, 1, 1, 2, 3, 5, 8, 13, 21, \ldots$$

This definition befits the framework of the proposition above by taking $X = \mathbb{N}$, $x = 0$, $f_1(a_0) = 1$ and $f_{n+1}(a_0, a_1, a_2, \ldots, a_n) = a_{n-1} + a_n$ for all $n \geq 1$.

1.5   Operations

Our study of algebra is mainly concerned with sets on which operations having certain properties can be defined in a natural way. An *operation* on a set X is a map from the Cartesian product $X \times X$ to X. Instead of denoting operations as we generally denote maps, we shall typically indicate the result of some operation $\circ$, applied to an ordered pair (x, y) of elements of X, by $x \circ y$; the operation $\circ$ is thus the map

$$X \times X \ni (x, y) \xmapsto{\circ} x \circ y \in X.$$

We shall also say that $x \circ y$ is the *composition* of x and y via the operation $\circ$.

An operation $\circ$ is said to be *associative* if $(x \circ y) \circ z = x \circ (y \circ z)$ for all $x, y, z \in X$. Given n elements $x_1, x_2, \ldots, x_n$ of X and an associative operation $\circ$, one can give an unambiguous meaning to the expression $x_1 \circ x_2 \circ \cdots \circ x_n$; indeed, pairing up elements arbitrarily will always yield the same final result.

An operation $\circ$ is said to be *commutative* if $x \circ y = y \circ x$ for all $x, y \in X$. If an operation is both commutative and associative then the composition $x_1 \circ x_2 \circ \cdots \circ x_n$ does not depend on the order of its terms.

A *neutral element* e for an operation $\circ$ on a set X is an element of X such that $e \circ x = x \circ e = x$ for all $x \in X$. One can easily prove that, if an operation admits a neutral element, then such an element is unique.

When an operation $\circ$ admits a neutral element e, we say that y is a *left inverse* of x if $y \circ x = e$. Similarly, a *right inverse* of x is an element y such that $x \circ y = e$ and, finally, an *inverse* of x is an element which is both a left inverse and a right inverse of x.

Given an operation $\circ$ on a set X and a subset Y of X, we say that Y is *closed* under $\circ$ if for all pairs of elements y_1, y_2 in Y we have $y_1 \circ y_2 \in Y$. If Y is closed under $\circ$, then we may *restrict* $\circ$ to an operation on Y by setting $Y \times Y \ni (y_1, y_2) \longmapsto y_1 \circ y_2 \in Y$. Usually, the operation restricted to Y is denoted by the same symbol as the operation on X.

Given two operations $\circ$ and $+$ on a set X, we say $\circ$ *distributes* over $+$ if for all $x, y, z \in X$ we have $(x+y) \circ z = (x \circ z) + (y \circ z)$ and $x \circ (y+z) = (x \circ y) + (x \circ z)$.

1.6 Numbers

Throughout the book we shall use several number sets, mostly as examples of algebraic structures. These can all be constructed from the set of natural numbers introduced in Sect. 1.4. For instance, the set $\mathbb{Z}$ of integers can be seen as "natural numbers endowed with a sign" and can be formally defined as $\mathbb{N} \times \mathbb{N}$ modulo the equivalence relation $\sim$ such that $(n, m) \sim (h, k)$ if and only if $n + k = m + h$. Indeed, the equivalence class of (n, m) under $\sim$ is the integer $n - m$. Addition and multiplication of natural numbers can be extended to integers. Moreover, while the equation $x + a = b$ can be solved within the set of natural numbers if and only if $a \leq b$, it always admits the solution $b - a$ in the set of integers.

The set $\mathbb{Q}$ of rational numbers can be constructed in a similar way, by taking $\mathbb{Z} \times (\mathbb{Z} \setminus \{0\})$ modulo the equivalence relation $\sim$, where we set $(n, m) \sim (h, k)$ if $nk = mh$; the equivalence class of (n, m) is the rational number n/m. Once more, the operations on $\mathbb{Z}$ can be extended to $\mathbb{Q}$. While the equation $ax = b$ is solvable in $\mathbb{Z}$ if and only if b is a multiple of a, it is always possible to find a solution in $\mathbb{Q}$ as long as $a \neq 0$: in particular, one has the rational solution b/a. There are, however, equations that do not admit rational solutions, such as $x^2 - 2 = 0$.

One may perform another extension and introduce the set $\mathbb{R}$ of real numbers. Constructing the set $\mathbb{R}$ proves rather more complicated. Although the Greeks had already understood several properties of real numbers, which they investigated within the scope of what they called the "theory of proportions", it was not until the end of the 19th century that a formal definition was given. Here we shall merely mention that several different routes are possible; for example, employing Cauchy sequences of rationals or Dedekind cuts. Both options involve "completing" $\mathbb{Q}$ by including all quantities that may be approximated with arbitrary precision by rational numbers, but do not belong to $\mathbb{Q}$. For example, $\mathbb{R}$ contains the number $\sqrt{2}$, which is a solution to the equation $x^2 - 2 = 0$ above. Even at this juncture the task of rendering all polynomial equations solvable is not yet complete: the equation $x^2 + 1 = 0$, for example, still has no solution even over the reals, since the square of a real number can never be negative.

We shall now spend a few words discussing the next step in this sequence of extensions, i.e. complex numbers. The definition does not pose any serious difficulty, once the set of real numbers has been constructed. Indeed, we shall call any ordered pair (a, b) of real numbers a complex number. Traditionally, the complex number (a, b) is denoted by $a + ib$; the symbol i is called the *imaginary unit*, the number a is called the *real part* and the number b the *imaginary part*. Addition and multiplication of complex numbers are defined as follows:

$$(a + ib) + (c + id) = (a + c) + i(b + d)$$
$$(a + ib)(c + id) = (ac - bd) + i(ad + bc).$$

By considering complex numbers of the form $a + i \cdot 0$, one immediately finds that the set of real numbers can be naturally identified with a subset of the set $\mathbb{C}$ of complex numbers. One can easily show that, under this identification, the operations on $\mathbb{C}$ extend the operations on real numbers. The definition given above for the product of two complex numbers is devised in order to have $i^2 = (0 + i \cdot 1)^2 = -1$: the number -1 now has a square root. In other words, the equation $x^2 + 1 = 0$ admits the two complex solutions $\pm i$.

A complex number of the form $0 + ib$ is called (purely) *imaginary*. It is sometimes useful to think of complex numbers as points in a Cartesian plane, called the *complex plane*, whose x-axis is that of real numbers and whose y-axis contains purely imaginary numbers: a complex number $a + ib$ has abscissa a and ordinate b in the complex plane. According to this representation, the origin corresponds to the number $0 = 0 + i \cdot 0$, which is the addition's neutral element. The notation for complex numbers we have been discussing so far is called *algebraic form*.

Given a complex number $z = a + ib$ we shall call the number $\bar{z} = a - ib$ the *complex conjugate* of z. The point corresponding to $\bar{z}$ is the symmetric point of z with respect to the real axis in the complex plane. The distance between the point z and the origin is given by the quantity $|z| = \sqrt{a^2 + b^2}$ and is called the *modulus* of z. Notice that we have $|z|^2 = z \cdot \bar{z}$; from this, one can obtain the formula $1/z = \bar{z}/|z|^2$ for the inverse of a nonzero complex number z. In particular, for a complex number whose modulus is equal to 1 (a point on the unit circle in the complex plane), $z^{-1} = \bar{z}$.

If $z \neq 0$, we denote by θ the *argument* of z, that is, the angle between the positive real axis and the segment joining the point z to the origin 0; we then have

$$z = |z|(\cos \theta + i \sin \theta).$$

Euler's formula

$$e^{i\theta} = \cos \theta + i \sin \theta,$$

is the fundamental relationship between the complex exponential map and the trigonometric functions. It allows us to express a complex number z in its *polar form*

$$z = |z|e^{i\theta}.$$

When $z = 0$ the argument θ is not defined, but whenever $z \neq 0$ the polar form for z is unique up to adding integer multiples of 2π to θ. Notice that, as θ varies, the complex number $|z|e^{i\theta}$ moves along the circle with centre 0 and radius $|z|$. In particular, by Euler's formula we know that the map $\theta \longmapsto e^{i\theta}$ parametrises the unit circle.

The polar form is especially well suited to compute powers of a complex number. Indeed, we can immediately obtain that, given a complex number $z = |z|e^{i\theta}$ and an integer n, we have

$$z^n = |z|^n e^{in\theta};$$

in other words, z^n is the complex number whose modulus is $|z|^n$ and whose argument is $n\theta$. Analogously we can find the nth roots of z, i.e. the complex numbers ζ such that $\zeta^n = z$, by

$$\sqrt[n]{|z|}e^{i\frac{\theta+2k\pi}{n}}, \qquad \text{with } k = 0, 1, 2, \ldots, n-1.$$

Notice that, when $z \neq 0$, the number of distinct nth roots of z is n. If we set $z = 1$, the formula above gives the *nth roots of unity*, which will be especially important for our purposes:

$$e^{2\pi ik/n} = \cos\frac{2\pi k}{n} + i\sin\frac{2\pi k}{n}, \qquad \text{with } k = 0, 1, 2, \ldots, n-1.$$

Also notice that, setting $\zeta_n = e^{2\pi i/n}$, all nth roots of unity can be obtained as $1, \zeta_n, \zeta_n^2, \ldots, \zeta_n^{n-1}$.

We shall conclude this brief introduction to number sets by remarking that every equation of the form

$$a_n x^n + a_{n-1}x^{n-1} + \cdots + a_1 x + a_0 = 0,$$

where $a_0, a_1, \ldots, a_{n-1}, a_n$ are complex numbers and $a_n \neq 0$, admits a solution in $\mathbb{C}$ if $n > 0$: there is no further need to extend the set of numbers to solve polynomial equations. Throughout the rest of the book we shall study polynomial equations in detail, focusing on the process—encountered several times in this section—of extending number sets by adding solutions to certain equations.

Finally, notice that the familiar addition and multiplication between integer, rational, real and complex numbers are associative and commutative. In all cases the neutral element for the addition operation is always the number 0 and the additive inverse of a number a is denoted by $-a$. The multiplication operation has 1 as neutral element and distributes over addition.

2 Combinatorics

A set X is *finite* if it has a finite number of elements; we call this number the *cardinality* of X and denote it by $|X|$. When X is not finite we say it is *infinite* and has infinite cardinality. Two finite sets have the same cardinality if and only if there is a bijection between them. In particular, X is finite with cardinality n if and only if there is a bijective map between $\{1, 2, \ldots, n\}$ and X; in that case, we can enumerate the elements of X and write $X = \{x_1, x_2, \ldots, x_n\}$.

With a finer definition, one could distinguish between different cardinalities for infinite sets; this, however, would go beyond the scope of this work: we shall be content with the ability to distinguish infinite sets from finite ones.

A first remark one can make about finite sets is the following:

Remark 2.1 A map $X \longrightarrow Y$ between finite sets with the same cardinality is injective if and only if it is surjective, if and only if it is bijective.

Given two finite sets $X = \{x_1, x_2, \ldots, x_n\}$ and $Y = \{y_1, y_2, \ldots, y_m\}$, the Cartesian product $X \times Y$ consists of all pairs of the form (x_i, y_j) with $i = 1, 2, \ldots, n$ and $j = 1, 2, \ldots, m$. We therefore have

Remark 2.2 If X and Y are finite sets then

$$|X \times Y| = |X| \cdot |Y|.$$

On the other hand, if one of the two sets is infinite and the other is nonempty, then their Cartesian product is also infinite.

A map $f : X \longrightarrow Y$, where X has cardinality n, can be described completely by an n-tuple of elements of Y; that is, there is a bijection between Y^X and Y^n. In particular,

Remark 2.3 Given two finite sets X and Y that are not both empty, the cardinality of the set Y^X of maps from X to Y is given by

$$|Y^X| = |Y|^{|X|}.$$

On the other hand, if one of the two sets is infinite and the other one is nonempty, then Y^X is also infinite.

Each subset A of X has a corresponding *characteristic function* $\chi_A : X \longrightarrow \{0, 1\}$, constructed so that $\chi_A(x) = 1$ if $x \in A$ and $\chi_A(x) = 0$ if $x \notin A$. Subsets of X are in bijection with the set of their corresponding characteristic functions, so that

Remark 2.4 If X is a finite set then the cardinality of its power set $\mathcal{P}(X)$ is

$$|\mathcal{P}(X)| = 2^{|X|};$$

on the other hand, if X is infinite then $\mathcal{P}(X)$ is also infinite.

The number of injective maps from a set X with n elements to a set Y with m elements is easy to find; one need only notice that an injective map corresponds to the choice of n distinct elements of Y, in some order.

Remark 2.5 Let $|X| = n$ and $|Y| = m$. If $n > m$ then there are no injective maps from X to Y. If $n \leq m$ then the number of injective maps from X to Y is $m(m - 1)(m - 2) \cdots (m - n + 1)$.

The content of this remark for $n > m$ is sometimes expressed as follows:

Remark 2.6 (Pigeonhole Principle) If n items are put into m boxes and $n > m$ then there is a box that contains more than one item.

The *factorial n!* of a natural number n is recursively defined by $0! = 1$ and, for all $n \geq 0$, $(n + 1)! = (n + 1) \cdot n!$. Clearly, this is equivalent to setting $n! = n \cdot (n - 1) \cdots 2 \cdot 1$. Thanks to Remark 2.1, a special case of the formula above for the number of injective functions is given by

Remark 2.7 The number of permutations of a set of n elements is $n!$.

As we will see later, permutations of a set with n elements form a fundamental object in group theory; if $X = \{1, 2, \ldots, n\}$, then the set $S(X)$ of permutations of X is simply denoted by S_n. Thanks to the remark we just made, we know that $|S_n| = n!$.

The power set $\mathcal{P}(X)$ of a set X consisting of n elements can be partitioned according to the cardinality of its elements, as follows:

$$\mathcal{P}(X) = \bigsqcup_{k=0}^{n} \{A \subseteq X \mid |A| = k\}.$$

Moreover, from the above we immediately obtain

Remark 2.8 If $|X| = n$ and $0 \leq k \leq n$, then the number of subsets of X having cardinality k is

$$\frac{n \cdot (n - 1) \cdots (n - k + 1)}{k!}.$$

This is a fact of great importance in combinatorics. Given $n \geq 0$ and $0 \leq k \leq n$ we shall define the *binomial coefficient* indexed by n, k as

$$\binom{n}{k} = \frac{n \cdot (n - 1) \cdots (n - k + 1)}{k!} = \frac{n!}{k!(n - k)!}.$$

Note that, in particular, $\binom{n}{0} = 1$: indeed, the only zero-element subset of X is the empty set. Moreover, we have $\binom{n}{n} = 1$, and indeed the only n-element subset of X is X itself. It is sometimes useful to extend the meaning of the symbol $\binom{n}{k}$ by defining it as 0 for all $k < 0$ and $k > n$.

Binomial coefficients satisfy several useful identities; two of the main ones are

Remark 2.9 For all $n \geq 0$ we have

$$\binom{n}{n-k} = \binom{n}{k}$$

and also

$$\binom{n+1}{k} = \binom{n}{k} + \binom{n}{k-1}.$$

The second identity can be used as a recursive definition for binomial coefficients if we first set $\binom{0}{0} = 1$ and $\binom{0}{k} = 0$ for all $k \neq 0$.

From the partition of $\mathcal{P}(X)$ according to cardinalities we obtain one more identity:

Remark 2.10 For all natural numbers n we have

$$\sum_{k=0}^{n} \binom{n}{k} = 2^n.$$

It is common to arrange binomial coefficients into a triangle whose rows consist of the values of $\binom{n}{k}$ for a fixed n. The first six rows of this triangle, called Pascal's triangle, are

$$
\begin{array}{ccccccccccc}
 & & & & & 1 & & & & & \\
 & & & & 1 & & 1 & & & & \\
 & & & 1 & & 2 & & 1 & & & \\
 & & 1 & & 3 & & 3 & & 1 & & \\
 & 1 & & 4 & & 6 & & 4 & & 1 & \\
1 & & 5 & & 10 & & 10 & & 5 & & 1
\end{array}
$$

Binomial coefficients appear in the algebraic expansion of the powers of binomials.

Theorem 2.11 (Binomial Theorem)   *Given two numbers a and b and a non-negative integer n, we have*

$$(a+b)^n = \sum_{k=0}^{n} \binom{n}{k} a^k b^{n-k}.$$

Actually, the formula above holds for all elements a, b of any commutative ring: we will mention this again in the chapter about rings.

Given two subsets X_1 and X_2 of a set X such that $X = X_1 \cup X_2$, we have

$$|X| = |X_1| + |X_2| - |X_1 \cap X_2|.$$

This is because elements of $X_1 \cap X_2$ belong to both subsets and are therefore counted twice in the sum $|X_1| + |X_2|$. The formula above is a special case of the following

Proposition 2.12 (Inclusion-Exclusion Principle) *If X is a finite set and $X_1, X_2, \ldots, X_k$ are subsets of X such that $X_1 \cup X_2 \cup \cdots \cup X_k = X$ then we have*

$$|X| = \sum (-1)^{h+1} |X_{i_1} \cap X_{i_2} \cap \cdots \cap X_{i_h}|,$$

where the sum is over $h = 1, \ldots, k$ and all h-tuples $(i_1, i_2, \ldots, i_h)$ with $1 \le i_1 < i_2 < \cdots < i_h \le n$.

For example, the case $k = 3$ of the inclusion-exclusion formula above is

$$|X| = |X_1| + |X_2| + |X_3| - |X_1 \cap X_2| - |X_1 \cap X_3| - |X_2 \cap X_3| + |X_1 \cap X_2 \cap X_3|.$$

Naturally, if we assume that $\{X_1, X_2, \ldots, X_k\}$ is a partition of the set X, then we have

$$|X| = |X_1| + |X_2| + \ldots + |X_k|.$$

3 Integers

3.1 Divisibility of Integers

At the foundation of integer arithmetic is the concept of *Euclidean division*, which we recall in the following

Proposition 3.1 (Euclidean Division) *Given an integer a and a positive integer m, there exist a unique integer q, called the* quotient, *and a unique non-negative integer r, called the* remainder, *such that $a = q \cdot m + r$ and $0 \le r < m$.*

If, when performing Euclidean division, we find that $r = 0$, then we have $a = q \cdot m$ and we write $m \mid a$: we shall say that a is a *multiple* of m or that m is *divisible* by a and we shall call m a *divisor* of a. If m does not divide a, then we write $m \nmid a$. Only the integers ± 1 divide 1, whereas every nonzero integer divides 0.

Given two integers a and b at least one of which is nonzero, we shall call *greatest common divisor* of a and b any positive integer m such that: m is a divisor of a and a divisor of b and, if n is a common divisor of a and b, then n divides m. Using Euclidean division, one can show

Proposition 3.2 *The greatest common divisor m of two integers at least one of which is nonzero exists and is unique. Moreover, there exist two integers x and y such that $m = xa + yb$; this identity is called Bézout's identity.*

We shall write (a, b) for the greatest common divisor of a and b. Remark that $(a, b) = (|a|, |b|)$, that is, we may always assume that a and b are non-negative. In order to compute the greatest common divisor, one can use

Proposition 3.3 (Euclid's Algorithm) *Suppose a and b are non-negative integers with $a \geq b$; set $r_0 = a$, $r_1 = b$. If for $k \geq 1$ we have $r_k > 0$ then define r_{k+1} recursively as the remainder in the Euclidean division of r_{k-1} by r_k. Since $r_0 > r_1 > r_2 > \ldots \geq 0$, after a finite number of steps (n, say) we have $r_n = 0$. Then $(a, b) = r_{n-1}$.*

This algorithm can be used to produce an explicit solution (x, y) of Bézout's identity. Iteratively replacing r_k with its expression in terms of preceding remainders until one gets back to the equation $a = qb + r_1$ yields an identity involving only a, b and the last nonzero remainder $r_{n-1} = (a, b)$.

Two integers are said to be *relatively prime* when $(a, b) = 1$; this, in turn, holds if and only if there exist integers x, y such that $xa + yb = 1$.

Remark 3.4 (Euclid's Lemma) If m divides the integer product $a \cdot b$ and integers m and a are relatively prime, then m divides b.

Thanks to Euclid's lemma, we can easily find all solutions of Bézout's identity:

Proposition 3.5 *Let a, b be two integers at least one of which is nonzero; let m be their greatest common divisor and let (x_0, y_0) be a solution of Bézout's identity for a, b. Then all pairs of integers satisfying Bézout's identity are given by*

$$(x_0 + k\frac{b}{m}, y_0 - k\frac{a}{m}), \quad k \in \mathbb{Z}.$$

We call *linear Diophantine equation* in the two integer indeterminates x and y an equation of the form $ax + by = c$ with integer coefficients a, b and c. More generally, any equation with integer coefficients for which we seek integer solution is called a Diophantine equation. Solving such equations is quite different from solving equations in the reals and usually turns out to be very difficult. We might almost say that the Mathematics of today is only adequate for solving linear and quadratic Diophantine equations; cubic equation already cross into a new intricate and fascinating algebraic world inhabited by such objects as elliptic curves.

In Preliminary Exercise 7, we deal with the linear case by means of Bézout's identity, which we use to prove

Proposition 3.6 *The Diophantine equation $ax + by = c$ has a solution if and only if the greatest common divisor $m = (a, b)$ divides the constant term c. In that case,* given a solution (x_0, y_0), all solutions are of the form

$$(x_0 + k\frac{b}{m}, y_0 - k\frac{a}{m}), \quad k \in \mathbb{Z}.$$

Let us now present the central definition in integer arithmetic. A positive integer p is said to be *prime* if it has exactly two positive divisors, that is, 1 and p. Note that 1 is not prime. Given an integer n and a prime p, the greatest common divisor (p, n) can only be p (if p divides n) or 1 (if p does not divides n). It immediately follows that

Remark 3.7 If a prime p divides an integer product $a \cdot b$ and p does not divide a, then p divides b.

A classical result from ancient Greek arithmetic states that each integer can be written as a product of primes in an essentially unique way; it is the

Theorem 3.8 (Fundamental Theorem of Arithmetic) *Given a positive integer $n > 1$ there exist (not necessarily distinct) primes $p_1, p_2, \ldots, p_r$, unique up to reordering, such that $n = p_1 p_2 \cdots p_r$.*

Given a prime p and an integer n, we say that a power p^e *exactly divides* n if p^e divides n and p^{e+1} does not divide n. In other words, p^e exactly divides n if and only if the prime p appears in the factorisation of n with exponent e.

Given two integers a and b at least one of which is nonzero, we shall define their *least common multiple*, denoted by $[a, b]$, as a common multiple of a and b that is divisible by every common multiple of a and b. Analogously to the greatest common divisor, the least common multiple exists and is unique; moreover, we have $(a, b)[a, b] = ab$ if $ab \geq 0$. The definition of least common multiple is the dual of that of greatest common divisor, thus the two objects share many similar properties.

3.2 Congruences

In his 1801 book "Disquisitiones Arithmeticae", Carl Friedrich Gauss introduced what would turn out to be one of the most important relations between integers in elementary arithmetic. We say that an integer a is *congruent* to an integer b modulo n, and write $a \equiv b \pmod{n}$, if $a - b$ is a multiple of n. It is easy to show that the relation of "congruence" is an equivalence relation.

The equivalence class of an integer a, that is, the set of all integers congruent to a, is denoted by $[a]_n$ and consists of the set $\{a + kn \mid k \in \mathbb{Z}\}$. In other words, $[a]_n$ is the set of all integers having the same remainder as a in the division by n. If the modulus n can be inferred from the context, we may alternatively denote the class $[a]_n$ by $\bar{a}$.

As is the case with all equivalence relations, congruence classes modulo n form a partition of $\mathbb{Z}$. A possible set of representatives is $\{0, 1, 2, \ldots, n - 1\}$: these are called *residues modulo n*. Note that, in particular, the number of equivalence classes is n; naturally, any set of n integers with different remainders in the division by n is a set of representatives for the congruence classes modulo n. Remark that any set of n consecutive integers is a set of representatives.

The quotient of $\mathbb{Z}$ by the congruence relation modulo n is denoted by $\mathbb{Z}/n\mathbb{Z}$. We shall find out later, when discussing groups, the reason behind this specific notation: for now, we shall interpret it as a reminder that we are identifying integers that differ by a multiple of n, that is, by an element of $n\mathbb{Z}$.

Here are some properties of the relation of congruence modulo n.

Proposition 3.9 *Let a and b be integers such that $a \equiv b \pmod{n}$. We have*

(i) $(a, n) = (b, n)$,
(ii) if $m \mid n$ then $a \equiv b \pmod{m}$,
(iii) if we also have $a \equiv b \pmod{m}$ then $a \equiv b \pmod{[n, m]}$.

The second property is equivalent to the following: if n is a multiple of m then the partition given by the congruence classes modulo n is a finer partition than that given by the congruence classes modulo m. Indeed, we have

$$[a]_m = \bigsqcup_{h=0,1,\ldots,\frac{n}{m}-1} [a + mh]_n.$$

Thus, when n is a multiple of m, there is a map from $\mathbb{Z}/n\mathbb{Z}$ to $\mathbb{Z}/m\mathbb{Z}$ sending the class $[a]_n$ to $[a]_m$, which renders the following diagramme commutative

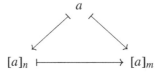

as remarked in the section about relations.

The relation of congruence modulo n is compatible with integer addition and multiplication.

Proposition 3.10 *Given integers a, b, a' and b' such that $a \equiv b \pmod{n}$ and $a' \equiv b' \pmod{n}$, we have $a + a' \equiv b + b' \pmod{n}$ and $a \cdot a' \equiv b \cdot b' \pmod{n}$.*

In particular, if we have $a \equiv b \pmod{n}$ and k is any integer, then we have $ka \equiv kb \pmod{n}$. Let us also explicitly note that the converse of our previous remark is false in general: for example, $2 \cdot 2 \equiv 2 \cdot 0 \pmod{4}$, but $2 \not\equiv 0 \pmod{4}$. A weaker but accurate converse statement is the following

Proposition 3.11 *Given integers a, b and k with $k \neq 0$,*

$$ka \equiv kb \pmod{n} \quad \text{implies} \quad a \equiv b \pmod{\frac{n}{(n, k)}}.$$

In particular, if the factor k and the modulus n are relatively prime, then

$$ka \equiv kb \pmod{n} \quad \text{if and only if} \quad a \equiv b \pmod{n}.$$

In particular, in the example above what we can correctly deduce from $2 \cdot 2 \equiv 2 \cdot 0$ (mod 4) is that $2 \equiv 0 \pmod{2}$, because $4/(4, 2) = 2$.

What has been recalled so far, though elementary, already has some consequences that are not completely trivial. For example, by using the base 10 representation of the integers together with congruences modulo 3, one can deduce the well-known divisibility rule for 3, which easily follows from $10 \equiv 1 \pmod{3}$. One can similarly deduce divisibility rules for 2, 4, 5, 9, 11 and 25, and more elaborate rules for 7 and 13.

A first nontrivial fact about residue classes modulo a prime is the following

Theorem 3.12 (Freshman's Dream)   *Given a prime p, for all pairs of integers a and b we have*

$$(a + b)^p \equiv a^p + b^p \pmod{p}.$$

The proof of this theorem is immediate if one notices that all binomial coefficients $\binom{p}{h}$, where p is prime and $1 \le h \le p - 1$, are divisible by p.

Using Theorem 3.12, one can prove by induction

Theorem 3.13 (Fermat's Theorem)   *Given a prime p, we have*

$$a^p \equiv a \pmod{p}$$

for all integers a.

An immediate corollary of the theorem above is among the first nontrivial results about prime numbers to appear after Euclid's time:

Corollary 3.14 (Fermat's Little Theorem)   *Given a prime p and an integer a such that a and p are relatively prime, we have $a^{p-1} \equiv 1 \pmod{p}$.*

Finding the solution of a linear congruence, that is, of an equation of the type $ax \equiv b \pmod{n}$, where x is the indeterminate and a, b are integers, amounts to solving the Diophantine equation $ax + ny = b$ in the indeterminates x and y. One immediately obtains

Proposition 3.15   *The linear congruence $ax \equiv b \pmod{n}$ has a solution if and only if $m = (a, n)$ divides b, in which case, given a solution x_0, all solutions are given by*

$$[x_0 + k\frac{n}{m}]_n \quad \text{for} \quad k = 0, 1, \ldots, m - 1.$$

In particular, there are m solutions modulo n.

In order to solve systems of linear congruences, the following theorem is of fundamental importance.

Theorem 3.16 (Chinese Remainder Theorem)   *Given   r   nonzero integers $n_1, n_2, \ldots, n_r$ that are   pairwise   relatively   prime   and   r   integers $a_1, a_2, \ldots, a_r$, the system of congruences*

$$\begin{cases} x \equiv a_1 \pmod{n_1} \\ x \equiv a_2 \pmod{n_2} \\ \quad \vdots \\ x \equiv a_r \pmod{n_r} \end{cases}$$

has a unique solution modulo $n_1 n_2 \cdots n_r$.

At the end of the next section, we will see how the solution of such a system can be computed.

3.3   *Modular Arithmetic*

The properties of compatibility for the congruence relation modulo a nonzero integer n that were discussed in the previous section allow us to define two operations on the quotient set $\mathbb{Z}/n\mathbb{Z}$ of residue classes modulo n.

We shall define the addition $+$ and the multiplication $\cdot$ on pairs of residue classes $[a]_n, [b]_n$ by

$$[a]_n + [b]_n = [a + b]_n \qquad [a]_n \cdot [b]_n = [ab]_n.$$

These operations are well defined: their result does not depend on the representatives a and b chosen in $\mathbb{Z}$ for the classes $[a]_n$ and $[b]_n$, but only on the residue classes themselves; this follows directly from Proposition 3.10.

From standard properties of the integers, one obtains the following corresponding results:

Theorem 3.17

(i) *The operations $+$ and $\cdot$ are associative, that is,*

$$([a]_n + [b]_n) + [c]_n = [a]_n + ([b]_n + [c]_n), \quad ([a]_n \cdot [b]_n) \cdot [c]_n = [a]_n \cdot ([b]_n \cdot [c]_n)$$

for all $[a]_n, [b]_n, [c]_n \in \mathbb{Z}/n\mathbb{Z}$.

(ii) *The operations $+$ and $\cdot$ are commutative, that is,*

$$[a]_n + [b]_n = [b]_n + [a]_n, \quad [a]_n \cdot [b]_n = [b]_n \cdot [a]_n$$

for all $[a]_n, [b]_n \in \mathbb{Z}/n\mathbb{Z}$.

(iii) *The addition's neutral element is the class* $[0]_n$ *and the multiplication's neutral element is the class* $[1]_n$. *Moreover, the inverse of the element* $[a]_n \in \mathbb{Z}/n\mathbb{Z}$ *with respect to the operation* $+$ *is the element* $[-a]_n \in \mathbb{Z}/n\mathbb{Z}$.

(iv) *The operation* $\cdot$ *distributes over the operation* $+$, *that is,*

$$[a]_n \cdot ([b]_n + [c]_n) = [a]_n \cdot [b]_n + [a]_n \cdot [c]_n$$

for all $[a]_n, [b]_n, [c]_n \in \mathbb{Z}/n\mathbb{Z}$.

As opposed to the addition, not all classes do have an inverse for the multiplication: for instance, $[2]_4$ does not have an inverse in $\mathbb{Z}/4\mathbb{Z}$. The classes $[a]_n$ such that their inverse for the multiplication does exist are called *invertible*. The class $[a]_n$ is invertible if there exists a class $[b]_n$ such that $[a]_n[b]_n = [1]_n$. The set of invertible classes modulo n is denoted by $(\mathbb{Z}/n\mathbb{Z})^*$.

The criterion for the existence of solutions of a linear congruence implies that a class $[a]_n$ is invertible if and only if $(a, n) = 1$. Therefore, it is clear that the product of two invertible classes is itself invertible. In other words, the multiplication can be restricted to the subset $(\mathbb{Z}/n\mathbb{Z})^*$ and induces an operation on $(\mathbb{Z}/n\mathbb{Z})^*$ which will still be denoted by $\cdot$. In the special case of congruences modulo p, where p is prime, we have $(\mathbb{Z}/p\mathbb{Z})^* = \mathbb{Z}/p\mathbb{Z} \setminus \{[0]_p\}$, that is, every nonzero class is invertible.

In order to compute the inverse of a class $[a]_n$ one can use Bézout's identity. Indeed, from $(a, n) = 1$ it follows that there are integers b and c, which one can compute explicitly by means of Euclid's algorithm, such that $ab + nc = 1$. Considering this equality modulo n yields $ab \equiv 1 \pmod{n}$, so $[b]_n$ is the inverse of $[a]_n$.

One may state the Chinese remainder theorem for systems of congruences in an equivalent form involving residue classes:

Theorem 3.18 *Given two nonzero, relatively prime integers* m *and* n, *the map*

$$\mathbb{Z}/mn\mathbb{Z} \ni [a]_{mn} \longmapsto ([a]_m, [a]_n) \in \mathbb{Z}/m\mathbb{Z} \times \mathbb{Z}/n\mathbb{Z}$$

is bijective.

The following is a direct corollary:

Corollary 3.19 *Let* m *and* n *be relatively prime, nonzero integers. The class* $[a]_{mn}$ *is invertible if and only if* $[a]_m$ *and* $[a]_n$ *are both invertible. Moreover, the map* $[a]_{mn} \longmapsto ([a]_m, [a]_n)$ *from* $(\mathbb{Z}/mn\mathbb{Z})^*$ *to* $(\mathbb{Z}/m\mathbb{Z})^* \times (\mathbb{Z}/n\mathbb{Z})^*$ *is bijective.*

The so-called *Euler's totient function* $n \longmapsto \phi(n)$ maps the integer n to the number $\phi(n)$ of integers between 1 and n that are coprime to n. This map plays a very important role in modular arithmetic. As seen above, $\phi(n)$ is equal to the number of invertible classes modulo n, that is, $\phi(n)$ also represents the cardinality of $(\mathbb{Z}/n\mathbb{Z})^*$. For instance, we have $\phi(p) = p - 1$.

A map f defined on $\mathbb{N}$ is said to be *multiplicative* if, for all pairs of relatively prime natural numbers n and m, we have $f(nm) = f(n)f(m)$. The corollary above implies that Euler's totient function is multiplicative. By considering the numbers between 1 and p^e that are not divisible by p, it is easy to see that $\phi(p^e) = (p - 1)p^{e-1}$ for all integers $e \geq 1$. We have thus shown the following formula:

Remark 3.20 If $n = p_1^{e_1} p_2^{e_2} \cdots p_r^{e_r}$ is the prime factorisation of n, with $p_i \neq p_j$ for $i \neq j$ and $e_i \geq 1$ for all i, then

$$\phi(n) = \prod_{i=1}^{r}(p_i - 1)p_i^{e_i-1}.$$

Corollary 3.14 can also be expressed as follows: if p is prime and a is not divisible by p, then $a^{\phi(p)} \equiv 1 \pmod{p}$. This form can be generalised to become

Theorem 3.21 (Euler's Theorem)   *Given an integer a which is coprime to the modulus n, we have $a^{\phi(n)} \equiv 1 \pmod{n}$.*

We shall conclude this discussion of congruences by illustrating in detail some methods one can use to solve linear systems of congruences whose equations have coprime moduli, as in the statement of the Chinese remainder theorem. Consider the system of two congruences

$$\begin{cases} x \equiv a_1 \pmod{n_1} \\ x \equiv a_2 \pmod{n_2} \end{cases}$$

where n_1 and n_2 are relatively prime. Solving this system amounts to finding integers u and v such that $x = a_1 + un_1$ and $x = a_2 + vn_2$. In other words, u and v must be solutions of the linear Diophantine equation

$$n_1 u - n_2 v = a_2 - a_1.$$

The solution of equations of this type is fully worked out in Preliminary Exercise 7. Once we have u and v, the solution of the original system is the residue class of $x_0 = a_1 + un_1 = a_2 + vn_2$ modulo $n_1 n_2$. We know that the solution is unique modulo $n_1 n_2$. Equivalently, all solutions are of the form $x_0 + hn_1 n_2$, for h in $\mathbb{Z}$.

Let us now move on to system consisting of r linear congruences, for some $r \geq 2$, of the form

$$\begin{cases} x \equiv a_1 \pmod{n_1} \\ x \equiv a_2 \pmod{n_2} \\ x \equiv a_3 \pmod{n_3} \\ \quad\vdots \\ x \equiv a_r \pmod{n_r} \end{cases}$$

where $n_1, n_2, \ldots, n_r$ are pairwise relatively prime. The subsystem given by the first two congruences can be solved with the method described above, which gives us a unique solution x_0 modulo $n_1 n_2$. We can then consider the equivalent system

$$\begin{cases} x \equiv x_0 \pmod{n_1 n_2} \\ x \equiv a_3 \pmod{n_3} \\ \quad \vdots \\ x \equiv a_r \pmod{n_r} \end{cases}$$

which consists of $r - 1$ congruences. We repeatedly solve the first two congruences and decrease the number of congruences in the system until we find the solution.

Let us now discuss a different method for solving the system above. We know that the solution is unique modulo $n_1 n_2 \cdots n_r$. In order to produce the solution, we shall first find integers $x_1, x_2, \ldots, x_r$ such that, for $i = 1, 2, \ldots, r$, we have $x_i \equiv 0$ $(\bmod\ n_j)$ for all $j \neq i$ and $x_i \equiv 1 \pmod{n_i}$. In other words, $x_1, x_2, \ldots, x_r$ are solutions of the systems

$$\begin{cases} x_1 \equiv 1 \pmod{n_1} \\ x_1 \equiv 0 \pmod{n_2} \\ \quad \vdots \\ x_1 \equiv 0 \pmod{n_r} \end{cases} \quad \begin{cases} x_2 \equiv 0 \pmod{n_1} \\ x_2 \equiv 1 \pmod{n_2} \\ \quad \vdots \\ x_2 \equiv 0 \pmod{n_r} \end{cases} \cdots \begin{cases} x_r \equiv 0 \pmod{n_1} \\ x_r \equiv 0 \pmod{n_2} \\ \quad \vdots \\ x_r \equiv 1 \pmod{n_r}. \end{cases}$$

Once we have solved each of these systems, it is clear that the solution of the original system is given by

$$x_0 \equiv a_1 x_1 + a_2 x_2 + \ldots + a_r x_r \pmod{n_1 n_2 \cdots n_r}.$$

In order to find the integer x_1 (and similarly $x_2, \ldots, x_r$) it is enough to remark that we must have $x_1 = y_1 n_2 \cdots n_r$ for some integer y_1. Moreover, $n_2 \cdots n_r$ is an invertible class modulo n_1, given our assumptions about the moduli $n_1, n_2, \ldots, n_r$. We can therefore obtain y_1 by solving the equation $y_1(n_2 \cdots n_r) \equiv 1 \pmod{n_1}$ or, equivalently, $y_1 \equiv (n_2 \cdots n_r)^{-1} \pmod{n_1}$. This congruence is again equivalent to a linear Diophantine equation which we can solve explicitly.

This second method for solving linear systems via the auxiliary solutions $x_1, x_2, \ldots, x_r$ can prove especially efficient when we wish to solve several systems whose equations have the same moduli $n_1, n_2, \ldots, n_r$ but different constant terms $a_1, a_2, \ldots, a_r$.

The remarks we just made about explicit ways of computing the solutions of a system of linear congruences will be used repeatedly, often without referencing them, in the solutions of the exercises.

4 Groups

4.1 Definition and Basic Properties

One of the fundamental structures in algebra is that of a group. Though simple enough to be defined in a few lines, groups truly play a crucial role: all of the more complex structures we will encounter throughout the book will be based on groups.

A nonempty set G endowed with an operation $\cdot$ is called a *group* if

(i) the operation $\cdot$ is associative,

(ii) there exists an element e in G, called the *neutral element*, such that $g \cdot e = g = e \cdot g$ for all g in G,

(iii) for each element g of G there exists an element $h \in G$, called the *inverse* of g, such that $g \cdot h = e = h \cdot g$.

Throughout the book we will say that $(G, \cdot)$ is a group to mean that $\cdot$ is an operation on the set G which makes it a group. When the operation is clear from the context we will simply say that G is a group. We shall occasionally also omit the group operation symbol $\cdot$ when composing elements and simply write gh for the composition $g \cdot h$ of g and h in G.

There are innumerable examples of groups: the set $\mathbb{Z}$ of integers, endowed with the addition operation; the set $\mathbb{Q}^*$ of nonzero rational numbers, endowed with multiplication; the set $\mathbb{R}^*$ of nonzero real numbers, as well as the set $\mathbb{C}^*$ of nonzero complex numbers, endowed with multiplication; the set S_n of all permutations of n elements, endowed with the composition operation.

It is very easy to show that the neutral element in a group is unique. It is also easy to show that, given g in a group, the inverse of g is unique: we will denote it by g^{-1}.

In general, given a positive integer n and an element g in a group, we define g^n as the composition of g with itself performed n times, and set $g^{-n} = (g^n)^{-1}$. The usual rules for powers still hold in this context, namely we have $g^n \cdot g^m = g^{n+m}$ and $(g^n)^m = g^{nm}$ for all natural numbers n and m.

Given two elements g and h in a group, we say they *commute* if $gh = hg$. If the operation on G is commutative, that is, if $gh = hg$ for all g and h in G (every two elements commute), we say that the group itself is *commutative* or *Abelian*. It is customary to denote the operation of an Abelian group by $+$, or to *use the additive notation*. When using the additive notation, we write $-g$ for the inverse of an element g and ng for the composition g^n. Clearly, we have $(n+m)g = ng + mg$ and $(nm)g = n(mg)$ for all natural numbers n and m.

The *order* of a group G is the cardinality of the set G and is denoted by $|G|$. The order of G is the number of elements in G if G is finite and is infinite otherwise. The *order* of a single element g is the smallest positive integer n such that $g^n = e$, if such an integer exists; if $g^n \neq e$ for all positive n then we say that g has *infinite* order. We denote the order of g by $\mathrm{ord}(g)$.

Note that the set $\mathbb{Z}/n\mathbb{Z}$ of all congruence classes modulo a nonzero natural number n is a group when endowed with the operation $+$ of addition between classes; this immediately follows from Theorem 3.17. Clearly, such a group is Abelian of order n. The set $(\mathbb{Z}/n\mathbb{Z})^*$ of invertible classes modulo n is itself an Abelian group when endowed with the operation $\cdot$ of multiplication between classes; its order is $\phi(n)$.

Our first remark about groups, which immediately follows from the definition of a group, is

Remark 4.1 (Cancellation Laws) If, given elements g, h, k in a group G, we have $gh = gk$, then $h = k$; similarly, if $hg = kg$ then $h = k$.

4.2 Subgroups

A subset H of a group G is a *subgroup* if the group operation $\cdot$ of G can be restricted to an operation on H that makes H into a group. We write $H \leq G$ to indicate that H is a subgroup of G. In order to check whether a nonempty subset H is a subgroup it is enough to check that for all h and k in H we have $h \cdot k \in H$ and that for all h in H we have $h^{-1} \in H$.

For instance, the subset $2\mathbb{Z}$ of even numbers is a subgroup of $\mathbb{Z}$, as is the subset $n\mathbb{Z}$ of all multiples of a fixed integer n. The subset $\{\pm 1\}$ is a subgroup of $\mathbb{Q}^*$, which is itself a subgroup of $\mathbb{R}^*$. The subset of permutations fixing the element 1 is a subgroup of S_n.

The subset $\{e\}$ is always a subgroup for any group G and is often called the *trivial* subgroup; the group G itself is also a subgroup of G, though not a *proper* subgroup. We will refer to subgroups of G other than these two as *proper nontrivial* subgroups.

Given a group G, the subset $Z(G)$ of elements z of G such that $zg = gz$ for all g in G is called the *centre* of G; an element z belongs to the centre if it commutes with all elements of the group.

Remark 4.2 The centre $Z(G)$ is a subgroup of G.

From now on, we will often mention the centre. The centre of a group is an indicator of "how non-Abelian" the group is; for instance, G is Abelian if and only if $Z(G) = G$.

Remark that the intersection of two subgroups is a subgroup, but the union of two subgroups is not necessarily a subgroup. Given a subset X of a group G, we shall denote by $\langle X \rangle$ the subgroup *generated* by X in G, that is, the intersection of all subgroups of G that contain X. Remark that such a subgroup always exists, because G always contains X. We say that X is a *set of generators* for the group $\langle X \rangle$. The subgroup $\langle X \rangle$ generated by X in G can also equivalently be defined as the smallest subgroup of G that contains X.

A group G is *cyclic* if there exists an element g in G such that $G = \langle g \rangle$. In this case, the element g is a generator of G. Clearly, if G is cyclic then it is Abelian.

Moreover, if its generator g has finite order n then $G = \{e, g, g^2, \ldots, g^{n-1}\}$, that is, the cyclic group generated by g has order n. Similarly, if the order of g is infinite then we have $G = \{\ldots, g^{-2}, g^{-1}, e, g, g^2, \ldots\}$.

For example, $\mathbb{Z}$ is an infinite cyclic group, because $\mathbb{Z} = \langle 1 \rangle$. The group $n\mathbb{Z} = \langle n \rangle$ is also an infinite cyclic group, and $\mathbb{Z}/n\mathbb{Z} = \langle [1]_n \rangle$ is a cyclic group of order n.

Using Euclidean division one can easily show

Remark 4.3 Any subgroup of a cyclic group is cyclic.

This remark yields a description of all subgroups of $\mathbb{Z}$:

Corollary 4.4 *If H is a subgroup of $\mathbb{Z}$ then $H = n\mathbb{Z}$ for some non-negative integer n.*

As we recalled above, $\mathbb{Z}/n\mathbb{Z}$ is also a cyclic group, and we will later see that it is, in some sense, the prototype for all finite cyclic groups. In order to classify all of its subgroups, the following remark, which follows from standard properties of congruences, will be useful.

Remark 4.5 For all integers a the order of $[a]_n$ in $\mathbb{Z}/n\mathbb{Z}$ is

$$\mathrm{ord}([a]_n) = \frac{n}{(a, n)}.$$

As we can immediately see from the formula above, $\mathrm{ord}([a]_n)$ always divides $n = |\mathbb{Z}/n\mathbb{Z}|$; this, as we will see in the next section, is an example of a more general phenomenon. Other important consequences are those of

Remark 4.6 For each divisor d of a positive integer n, the group $\mathbb{Z}/n\mathbb{Z}$ contains exactly $\phi(d)$ elements of order d; moreover, there is a unique subgroup of order d in $\mathbb{Z}/n\mathbb{Z}$, generated by the class $[n/d]_n$. The above is an exhaustive description of all subgroups of $\mathbb{Z}/n\mathbb{Z}$.

Another interesting consequence follows if we list elements of $\mathbb{Z}/n\mathbb{Z}$ according to their order:

Remark 4.7 If n is a positive integer, then $\sum_{d \mid n} \phi(d) = n$.

4.3 Product of Subgroups

Given two subsets H and K of a group G, let HK be the set of all products hk, where h is an element of H and k an element of K. Even when H and K are subgroups, the subset HK of G is not necessarily a subgroup of G. If G is Abelian, then HK is indeed a subgroup. More generally, we have

Proposition 4.8 *The product HK of two subgroups H and K of a group G is a subgroup of G if and only if $HK = KH$.*

In general, even when HK is not a subgroup, we do have some information about its cardinality. Indeed, it is easy to see that each element of HK can be written in the form hk for exactly $|H \cap K|$ pairs (h, k) in $H \times K$. Therefore, we have

Remark 4.9 Let H and K be finite subgroups of a group G. The order of HK is $|H||K|/|H \cap K|$.

4.4 Cosets of a Subgroup

Given a subgroup H of a group G, we introduce a relation $\sim_H$ on G as follows: $g \sim_H k$ if and only if $g^{-1}k \in H$, in which case we shall say that g is *congruent* to k modulo H, or that g and k are *congruent* modulo H. It is easy to prove that $\sim_H$ is an equivalence relation. In particular, note that for all nonzero natural numbers n, the relation of congruence modulo the subgroup $n\mathbb{Z}$ of the group $\mathbb{Z}$ is the same as congruence of integers as defined previously.

We call the equivalence classes for $\sim_H$ *left cosets* of H in G; the name comes from the fact that the equivalence class of g is the subset $gH = \{gh \mid h \in H\}$. Clearly, we can define a "right" version of the relation by letting $g \, _H\!\sim k$ if and only if $gk^{-1} \in H$. Equivalence classes for this relation are the *right cosets* $Hg = \{hg \mid h \in H\}$. Evidently, the two relations are exactly the same in an Abelian group, where a subset is a right coset if and only if it is a left coset.

The quotient of G with respect to the relation $\sim_H$ is denoted by G/H. When quotienting by $_H\!\sim$, we write $H\backslash G$ instead. Note that $gH \longmapsto Hg$ is a bijective correspondence between G/H and $H\backslash G$.

We define the *index* $[G : H]$ of the subgroup H in G as the cardinality of the quotient G/H. By our previous remarks, this is the same as the cardinality of $H\backslash G$. We shall mention the index of a subgroup almost exclusively when it is finite.

We shall now deduce an important property of the subgroups of a finite group. The map $h \longmapsto gh$ is a bijection between H and its left coset gH; in particular, every left coset has the same cardinality as H. Since $\sim_H$ induces a partition of G, we immediately have

Theorem 4.10 (Lagrange's Theorem) *The order of every subgroup of a finite group divides of the order of the group.*

This yields

Corollary 4.11 *The order of every element in a group divides the order of the group. In particular, if a finite group G has order n then $g^n = e$ holds for all $g \in G$.*

This statement directly implies Euler's theorem (Theorem 3.21) if we remark that the order of $(\mathbb{Z}/n\mathbb{Z})^*$ is $\phi(n)$.

Another immediate consequence concerns groups whose order is prime. Given an element g which is not the neutral element in a group G with prime order p, the order of g must divide p, and thus must be p itself. We therefore have

Corollary 4.12 *Any group whose order is prime is a cyclic group.*

Finally, given a finite group G and a subgroup H of G, we have that $[G : H] = |G|/|H|$; in particular, the index of a subgroup of a finite group is also a divisor of the order of the group.

4.5 Normal Subgroups

We have remarked in the previous section that, given a subgroup of an Abelian group, its right cosets coincide with its left cosets. This is a key property: subgroups enjoying this property are of fundamental importance, as the mathematician Évariste Galois was the first to realise.

Given a group G and an element h of G, elements of the form ghg^{-1} are called *conjugates* of h. Given a subset H of G we denote by gHg^{-1} the set of elements of the form ghg^{-1} for some h in H. A subgroup H is called *normal* if $gHg^{-1} = H$ for all g in G. Clearly, all subgroups of an Abelian group are normal, since $ghg^{-1} = h$ for all h and g. Remark that we can also write the normality condition for H as $gH = Hg$ for all $g \in G$. Therefore, a subgroup is normal if and only if each of its right cosets is a left coset.

The trivial subgroup $\{e\}$ and the subgroup G are both normal. Another example of a normal subgroup is the centre: indeed, $gZ(G) = Z(G)g$ because all elements of $Z(G)$ commute with all elements of G, in particular with the element g.

Moreover, if H is a normal subgroup then for all pairs of elements g_1, g_2 in G we have $g_1 H g_2 H = g_1 g_2 H H = g_1 g_2 H$; that is, the product of any two left cosets, seen as subsets of G, is again a left coset.

This suggests introducing an operation on the quotient G/H by setting $(g_1 H) \cdot (g_2 H) = (g_1 g_2) H$. The result of the operation on cosets $g_1 H$, $g_2 H$ only depends on the cosets and not on the chosen representatives g_1 and g_2. The requirement that this operation is well posed is precisely equivalent to the condition that H is a normal subgroup of G. But we can say something much stronger:

Theorem 4.13 *Given a normal subgroup H of G, the operation $g_1 H \cdot g_2 H = (g_1 g_2) H$ endows the quotient set G/H with a group structure. The resulting group has order $[G : H]$.*

We shall say that the group structure on G/H is *induced* by the group structure of G. Let us go back to the Abelian group $\mathbb{Z}$ and its subgroup $n\mathbb{Z}$, which is clearly normal. The addition operation defined on residue classes is exactly the structure induced by the usual addition on $\mathbb{Z}$ on the quotient $\mathbb{Z}/n\mathbb{Z}$. This finally gives a complete motivation for the notation $\mathbb{Z}/n\mathbb{Z}$ we introduced for the set of residue classes.

Endowing the quotient set with a group structure is a very useful procedure. Though on the one hand information on G is lost when constructing G/H, in the sense that elements of G that differ by an element of H are identified, on the other hand G/H can be "simpler" than G and easier to study. The value of the construction is that we can sometimes deduce useful information on G from information on G/H.

We remarked before that all subgroups of an Abelian group are normal; the converse, however, is not true. As an example of a non-Abelian group all of whose subgroups are normal, we can construct the group $\mathbf{Q_8}$ of *unit quaternions*, which is defined as follows. The elements of $\mathbf{Q_8}$ are ± 1, $\pm i$, $\pm j$ and $\pm k$, where 1 is the neutral element; multiplication by -1 switches the sign of an element, and $i^2 = j^2 = k^2 = -1$, $ij = k = -ji$, $jk = i = -kj$, $ki = j = -ik$. It is easy to show that $\mathbf{Q_8}$ has the following subgroups

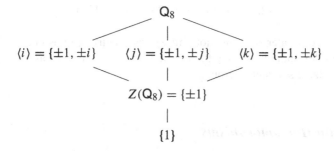

where two subgroups are connected if the lower one is a subgroup of the one above.

Normality of the subgroups of $\mathbf{Q_8}$ follows from general principles: we need only realise that any nontrivial proper subgroup either has index 2 or is the centre. But, by our previous remark, the centre is always a normal subgroup, and moreover

Remark 4.14 Any subgroup of index 2 is normal.

4.6 The Symmetric Group

Given a natural number n, the set $\mathbf{S}_n$ of permutations of $\{1, 2, \ldots, n\}$, endowed with the operation of composition of maps, is a group called the *symmetric group* on n elements. Indeed, the composite of bijective maps is bijective, the composition operation is associative, the identity map is its neutral element, and every bijective map is invertible. Moreover, we have remarked before that $\mathbf{S}_n$ has $n!$ elements.

Given a permutation $\sigma \in \mathbf{S}_n$ and elements $i_1, i_2, \ldots, i_n$ such that $\sigma(k) = i_k$ for $k = 1, 2, \ldots, n$, we denote σ as

$$\begin{pmatrix} 1 & 2 & \cdots & n \\ i_1 & i_2 & \cdots & i_n \end{pmatrix}.$$

Given distinct integers k_1, k_2, ..., k_ℓ from the set $\{1, 2, \ldots, n\}$, the permutation σ for which

$$\sigma(k_t) = k_{t+1} \text{ for } t = 1, 2, \ldots, \ell - 1,$$
$$\sigma(k_\ell) = k_1,$$
$$\sigma(j) = j \text{ for all } j \in \{1, 2, \ldots, n\} \setminus \{k_1, k_2, \ldots, k_\ell\}$$

is called a *cycle* of *length* ℓ, or an ℓ-cycle. We shall denote this cycle σ by $(k_1, k_2, \ldots, k_\ell)$.

Notice that the order of a cycle is the same as its length. For instance, if $n \geq 3$ then the cycle $(1, 2, 3)$ in S_n has order 3. A cycle (i, j) of length 2 is called a *transposition*: it exchanges i and j and fixes all other elements of $\{1, 2, \ldots, n\}$.

The symmetric group S_n is *not* Abelian for any $n \geq 3$. Indeed, we have for example

$$(123)(12) = (13) \neq (23) = (12)(123).$$

As we shall see in later sections, not only are symmetric groups not Abelian, but they are complex enough that any finite group can be constructed as a subgroup of a sufficiently large symmetric group.

4.7 Group Homomorphisms

We shall now introduce group homomorphisms, that is, maps that preserve the group structure. Homomorphisms allow us to compare groups by relating them to one another, which will prove very fruitful as a general approach.

Let G, H be groups whose operations are $\cdot$ and $\circ$, respectively. A map $f : G \longrightarrow H$ from one to the other is a *homomorphism* if $f(g_1 \cdot g_2) = f(g_1) \circ f(g_2)$ for all elements g_1, g_2 in G.

It is immediate from the definition that a homomorphism will send the neutral element of G to the neutral element of H, that is, $f(e_G) = e_H$. Moreover, $f(g^{-1}) = f(g)^{-1}$ and $\mathrm{ord}(f(g)) \mid \mathrm{ord}(g)$. Subgroups are sent to subgroups, as stated by the following

Proposition 4.15 *Let $G \xrightarrow{f} H$ be a group homomorphism. If G' is a subgroup of G then $f(G')$ is a subgroup of H; if H' is a subgroup of H then $f^{-1}(H')$ is a subgroup of G.*

In particular, the image $f(G)$ of f is a subgroup of H; it is called a *homomorphic image* of G. The inverse image of the trivial subgroup $\{e_H\}$ of H is of fundamental importance; it is called the *kernel* of f and denoted by $\mathrm{Ker}(f)$. In other words, we have

$$\mathrm{Ker}(f) = \{g \in G \mid f(g) = e_H\}.$$

The kernel is a measure of the extent to which the homomorphism fails to be injective. Indeed,

Proposition 4.16 *The kernel of a group homomorphism is a normal subgroup of G. Moreover, for all h in* $\text{Im}(f)$ *we have* $f^{-1}(h) = g\,\text{Ker}(f)$ *where g is any element of* $f^{-1}(h)$. *In particular, f is injective if and only if* $\text{Ker}(f)$ *is trivial.*

Not only is the kernel of a homomorphism a normal subgroup, but every normal subgroup is the kernel of some homomorphism: if H is a normal subgroup of G, then H is the kernel of the projection homomorphism $G \longrightarrow G/H$. We express this remark in the following

Proposition 4.17 *A subgroup of a group is normal if and only if it is the kernel of a group homomorphism.*

The most important theorem concerning group homomorphisms is the

Theorem 4.18 (Fundamental Homomorphism Theorem) *Given a group homomorphism* $G \xrightarrow{f} H$, *and letting* $G \xrightarrow{\pi} G/\text{Ker}(f)$ *be the quotient homomorphism, there exists a unique homomorphism* $\overline{f}$, *necessarily injective, that renders the following diagramme commutative*

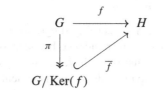

A bijective homomorphism is called an *isomorphism*. If there exists an isomorphism between groups G and H then we say the two groups are *isomorphic* and write $G \simeq H$. In particular, the theorem above implies

Corollary 4.19 *If f is surjective then* $\overline{f}$ *is an isomorphism between* $G/\text{Ker}(f)$ *and H.*

This corollary allows us to conclude that the homomorphic images of the group G are quotients of G and can therefore be constructed using only G itself. Moreover, every group homomorphism $f : G \longrightarrow H$ factors as follows: first, the projection $\pi : G \longrightarrow G/\text{Ker}(f)$, followed by the isomorphism $\overline{f} : G/\text{Ker}(f) \longrightarrow \text{Im}(f)$ and finally the inclusion $\text{Im}(f) \longhookrightarrow H$. That is, we have the following commutative diagramme:

$$
\begin{array}{ccc}
G & \xrightarrow{\ f\ } & H \\[4pt]
{\scriptstyle \pi}\big\downarrow & & \big\uparrow{\scriptstyle i} \\[4pt]
G/\text{Ker}(f) & \xrightarrow[\sim]{\ \overline{f}\ } & \text{Im}(f)
\end{array}
$$

Let G be a group; the isomorphisms from G to itself are called *automorphisms* and the set of automorphisms of G is denoted by $\mathrm{Aut}(G)$. Clearly, the identity is an automorphism, the composition of two automorphisms is an automorphism, and the inverse of an automorphism is an automorphism. In other words, the set $\mathrm{Aut}(G)$ endowed with the composition of maps is a group.

The quotient homomorphism π corresponding to a normal subgroup H of G can be used to further refine Proposition 4.15.

Proposition 4.20 *There is a bijective correspondence between the set of all subgroups of G/H and the set of subgroups of G that contain H, given by the maps $K' \longmapsto \pi^{-1}(K')$ and $G' \longmapsto \pi(G')$. Moreover, normal subgroups of G/H correspond to normal subgroups of G.*

It is now easy to obtain the details of the structure of any cyclic group. Indeed, if $G = \langle g \rangle$ is cyclic then $\mathbb{Z} \ni k \longmapsto g^k \in G$ is a surjective homomorphism. By applying the results above we obtain

Theorem 4.21 (Structure Theorem for Cyclic Groups) *Given a cyclic group G, if G is infinite then we have $G \simeq \mathbb{Z}$, whereas if its order $|G|$ is finite and equal to n then $G \simeq \mathbb{Z}/n\mathbb{Z}$. Moreover, if $G = \langle g \rangle$ is infinite then its subgroups are all those of the form $\langle g^k \rangle$, where k is any positive integer. If $|G| = n < \infty$ then G has a unique subgroup of order d for each divisor d of n.*

Note that the part of the theorem concerning subgroups of a finite cyclic group is a consequence of the bijective correspondence between subgroups of $\mathbb{Z}/n\mathbb{Z}$ and subgroups of $\mathbb{Z}$ containing $n\mathbb{Z}$. That is, we have a new way of proving Remark 4.6.

We know that in a finite group the order of each element divides the order of the group. It is not necessarily true, however, that for each divisor d of the order of G there is an element of order d in G. This does hold when d is a prime divisor and, if G is an Abelian group, it can be sown by applying the fundamental homomorphism theorem and setting up a simple induction.

Theorem 4.22 (Cauchy's Theorem) *Given a finite group G and a prime p that divides the order of G, there exists an element of order p in G.*

The order of any subgroup is also a divisor of the order of the group; however, as for single elements, we cannot say in general whether there exists a subgroup whose order is a certain divisor of the order of the group. We do know that a cyclic group has exactly one subgroup for each divisor of its order, but this case is very special: indeed, we have

Remark 4.23 Given a finite group G which has exactly one subgroup of order d for each divisor d of $|G|$, the group G is cyclic.

4.8 Direct Product of Groups

Given two groups G and H whose operations are $\cdot$ and $\circ$, respectively, we can endow the Cartesian product $G \times H$ with an operation given by $(g_1, h_1)(g_2, h_2) = (g_1 \cdot g_2, h_1 \circ h_2)$ for all $g_1, g_2 \in G$ and $h_1, h_2 \in H$. It is easy to show that this operation makes the set $G \times H$ into a group, which we call the *direct product* of the groups G and H.

We shall see that many properties of the group $G \times H$ can easily be inferred from the properties of the groups G and H. We shall often find that some group, however defined, is isomorphic to a direct product of groups, and thus will obtain information about the group from information about its factors.

The cardinality of the set $G \times H$ is the product of the cardinalities of G and H: if G and H are both finite, then $G \times H$ has order $|G| \cdot |H|$, whereas if at least one of the two groups is infinite then $G \times H$ is infinite. As for the order of its elements, we have the following

Remark 4.24 Given $g \in G$ of finite order m and $h \in H$ of finite order n, the order of the element (g, h) of $G \times H$ is the least common multiple of m and n.

The centre of $G \times H$ is easy to describe in terms of G and H: one can check that $Z(G \times H) = Z(G) \times Z(H)$. In particular,

Remark 4.25 The group $G \times H$ is Abelian if and only if G and H are both Abelian.

Given a subgroup G' of a group G and a subgroup H' of a group H, the group $G' \times H'$ is a subgroup of $G \times H$ in a natural way; note, however, that a subgroup of $G \times H$ need *not* be the direct product of two subgroups. For instance, the diagonal subgroup $\{(g, g) \mid g \in G\}$ of $G \times G$ is not a product of two subgroups if G has more than one element.

We shall now give a consequence of our remark about the order of elements in a direct product of groups. We immediately obtain

Remark 4.26 A direct product of cyclic groups of order m and n is itself cyclic if and only if m and n are relatively prime.

In particular, for the cyclic groups given by the residue classes $\mathbb{Z}/m\mathbb{Z}$ and $\mathbb{Z}/n\mathbb{Z}$, the homomorphism

$$\mathbb{Z}/mn\mathbb{Z} \ni [a]_{mn} \longmapsto ([a]_m, [a]_n) \in \mathbb{Z}/m\mathbb{Z} \times \mathbb{Z}/n\mathbb{Z}$$

is an isomorphism if and only if m and n are relatively prime. Moreover, this homomorphism can be restricted to $(\mathbb{Z}/mn\mathbb{Z})^*$, yielding a homomorphism between multiplicative structures

$$(\mathbb{Z}/mn\mathbb{Z})^* \ni [a]_{mn} \longmapsto ([a]_m, [a]_n) \in (\mathbb{Z}/m\mathbb{Z})^* \times (\mathbb{Z}/n\mathbb{Z})^*.$$

Again, this is an isomorphism if and only if m and n are relatively prime. Note that this discussion further specifies the content of Theorem 3.18 and of its Corollary 3.19.

5 Rings

5.1 Definition and Basic Properties

Rings are sets endowed with two operations whose properties are analogous to those of integer addition and multiplication. A set A endowed with two operations $+$ and $\cdot$ is a *ring* if

(i) A, endowed with the operation $+$, is an Abelian group,
(ii) the operation $\cdot$ is associative,
(iii) the operation $\cdot$ distributes over the operation $+$.

We shall call the operation $+$ *addition* and the operation $\cdot$ *multiplication*. The addition's neutral element is denoted by 0 and called the *zero* of the ring A. Remark that a neutral element for $\cdot$ need not exist; if it does, then we say that A is a ring *with identity* or a *unitary ring*. In a unitary ring, the identity 1 is necessarily unique, and is often called the *one* of the ring. Note that we can have $0 = 1$, in which case it is easy to prove that $A = \{0\}$ and A is called the *zero* ring. If the multiplication of A is commutative then the ring itself is called *commutative*.

The basic rules of integer arithmetic still hold for rings; indeed, we have

Remark 5.1 Given a ring A, for all a and b in A we have: $a0 = 0a = 0$, $a(-b) = (-a)b = -(ab)$, $(-a)(-b) = ab$. Moreover, if A is a unitary ring then $(-1)a = -a$ and $(-1)(-1) = 1$.

The most standard example of a ring is of course $\mathbb{Z}$, which is a commutative unitary ring. Thanks to Theorem 3.17, the set $\mathbb{Z}/n\mathbb{Z}$ of residue classes modulo a positive integer n is also a commutative unitary ring whose zero is $0 + n\mathbb{Z}$ and whose one is $1 + n\mathbb{Z}$.

An element a of a commutative ring A is a *zero divisor* if there exists an element $b \neq 0$ in A such that $ab = 0$. We denote the set of zero divisors of A by $D(A)$. Naturally, zero is a zero divisor in any nonzero commutative ring. A commutative ring with no zero divisors other than 0 is called an *integral domain*. The ring of integers is an integral domain, whereas $\mathbb{Z}/n\mathbb{Z}$ is an integral domain if and only if n is prime. This is a consequence of Proposition 3.11, which tells us that the zero divisors in $\mathbb{Z}/n\mathbb{Z}$ are the classes $a + n\mathbb{Z}$ such that $(a, n) \neq 1$.

An element a of a ring is *nilpotent* if there exists a positive integer k such that $a^k = 0$. For instance, assuming the prime factorisation of n is $n = p_1^{e_1} p_2^{e_2} \cdots p_r^{e_r}$, a class $\bar{a}$ in $\mathbb{Z}/n\mathbb{Z}$ is nilpotent if and only if $p_1 p_2 \cdots p_r$ divides a in $\mathbb{Z}$.

Throughout the rest of the book, we shall often make use of the following remark, which is essentially equivalent to the definition of an integral domain.

Remark 5.2 (Principle of Zero Products) Given two elements a, b in an integral domain, if $ab = 0$ then $a = 0$ or $b = 0$.

An element a of a unitary ring A is *invertible* if there exists an element $b \in A$ such that $ab = ba = 1$. We shall write A^* for the set of all invertible elements in A. For the ring $\mathbb{Z}$ we have $\mathbb{Z}^* = \{1, -1\}$; the sets $\mathbb{Q}$, $\mathbb{R}$ and $\mathbb{C}$ are also commutative unitary rings, and we have $\mathbb{Q}^* = \mathbb{Q} \setminus \{0\}$, $\mathbb{R}^* = \mathbb{R} \setminus \{0\}$ and $\mathbb{C}^* = \mathbb{C} \setminus \{0\}$. One immediately obtains from Proposition 3.15 that the class $a + n\mathbb{Z}$ is invertible in $\mathbb{Z}/n\mathbb{Z}$ if and only if $(a, n) = 1$.

Remark 5.3 Given a unitary ring A, the set A^* of its invertible elements endowed with multiplication is a group; it is an Abelian group if A is commutative. Moreover, we always have $A^* \cap D(A) = \varnothing$.

In the case of a finite ring we can say something more:

Remark 5.4 Given a finite unitary ring A, we have $A = A^* \sqcup D(A)$; that is, every element of A is either invertible or a zero divisor.

If $\mathbb{K}$ is a nonzero commutative unitary ring whose nonzero elements are all invertible, we say that $\mathbb{K}$ is a *field*. In other words, $\mathbb{K} \neq \{0\}$ is a field if and only if $\mathbb{K}^* = \mathbb{K} \setminus \{0\}$; in this case, the set $\mathbb{K}^* = \mathbb{K} \setminus \{0\}$ endowed with multiplication is an Abelian group. Remark that every field is an integral domain, so the principle of zero products does hold for fields. Some examples of fields are $\mathbb{Q}$, $\mathbb{R}$ and $\mathbb{C}$. One can also construct fields with a finite number of elements: given a prime p, the ring $\mathbb{Z}/p\mathbb{Z}$ is actually a field, because every nonzero class is invertible modulo p.

In a commutative ring, one can develop powers of a binomial in a similar way as one does in $\mathbb{Z}$.

Remark 5.5 (Binomial Theorem) Given two elements a and b of a commutative ring A, for all positive integers n we have

$$(a + b)^n = \sum_{k=0}^{n} \binom{n}{k} a^k b^{n-k}.$$

5.2 Subrings, Ideals and Quotients

A subset B of a ring A is a *subring* if the operations $+$ and $\cdot$ can be restricted to B and their restrictions make B itself a ring. Clearly, in order to check whether a nonempty subset B is a subring it is enough to check that $b_1 + b_2$ and $b_1 \cdot b_2$ belong to B for all b_1 and b_2 in B, and that $-b \in B$ for all $b \in B$. An equivalent definition would be that a subset B is a subring if it is a subgroup under addition and is closed under multiplication.

Now, let A' be a ring containing the ring A and let X be a subset of A'. It is evident that an intersection of subrings of A' is itself a subring. We can thus define the ring *generated* by X over A as the intersection of all subrings of A' that contain $A \cup X$. We denote such a ring by $A[X]$: it is the smallest subring of A' that contains $A \cup X$. Given a commutative ring A, it is easy to show that the ring $A[X]$ is the set of all possible sums

$$a_1 y_1 + a_2 y_2 + \cdots + a_k y_k$$

where k is any natural number, $a_1, a_2, \ldots, a_k$ are elements of A and $y_1, y_2, \ldots, y_r$ are products of elements of X. If the set X is finite, say $X = \{x_1, x_2, \ldots, x_r\}$, then we may write $A[x_1, x_2, \ldots, x_r]$ for $A[X]$. In particular, if $X = \{x\}$ and A is commutative, then $A[x]$ is the set of all sums of the form

$$\sum_{h=0}^{k} a_k x^k$$

where k is a natural number and $a_0, a_1, \ldots, a_k$ are elements of A.

A *ring homomorphism* is a map $A \xrightarrow{f} B$ from a ring A to a ring B such that $f(a_1 + a_2) = f(a_1) + f(a_2)$ and $f(a_1 a_2) = f(a_1) f(a_2)$ for all $a_1, a_2 \in A$. Moreover, in the case of A and B being unitary rings, we require that a ring homomorphism f from A to B send the identity 1_A of A to the identity 1_B of B, that is, we impose $f(1_A) = 1_B$. A *ring isomorphism* is a bijective ring homomorphism. As we did for groups, we write $A \simeq B$ if there exists a ring isomorphism from A to B; in this case, we say that A and B are *isomorphic*.

Given a commutative unitary ring A, there is exactly one way to extend the pairing $\mathbb{Z} \ni 1 \longmapsto 1_A \in A$ to a ring homomorphism. In particular, we can always think of the integers as elements of A. Note, however, that the resulting homomorphism is not necessarily injective.

A ring homomorphism is also a group homomorphism between the additive groups $(A, +)$ and $(B, +)$. We can therefore define the *kernel* $\mathrm{Ker}(f)$ of a ring homomorphism f as the kernel of the corresponding homomorphism between additive groups. The kernel $\mathrm{Ker}(f)$ is the set of elements of A whose image under f is 0_B:

$$\mathrm{Ker}(f) = \{a \in A \mid f(a) = 0_B\}.$$

The set $\mathrm{Ker}(f)$ is not only a subring of A, but also such that $a_1 \cdot a_2 \in \mathrm{Ker}(f)$ if a_1 belongs to $\mathrm{Ker}(f)$ or a_2 belongs to $\mathrm{Ker}(f)$: we say that $\mathrm{Ker}(f)$ *absorbs* multiplication. More generally, any additive subgroup I of A that absorbs multiplication is called an *ideal*.

An ideal is to a ring what a normal subgroup is to a group. Indeed, since $(A, +)$ is an Abelian group, an ideal I is a normal subgroup of $(A, +)$. Moreover, if we consider the quotient set A/I given by all additive cosets $a + I$ of the ideal I, that is,

$$A/I = \{a + I \mid a \in A\},$$

this is of course an Abelian group if endowed with the addition of cosets induced
by the operation $+$ of A. But we can also endow A/I with a multiplication given by
$(a_1 + I) \cdot (a_2 + I) = a_1 a_2 + I$ and thus make it into a ring. In other words, given an
ideal I of A, the operations $+$ and $\cdot$ of A pass to the quotient A/I and make it into
a ring. The quotient map

$$A \ni a \longmapsto a + I \in A/I$$

is a surjective ring homomorphism. So, similarly to subgroups, ideals are exactly
the kernels of ring homomorphisms.

We now give another definition which will prove very important. An ideal $M \neq
A$ is *maximal* if it is not properly contained in any ideal other than A itself.

Let A be a ring and let X be a subset of A. The ideal *generated* by X is the
intersection of all ideals of A that contain X. We denote this ideal by (X): it is the
smallest ideal of A that contains X. In a commutative ring A, the ideal X is the set
of all sums

$$a_1 x_1 + a_2 x_2 + \cdots + a_k x_k$$

where k is any natural number, $a_1, a_2, \ldots, a_k$ belong to A and $x_1, x_2, \ldots, x_k$
belong to X. If the set X is finite, say $X = \{x_1, x_2, \ldots, x_r\}$, then we may write
$(x_1, x_2, \ldots, x_r)$ for (X). In particular, if $X = \{x\}$ and A is commutative then the
ideal (x) generated by x is the subset $A \cdot x$ of A consisting of all multiples of the
form ax for some $a \in A$.

5.3 Polynomial Rings

We now give the definition of a polynomial with coefficients in a commutative ring
A. We choose to take an informal approach and rely on the reader's intuition of what
a polynomial is rather than give a rigorous but heavy account involving for instance
the machinery of eventually zero sequences.

Let x be a symbol which we call *indeterminate*; we shall also use the symbols
$1 = x^0, x = x^1, x^2, x^3, \ldots$, which we call *powers* of the indeterminate x. Given a
natural number n and elements $a_0, a_1, a_2, \ldots, a_n$ of a ring A, the formal sum

$$f(x) = a_0 + a_1 x + a_2 x^2 + \cdots + a_n x^n$$

is a *polynomial* in the indeterminate x with *coefficients* $a_0, a_1, \ldots, a_n$ in the ring A.
For convenience, we shall think of $f(x)$ as having coefficients $a_{n+1}, a_{n+2}, \ldots$ equal

to zero. The coefficient a_0 is called the *constant coefficient* of the polynomial. The *zero polynomial*, which we denote by 0, corresponds to the choice of $n = 0$ and $a_0 = 0$; all of its coefficients are zero. Two polynomials $a_0 + a_1x + \cdots + a_nx^n$ and $b_0 + b_1x + \cdots + b_mx^m$ are *equal* if their corresponding coefficients are equal, that is, if $a_0 = b_0$, $a_1 = b_1$ and so on.

Given a nonzero polynomial $f(x) = a_0 + a_1x + \cdots + a_nx^n$, the *degree* of $f(x)$, denoted by $\deg(f)$, is the smallest integer r such that the coefficients $a_{r+1}, a_{r+2}, \ldots$ are all equal to zero. Remark that we are purposefully not defining the degree of the zero polynomial. For a polynomial $f(x)$ of degree r, the coefficient a_r is called the *leading coefficient* of $f(x)$. If A is a unitary ring, a polynomial with coefficients in A is called *monic* if its leading coefficient is 1. A *constant* polynomial is a polynomial which is zero or has degree 0: the polynomial $f(x)$ is constant if and only if $f(x) = a_0$ for some $a_0 \in A$.

The set of polynomials in the indeterminate x with coefficients in A is denoted by $A[x]$. We will now discuss how to use the operations of A to construct operations on $A[x]$ that make $A[x]$ into a ring. Given polynomials $f(x) = a_0 + a_1x + a_2x^2 + \cdots$ and $g(x) = b_0 + b_1x + b_2x^2 + \cdots$, we define the sum of $f(x)$ and $g(x)$ as

$$f(x) + g(x) = (a_0 + b_0) + (a_1 + b_1)x + (a_2 + b_2)x^2 + \cdots$$

It is easy to see that this addition operation makes $A[x]$ into an Abelian group where the neutral element is the zero polynomial and the additive inverse of $a_0 + a_2x + a_2x^2 + \cdots$ is the polynomial $-a_0 - a_1x - a_2x^2 - \cdots$.

In order to construct the multiplication of polynomials, we first describe products of powers of the indeterminate: we set $x^n \cdot x^m = x^{n+m}$ for all natural numbers n and m. We then extend this operation *in a bilinear way*: given the polynomials $f(x)$ and $g(x)$ above, we set

$$\begin{aligned} f(x) \cdot g(x) &= (a_0 + a_1x + a_2x^2 + \cdots) \cdot (b_0 + b_1x + b_2x^2 + \cdots) \\ &= a_0b_0 + (a_0b_1 + a_1b_0)x + (a_0b_2 + a_1b_1 + a_2b_0)x^2 + \cdots \\ &= \sum_k (\sum_{h=0}^{k} a_h b_{k-h})x^k. \end{aligned}$$

It is very easy to show that the operation $\cdot$ is commutative and associative and that, if A is a unitary ring, then the polynomial 1 is the multiplication's neutral element. Moreover, as we can immediately see from the definition, the multiplication distributes over the addition. We therefore have

Proposition 5.6 *Given a commutative ring A, the set $A[x]$ of polynomials with coefficients in A is a commutative ring. Moreover, if A is a unitary ring then $A[x]$ is as well.*

The degree of polynomials has two important properties in relation to addition and multiplication.

Remark 5.7 Given two polynomials $f(x)$ and $g(x)$ in $A[x]$, we have

(i) if $f(x) + g(x)$ is not the zero polynomial then

$$\deg(f + g) \leq \max\{\deg(f), \deg(g)\},$$

(ii) if A is an integral domain and $f(x), g(x) \neq 0$, then $f(x) \cdot g(x) \neq 0$ and $\deg(f \cdot g) = \deg(f)\deg(g)$; in particular, $A[x]$ is an integral domain.

Thanks to the remark above we can immediately identify all invertible elements of $A[x]$ when A is a unitary integral domain.

Corollary 5.8 *If A is a unitary integral domain, then $A[x]^* = A^*$.*

Throughout the rest of the book, we shall often evaluate polynomials at elements of a ring; we now explain what this procedure entails. Let a be a fixed element of A. There exists a unique ring homomorphism

$$A[x] \ni f(x) \overset{v_a}{\longmapsto} f(a) \in A$$

called *evaluation* at a, which replaces all occurrences of x in $f(x)$ with a. A *root* of $f(x)$ is an element a of A such that $f(a) = 0$. Clearly, given a subring A of a ring B, the ring $A[x]$ is a subring of $B[x]$. We can thus evaluate polynomials in $A[x]$ at elements of B. In particular, we can look for roots of a polynomial among the elements of a ring that contains the ring of its coefficients. It is clear that the image of $A[x]$ in B via the map of evaluation at a is the subring $A[a]$ of B generated by a over A.

Given a ring homomorphism $A \overset{f}{\longrightarrow} B$, we can introduce a map between the corresponding polynomial rings by setting

$$A[x] \ni a_0 + a_1 x + \cdots + a_n x^n \longmapsto f(a_0) + f(a_1)x + \cdots + f(a_n)x^n \in B[x].$$

It immediately follows from the definition of the addition and multiplication of polynomials that this map is a ring homomorphism. Remark that this homomorphism never increases the degree of a polynomial, and in particular, if the leading coefficient a_n of a polynomial $p(x)$ does not belong to the kernel of f, then the homomorphism leaves the degree of $p(x)$ unchanged.

5.4 Divisibility of Polynomials

The next few sections will be devoted to the study of the ring $\mathbb{K}[x]$ of polynomials with coefficients in a field $\mathbb{K}$. The ring $\mathbb{K}[x]$ bears a strong resemblance to the ring of integers: for instance, we have a version of Euclidean division for polynomials in $\mathbb{K}[x]$ where the degree plays the role of the absolute value.

Proposition 5.9 (Euclidean Division of Polynomials)   *Given two polynomials* $f(x)$, $g(x)$ *in* $\mathbb{K}[x]$, *where* $f(x)$ *is nonzero, there exist two unique polynomials* $q(x), r(x) \in \mathbb{K}[x]$ *such that* $g(x) = q(x)f(x) + r(x)$ *and* $r(x) = 0$ *or* $\deg(r) <$ $\deg(f)$.

As with integers, $q(x)$ is called the *quotient* of the division of $g(x)$ by $f(x)$, and $r(x)$ is called the *remainder*. If $r(x) = 0$, in which case $g(x) = q(x)f(x)$, then we say that $f(x)$ *divides* $g(x)$, or that $g(x)$ is a *multiple* of $f(x)$. We denote this by $f(x) \mid g(x)$.

Theorem 5.10 (Ruffini's Theorem)   *The element* $a \in \mathbb{K}$ *is a root of* $f(x) \in \mathbb{K}[x]$ *if and only if* $x - a$ *divides* $f(x)$.

Using Ruffini's theorem, we can define the *multiplicity* of a root a of the polynomial $f(x)$ as the natural number k such that $(x - a)^k$ divides $f(x)$ but $(x - a)^{k+1}$ does not divide $f(x)$. A root of multiplicity 1 is called a *simple* root, whereas a root of multiplicity strictly greater than 1 is called a *multiple* root.

If the roots of the polynomial $f(x)$ are $a_1, a_2, \ldots, a_r$ with multiplicities given by $k_1, k_2, \ldots, k_r$ respectively, then the number of roots of $f(x)$ counted with multiplicity is $k_1 + k_2 + \cdots + k_r$. By comparing the degree of the polynomial with the number of factors of the form $x - a$, where a is a root, we obtain

Corollary 5.11   *A nonzero polynomial* $f(x) \in \mathbb{K}[x]$ *has at most* $\deg(f)$ *roots in* $\mathbb{K}$, *counted with multiplicity.*

Let $f(x)$ and $g(x)$ be polynomials in $\mathbb{K}[x]$, not both zero; we call *greatest common divisor* of $f(x)$ and $g(x)$ any polynomial $p(x)$ with the following properties:

(i) $p(x)$ divides both $f(x)$ and $g(x)$,
(ii) if $q(x)$ is a polynomial that divides both $f(x)$ and $g(x)$ then $q(x)$ divides $p(x)$.

In exactly the same way as for the integers, we can show that the greatest common divisor does exist and prove Bézout's identity.

Proposition 5.12   *Given polynomials* $f(x)$, $g(x)$, *not both zero, there exists a greatest common divisor* $m(x)$ *of* $f(x)$ *and* $g(x)$ *in* $\mathbb{K}[x]$. *Moreover, there exist* $h(x), k(x) \in \mathbb{K}[x]$ *such that* $m(x) = h(x)f(x) + k(x)g(x)$.

We can immediately see that Euclid's algorithm works for polynomials exactly as it did for integers, by simply replacing integer division with Euclidean division of polynomials. This allows us to compute a greatest common divisor of two polynomials explicitly and to obtain Bézout's identity.

Note that the greatest common divisor of two polynomials is *not* unique; however, if $h(x)$ and $k(x)$ are greatest common divisors of $f(x)$ and $g(x)$, then there exists $\lambda \in \mathbb{K}^*$ such that $h(x) = \lambda \cdot k(x)$. We can therefore say that the greatest common divisor is unique *up to nonzero constant factors*. In particular, there is a unique monic greatest common divisor, which is often referred to as *the* greatest common divisor and denoted by $(f(x), g(x))$.

As with integers, we say that two polynomials are *relatively prime* if their greatest common divisor is 1.

5.5 Polynomial Factorisation

We continue to explore the analogy between the ring of integers and the ring $\mathbb{K}[x]$ of polynomials with coefficients in a field $\mathbb{K}$. A non-constant polynomial $f(x)$ is *irreducible* if $f(x) = g(x)h(x)$ implies that $g(x)$ is constant or $h(x)$ is constant. It will soon become clear that irreducible polynomials in $\mathbb{K}[x]$ are analogous to primes in $\mathbb{Z}$, and enjoy similar properties. For example, by arguments similar to those used for integers and by replacing the absolute value with the degree, we find

Theorem 5.13 *Every non-constant polynomial has a factorisation in irreducible polynomials; such a factorisation is unique up to the order of the factors, as long as polynomials that differ by a nonzero constant factor are identified.*

Let us now discuss how to factor polynomials with coefficients in some particular fields. We start with polynomials with complex coefficients, for which we have

Theorem 5.14 (Fundamental Theorem of Algebra) *Every non-constant polynomial with complex coefficients has a complex root.*

In spite of its name, there is no purely algebraic proof of this theorem: the completeness of the reals, or some other analytic or topological property, is essential for any proof of this fact.

We shall now list some immediate consequences of the theorem that concern irreducible polynomials with complex or real coefficients.

Corollary 5.15

(i) *A polynomial in $\mathbb{C}[x]$ is irreducible if and only if its degree is 1;*
(ii) *every polynomial in $\mathbb{C}[x]$ has a factorisation as a product of degree 1 polynomials;*
(iii) *any nonzero polynomial $f(x) \in \mathbb{C}[x]$ has exactly $\deg(f)$ roots, counted with multiplicity;*
(iv) *a polynomial is irreducible in $\mathbb{R}[x]$ if and only if it has degree 1 or it has degree 2 and is of the form $x^2 + ax + b$ with $a^2 - 4b < 0$.*

Let us now consider how to factor polynomials with rational coefficients; note that establishing whether or not a polynomial with coefficients in $\mathbb{Q}$ is irreducible is much harder than it is for polynomials with coefficients in $\mathbb{R}$ or $\mathbb{C}$. First of all, let us see how we can relate polynomials with rational coefficients to polynomials with integer coefficients.

Given a nonzero polynomial $f(x) = a_0 + a_1 x + \cdots + a_n x^n$ with integer coefficients, the *content* $c(f)$ of $f(x)$ is the greatest common divisor of the coefficients $a_0, a_1, \ldots, a_n$. A nonzero polynomial with content 1 is called *primitive*.

Naturally, every polynomial with integer coefficients can be expressed as $f(x) = c(f) \cdot f_1(x)$, where $f_1(x)$ is primitive. The key fact that will allow us to pass from $\mathbb{Q}$ to $\mathbb{Z}$ is the following

Lemma 5.16 (Gauss's Lemma) *The product of two primitive polynomials is primitive.*

Gauss's lemma yields the corollaries

Corollary 5.17 *Given $f(x), g(x) \in \mathbb{Z}[x] \setminus \{0\}$ we have $c(f \cdot g) = c(f)c(g)$.*

Corollary 5.18 *Given $f(x), g(x) \in \mathbb{Z}[x]$, assume that $f(x)$ is primitive and that it divides $g(x)$ in $\mathbb{Q}[x]$. Then $f(x)$ divides $g(x)$ in $\mathbb{Z}[x]$.*

Now remark that, given $f(x) \in \mathbb{Q}[x]$, we can always write $f(x) = cf_1(x)$ for some rational number c and some primitive polynomial $f_1(x) \in \mathbb{Z}[x]$. So, when attempting to factor $f(x)$, it is enough to find a factorisation in $\mathbb{Q}[x]$ of the primitive polynomial $f_1(x)$, which has integer coefficients. The following result will guarantee that we can look for a factorisation of $f_1(x)$ in $\mathbb{Z}[x]$ rather than $\mathbb{Q}[x]$.

Corollary 5.19 *A primitive polynomial in $\mathbb{Z}[x]$ is irreducible in $\mathbb{Q}[x]$ if and only if it is irreducible in $\mathbb{Z}[x]$.*

Let us now discuss some criteria for whether or not a polynomial is irreducible in $\mathbb{Z}[x]$. The first allows us to check whether or not the polynomial has roots: of course, by Ruffini's theorem, a nonzero polynomial with a root is irreducible if and only if its degree is one.

Remark 5.20 Let $f(x) = a_0 + a_1 x + a_2 x^2 + \cdots + a_n x^n$ be a polynomial with integer coefficients; any rational root of $f(x)$ can be expressed as a/b for some integer divisor a of a_0 and some integer divisor b of a_n.

This criterion can be used to determine whether or not $f(x)$ has a rational root within a finite number of steps: it is enough to check all rational numbers a/b satisfying the requirements of the remark, of which there is only a finite number.

The next criterion involves reducing the coefficients of the polynomial modulo a prime. Let $f(x)$, as above, be a polynomial in $\mathbb{Z}[x]$ and let p be a prime that does not divide the leading coefficient of $f(x)$. The quotient homomorphism $\mathbb{Z} \longrightarrow \mathbb{Z}/p\mathbb{Z}$ induces a homomorphism $\pi : \mathbb{Z}[x] \longrightarrow (\mathbb{Z}/p\mathbb{Z})[x]$ for which $f(x)$ is not in the kernel.

Remark 5.21 If $\pi(f)$ is irreducible in $(\mathbb{Z}/p\mathbb{Z})[x]$ then $f(x)$ is irreducible in $\mathbb{Z}[x]$.

This remark is especially useful because $\mathbb{Z}/p\mathbb{Z}$ has p elements: in particular, it is finite, hence the number of polynomials of any fixed degree with coefficients in $\mathbb{Z}/p\mathbb{Z}$ is finite. This can make it easy to prove that a polynomial is irreducible in $(\mathbb{Z}/p\mathbb{Z})[x]$. Do note, however, that there exist polynomials that are irreducible in $\mathbb{Z}[x]$ but are reducible modulo p for all primes p, so the remark above does not have a viable converse statement.

A useful criterion for showing the irreducibility of a polynomial with integer coefficients is the following

Proposition 5.22 (Eisenstein's Criterion) *Let* $f(x) = a_0 + a_1 x + \cdots + a_n x^n$ *be a polynomial with integer coefficients; if there exists a prime* p *that does not divide* a_n *but divides* $a_0, a_1, \ldots, a_{n-1}$ *and such that* p^2 *does not divide* a_0, *then* $f(x)$ *is irreducible in* $\mathbb{Z}[x]$, *hence it is irreducible in* $\mathbb{Q}[x]$.

Eisenstein's criterion applies in particular to polynomials of the form $x^n - p$ for some natural number n and some prime p. As a consequence, we have

Remark 5.23 For all natural numbers n there is an infinite number of irreducible polynomials of degree n in $\mathbb{Q}[x]$.

Here is another application of Eisenstein's criterion. Fix a prime p and consider the polynomial

$$f(x) = x^{p-1} + x^{p-2} + \cdots + x + 1 = \frac{x^p - 1}{x - 1},$$

which is called the *pth cyclotomic polynomial*. Remark that the polynomial $g(x) = f(x + 1) = \sum_{k=0}^{p-1} \binom{p}{k+1} x^k$ has constant coefficient p and that all of its coefficients are divisible by p except for the leading coefficient, which is 1. We can thus apply Eisenstein's criterion to obtain that $g(x)$ is irreducible. But then $f(x)$ must also be irreducible, because the map

$$\mathbb{Z}[x] \ni h(x) \longmapsto h(x + 1) \in \mathbb{Z}[x]$$

is a ring isomorphism. We have thus shown

Remark 5.24 For all primes p the polynomial $x^{p-1} + x^{p-2} + \cdots + x + 1$ is irreducible in $\mathbb{Q}[x]$.

5.6 Quotients of Polynomial Rings

Let $f(x)$ be a polynomial in $\mathbb{K}[x]$, where $\mathbb{K}$ is a field, and let $(f(x)) = \mathbb{K}[x] \cdot f(x)$ be the ideal generated by $f(x)$ in $\mathbb{K}[x]$. We wish to study the quotient ring $\mathbb{K}[x]/(f(x))$. In order to accomplish this goal, for the sake of completeness, we need to remind the reader of some definitions and results concerning vector spaces.

A *vector space* over the field $\mathbb{K}$ is an Abelian group $(V, +)$ endowed with a map

$$\mathbb{K} \times V \ni (\lambda, v) \longmapsto \lambda \cdot v \in V,$$

which we call *multiplication by scalars*, with the following properties:

(i) $(\lambda + \mu) \cdot v = \lambda \cdot v + \mu \cdot v$ for all $\lambda, \mu \in \mathbb{K}$ and $v \in V$,
(ii) $\lambda \cdot (u + v) = \lambda \cdot u + \lambda \cdot v$ for all $\lambda \in \mathbb{K}$ and $u, v \in V$,
(iii) $(\lambda\mu) \cdot v = \lambda \cdot (\mu \cdot v)$ for all $\lambda, \mu \in \mathbb{K}$ and $v \in V$,
(iv) $1 \cdot v = v$ for all $v \in V$.

In the context of vector spaces, elements of the field $\mathbb{K}$ are called *scalars* and elements of V are called *vectors*. In particular, the neutral element of V for the operation $+$ is called the *zero vector* and is denoted, as is customary in Abelian groups, by 0.

A natural example of a vector space is the set of *column vectors*, that is, the set $\mathbb{K}^n$, where n is some fixed positive integer, endowed with the addition and the multiplication by scalars given by

$$
\begin{pmatrix} x_1 \\ x_2 \\ \vdots \\ x_n \end{pmatrix} + \begin{pmatrix} y_1 \\ y_2 \\ \vdots \\ y_n \end{pmatrix} = \begin{pmatrix} x_1 + y_1 \\ x_2 + y_2 \\ \vdots \\ x_n + y_n \end{pmatrix}, \qquad \lambda \cdot \begin{pmatrix} x_1 \\ x_2 \\ \vdots \\ x_n \end{pmatrix} = \begin{pmatrix} \lambda x_1 \\ \lambda x_2 \\ \vdots \\ \lambda x_n \end{pmatrix}.
$$

Given k vectors $v_1, v_2, \ldots, v_k$ in a vector space V, a *linear combination* of them is a vector of the form

$$
\lambda_1 v_1 + \lambda_2 v_2 + \cdots + \lambda_k v_k
$$

where $\lambda_1, \lambda_2, \ldots, \lambda_k$ are scalars. If the only way to obtain the zero vector as a linear combination of $v_1, v_2, \ldots, v_k$ is to have $\lambda_1 = \lambda_2 = \cdots = \lambda_k = 0$, then the vectors are *linearly independent*; otherwise, we call them *linearly dependent*.

The vectors $v_1, v_2, \ldots, v_k$ are a set of *generators* for the vector space V if every vector of V is a linear combination of $v_1, v_2, \ldots, v_k$.

A *basis* of a vector space V is a set of generators that are linearly independent. It follows from the definitions that a set $v_1, v_2, \ldots, v_k$ is a basis of V if and only if every vector $v \in V$ can be expressed as a linear combination of $v_1, v_2, \ldots, v_k$ in a unique way. One can show

Theorem 5.25 *Every vector space has a basis. Moreover, any two bases of the same vector space have the same cardinality.*

The common cardinality of all bases of V is called the *dimension* of V and is denoted by dim V. For instance, it is easy to prove that the dimension of $\mathbb{K}^n$ is n.

We shall now make a general remark. Given a ring A containing a field $\mathbb{K}$ as a subring, if we consider the ring addition of A and the restriction to $\mathbb{K} \times A \longrightarrow A$ of the ring multiplication of A, then the ring A becomes a vector space over $\mathbb{K}$. In particular, if we take A to be the quotient ring $\mathbb{K}[x]/(f(x))$, we have

Proposition 5.26 *The quotient set $\mathbb{K}[x]/(f(x))$ is a commutative unitary ring. A set of representatives for the classes modulo $(f(x))$ is given by 0 together with*

all polynomials of degree strictly less than $\deg(f)$. *Moreover,* $\mathbb{K}[x]/(f(x))$ *is a vector space over* $\mathbb{K}$ *of dimension* $n = \deg(f)$. *A basis of* $\mathbb{K}[x]/(f(x))$ *is given by* $1 + (f(x)), x + (f(x)), \ldots, x^{n-1} + (f(x))$.

It is possible to give an explicit description of zero divisors, invertible elements and nilpotent elements in $\mathbb{K}[x]/(f(x))$, exactly as we did in the rings $\mathbb{Z}/n\mathbb{Z}$.

Remark 5.27 The class $g(x) + (f(x))$ is a zero divisor in the ring $\mathbb{K}[x]/(f(x))$ if and only if $g(x)$ and $f(x)$ are not relatively prime. It is invertible if and only if $g(x)$ and $f(x)$ are relatively prime, and it is nilpotent if and only if $g(x)$ is divisible by every irreducible factor of $f(x)$ in $\mathbb{K}[x]$.

We conclude this section with one last analogy between rings of the form $\mathbb{K}[x]/(f(x))$ and quotients $\mathbb{Z}/n\mathbb{Z}$.

Corollary 5.28 *The ring* $\mathbb{K}[x]/(f(x))$ *is a field if and only if* $f(x)$ *is irreducible in* $\mathbb{K}[x]$.

6 Fields

6.1 Characteristic of a Field

Given a field $\mathbb{K}$, consider the unique ring homomorphism $\mathbb{Z} \xrightarrow{\varphi} \mathbb{K}$ such that $\mathbb{Z} \ni 1 \longmapsto 1 \in \mathbb{K}$. It is easy to show that there are only two possibilities for its kernel: either φ is injective or there is a prime p such that $\mathrm{Ker}(\varphi) = p\mathbb{Z}$.

In the first case, $\mathbb{Z}$ is isomorphic to a subring of $\mathbb{K}$ and so $\mathbb{K}$ (being a field) contains an isomorphic copy of $\mathbb{Q}$: we say that $\mathbb{K}$ has *characteristic zero*.

If $\mathrm{Ker}(\varphi) = p\mathbb{Z}$ then $\mathbb{K}$ contains an isomorphic copy of the field $\mathbb{Z}/p\mathbb{Z}$: we say that $\mathbb{K}$ has *characteristic* p or has *positive characteristic*. Remark that in this case

$$p \cdot a = \underbrace{a + a + \ldots + a}_{p \text{ times}} = 0$$

for all a in $\mathbb{K}$. It is clear that a field with a finite number of elements has characteristic p for some prime p, since it does not contain the infinite set $\mathbb{Q}$ and therefore cannot have characteristic 0.

For a field of characteristic p, the binomial theorem takes an especially simple form when applied with exponent p.

Theorem 6.1 (Freshman's Dream) *Given a field* $\mathbb{K}$ *of positive characteristic* p, *we have*

$$(a + b)^p = a^p + b^p$$

for all $a, b \in \mathbb{K}$.

In the special case of the field $\mathbb{Z}/p\mathbb{Z}$, this is Theorem 3.12 about congruences modulo p.

An immediate consequence of the theorem is the fact that, given a finite field $\mathbb{K}$ of characteristic p, the map

$$\mathbb{K} \ni a \overset{F}{\longmapsto} a^p \in \mathbb{K}$$

is an *automorphism* of $\mathbb{K}$, that is, an isomorphism from $\mathbb{K}$ to itself; it is called the Frobenius automorphism.

6.2   Multiplicative Group

Given a field $\mathbb{K}$, the set $\mathbb{K}^* = \mathbb{K} \setminus \{0\}$ endowed with the field multiplication is a group with a very special structure. Remark that, given $a \in \mathbb{K}^*$ whose order is a divisor of some positive integer n, the element a is a root of the polynomial $x^n - 1$. This implies that $\mathbb{K}^*$ cannot contain more than n elements whose order is a divisor of n. It follows immediately from Remark 4.23 that we have

Proposition 6.2 *Any finite subgroup of the multiplicative group $\mathbb{K}^*$ of a field $\mathbb{K}$ is cyclic.*

In particular, for finite fields we have

Corollary 6.3 *The multiplicative group $\mathbb{K}^*$ of a finite field $\mathbb{K}$ is a cyclic group of order $|\mathbb{K}| - 1$.*

6.3   Field Extensions

Given two fields $\mathbb{K}$ and $\mathbb{F}$, we say that $\mathbb{F}$ is an *extension* of $\mathbb{K}$ if $\mathbb{K} \subseteq \mathbb{F}$, that is, if $\mathbb{K}$ is a subfield of $\mathbb{F}$. We will sometimes write $\mathbb{F}/\mathbb{K}$ to mean that $\mathbb{F}$ is an extension of $\mathbb{K}$. As we remarked before, given a field extension $\mathbb{F}/\mathbb{K}$ we have that $\mathbb{F}$ is a vector space over the field $\mathbb{K}$. The *degree* of the extension $\mathbb{F}/\mathbb{K}$, denoted by $[\mathbb{F} : \mathbb{K}]$, is the dimension $\dim_\mathbb{K} \mathbb{F}$ of $\mathbb{F}$ as a vector space over $\mathbb{K}$. We call the extension $\mathbb{F}/\mathbb{K}$ *finite* if it has finite degree.

A *tower of extensions* is a sequence of successive field extensions. In a tower of extensions, the degree is multiplicative:

Proposition 6.4 *Given field extensions $\mathbb{L}/\mathbb{F}$ and $\mathbb{F}/\mathbb{K}$ we have*

$$[\mathbb{L} : \mathbb{K}] = [\mathbb{L} : \mathbb{F}] \cdot [\mathbb{F} : \mathbb{K}].$$

Let $\mathbb{F}/\mathbb{K}$ be a field extension and let X be a subset of $\mathbb{F}$. Since, as for rings, an intersection of subfields of $\mathbb{F}$ is still a subfield, we can define $\mathbb{K}(X)$ as the intersection of all subfields of $\mathbb{F}$ that contain $\mathbb{K} \cup X$. The result is the smallest subfield of $\mathbb{F}$ that contains $\mathbb{K} \cup X$, that is, the set of all quotients of elements in the subring $\mathbb{K}[X]$ of $\mathbb{F}$ generated by X over $\mathbb{K}$. When the set X is finite, say $X = \{a_1, a_2, \ldots, a_k\}$, we may write $\mathbb{K}(a_1, a_2, \ldots, a_k)$ for $\mathbb{K}(X)$.

In particular, given $a \in \mathbb{F}$, the subfield $\mathbb{K}(a)$ generated by a over $\mathbb{K}$ is the set of all quotients of the form

$$\frac{f(a)}{g(a)}$$

for some $f(x)$ and $g(x)$ in $\mathbb{K}[x]$, subject to the condition that $g(a) \neq 0$.

An element a of an extension $\mathbb{F}$ of $\mathbb{K}$ is *algebraic* over $\mathbb{K}$ if there exists a nonzero polynomial $f(x)$ with coefficients in $\mathbb{K}$ such that $f(a) = 0$. A non-algebraic element is called *transcendental*. For instance, the number $\sqrt{2}$ is an element of $\mathbb{R}$ that is algebraic over $\mathbb{Q}$, because it is a root of $x^2 - 2 \in \mathbb{Q}[x]$. In 1882, the German mathematician Ferdinand von Lindemann gave the first proof that π is transcendental over $\mathbb{Q}$.

Given a field $\mathbb{L}$, a subfield $\mathbb{K}$ of $\mathbb{L}$ and two field extensions $\mathbb{E}, \mathbb{F}$ of $\mathbb{K}$ in $\mathbb{L}$, the *composite* of $\mathbb{E}$ and $\mathbb{F}$ is $\mathbb{E} \cdot \mathbb{F} = \mathbb{E}(\mathbb{F}) = \mathbb{F}(\mathbb{E})$; in other words, $\mathbb{E} \cdot \mathbb{F}$ is the smallest extension of $\mathbb{K}$ in $\mathbb{L}$ that contains both $\mathbb{E}$ and $\mathbb{F}$.

A field extension $\mathbb{F}/\mathbb{K}$ is *algebraic* if all elements of $\mathbb{F}$ are algebraic over $\mathbb{K}$. It is easy to show that $\mathbb{Q}(\sqrt{2})/\mathbb{Q}$ is an algebraic extension; on the contrary, the extension $\mathbb{R}/\mathbb{Q}$ is clearly not algebraic, since $\pi \in \mathbb{R}$ is not algebraic over $\mathbb{Q}$.

We can determine whether an element is algebraic or transcendental by studying the evaluation homomorphism at that element. We have

Remark 6.5 Let $\mathbb{F}/\mathbb{K}$ be a field extension and consider $a \in \mathbb{F}$. The element a is algebraic over $\mathbb{K}$ if and only if the evaluation homomorphism

$$\mathbb{K}[x] \ni f(x) \xmapsto{v_a} f(a) \in \mathbb{F}$$

has a nontrivial kernel. Equivalently, a is transcendental if and only if the above homomorphism is injective.

Given an element $a \in \mathbb{F}$ that is algebraic over $\mathbb{K}$, let $\mu(x)$ be a monic polynomial of minimal degree in the kernel $\mathrm{Ker}(v_a)$ of the evaluation at a.

Proposition 6.6 *If $a \in \mathbb{F}$ is algebraic over $\mathbb{K}$ then μ is irreducible in $\mathbb{K}[x]$ and generates the kernel $\mathrm{Ker}(v_a)$ of the evaluation at a. Moreover, μ is the only irreducible monic polynomial to have evaluation zero at a.*

Because of this proposition, we call $\mu(x)$ the *minimal polynomial* of a over $\mathbb{K}$. We know from the definition that $\mathrm{Ker}(v_a) = (\mu(x)) = \mu(x) \cdot \mathbb{K}[x]$, and therefore

Remark 6.7 Given an element $a \in \mathbb{F}$ that is algebraic over $\mathbb{K}$, the ring $\mathbb{K}[a]$ is isomorphic to $\mathbb{K}[x]/(\mu(x))$.

Moreover, since $\mu(x)$ is irreducible, we can use Bézout's identity to obtain the inverse of every nonzero element of $\mathbb{K}[a]$ and conclude that $\mathbb{K}[a]$ is a field. Therefore,

Corollary 6.8 *If $a \in \mathbb{F}$ is algebraic over $\mathbb{K}$ then $\mathbb{K}(a) = \mathbb{K}[a]$.*

Another consequence of Remark 6.5 is the way that the degree of an extension depends on whether its elements are algebraic or transcendental:

Corollary 6.9 *The element $a \in \mathbb{F}$ is algebraic over $\mathbb{K}$ if and only if $\mathbb{K}(a)/\mathbb{K}$ is a finite extension. If this is the case then the degree $[\mathbb{K}(a) : \mathbb{K}]$ is equal to the degree of the minimal polynomial of a over $\mathbb{K}$.*

It is therefore clear that a finite extension does not contain any transcendental elements:

Corollary 6.10 *Any finite extension $\mathbb{F}/\mathbb{K}$ is algebraic.*

Another important property of algebraic extensions, which is essentially a consequence of the corollary above and of the fact that the degree is multiplicative in towers, is the following

Proposition 6.11 *Given algebraic field extensions $\mathbb{L}/\mathbb{F}$ and $\mathbb{F}/\mathbb{K}$, the extension $\mathbb{L}/\mathbb{K}$ is algebraic.*

In particular, we have

Remark 6.12 Given elements $a_1, a_2, \ldots, a_k$ in $\mathbb{F}$ that are algebraic over $\mathbb{K}$, the ring $\mathbb{K}[a_1, a_2, \ldots, a_k]$ is a field of finite degree over $\mathbb{K}$.

Let us now discuss how the greatest common divisor of polynomials is affected by field extensions. Given a field extension $\mathbb{F}/\mathbb{K}$ and two polynomials $f(x)$ and $g(x)$ in $\mathbb{K}[x]$, we can compute the greatest common divisor of $f(x)$ and $g(x)$ both in $\mathbb{K}[x]$ and in $\mathbb{F}[x]$. We have

Remark 6.13 The greatest common divisor of $f(x)$ and $g(x)$ in $\mathbb{K}[x]$ is the same as their greatest common divisor in $\mathbb{F}[x]$.

We can therefore conclude that polynomials that are relatively prime in $\mathbb{K}[x]$ remain relatively prime in $\mathbb{F}[x]$.

6.4   Splitting Field

A field $\mathbb{K}$ is *algebraically closed* if all non-constant polynomials in $\mathbb{K}[x]$ have a root in $\mathbb{K}$. For example, $\mathbb{C}$ is algebraically closed but $\mathbb{R}$ is not, because the polynomial $x^2 + 1$ does not have any roots in $\mathbb{R}$. Remark that if $\mathbb{K}$ is algebraically closed, then by Ruffini's theorem the existence of one root of $f(x)$ implies that $f(x)$ factors as a product of degree one polynomials in $\mathbb{K}[x]$.

An extension Ω of a field $\mathbb{K}$ is an *algebraic closure* of $\mathbb{K}$ if $\Omega/\mathbb{K}$ is an algebraic extension and Ω is algebraically closed. For example, $\mathbb{C}$ is the algebraic closure of $\mathbb{R}$, but it is not an algebraic closure of $\mathbb{Q}$ because $\mathbb{C}/\mathbb{Q}$ is not an algebraic extension: not even the subextension $\mathbb{R}/\mathbb{Q}$ is algebraic.

In order to prove the following fact about algebraic closures, some rather advanced notions in mathematical logic are essential.

Theorem 6.14 *Every field has an algebraic closure. Any two algebraic closures of the same field $\mathbb{K}$ are isomorphic, and there is an isomorphism between them that fixes every point of $\mathbb{K}$.*

We now wish to state a criterion for a polynomial to have a multiple root, and in order to do this we need to introduce the concept of derivative. We shall give a definition that does not rely on the idea of limit we are familiar with from Analysis, and thus does not require a topology on the field $\mathbb{K}$ of the coefficients. Given a polynomial $f(x) = a_0 + a_1 x + \cdots + a_n x^n$ in $\mathbb{K}[x]$, its *derivative* is $f'(x) = a_1 + 2a_2 x + \cdots + n a_n x^{n-1}$. This definition coincides with the standard definition from Analysis for polynomials with real coefficients. It therefore should not be surprising that the following properties hold:

(i) $(f(x) + g(x))' = f'(x) + g'(x)$ for all $f(x), g(x) \in \mathbb{K}[x]$,
(ii) $(\lambda f(x))' = \lambda f'(x)$ for all $\lambda \in \mathbb{K}$ and $f(x) \in \mathbb{K}[x]$,
(iii) $(f(x)g(x))' = f'(x)g(x) + f(x)g'(x)$ for all $f(x), g(x) \in \mathbb{K}[x]$.

The last property is Leibniz's rule for the derivative of a product. It implies that, if an element a in the algebraic closure Ω of a field $\mathbb{K}$ is a multiple root of a polynomial $f(x) \in \mathbb{K}[x]$, then $(x - a)$ is a common factor of $f(x)$ and $f'(x)$ in $\Omega[x]$. But, since the greatest common divisor is unaffected by field extensions, $f(x)$ and $f'(x)$ must also have a common factor in $\mathbb{K}[x]$.

Remark 6.15 (Derivative Criterion for Multiple Roots)   A polynomial $f(x)$ in $\mathbb{K}[x]$ has a multiple root in the algebraic closure of $\mathbb{K}$ if and only if $f(x)$ and its derivative are not relatively prime in $\mathbb{K}[x]$.

Now, let $f(x) \in \mathbb{K}[x]$ be a polynomial and let Ω be an algebraic closure of $\mathbb{K}$. In particular, Ω contains all roots $a_1, a_2, \ldots, a_k$ of $f(x)$. The *splitting field* of $f(x)$ in Ω is the field $\mathbb{K}(a_1, a_2, \ldots, a_k)$, whose degree over $\mathbb{K}$ is at most $\deg(f)!$.

The splitting field of $f(x)$ is the smallest subfield of Ω in which the polynomial $f(x)$ factors as a product of linear polynomials. Once the algebraic closure Ω is fixed, the splitting field is uniquely determined by the polynomial $f(x)$. Moreover, the choice of the field Ω is just a technical step: splitting fields constructed from different algebraic closures are isomorphic.

As an example, we will describe the splitting field of the polynomial $x^n - 1$ over $\mathbb{Q}$. The complex roots of this polynomial are of course the nth roots of unity: if we set $\zeta_n = e^{2\pi i/n} = \cos(2\pi/n) + i \sin(2\pi/n)$, the roots are $1, \zeta_n, \zeta_n^2, \ldots, \zeta_n^{n-1}$. Note that this set of roots is a cyclic subgroup of order n of the group $\mathbb{C}^*$; a generator of this group – for example, ζ_n—is called a *primitive* nth root of unity. Clearly, there are $\phi(n)$ primitive nth roots of unity, all of them of the form ζ_n^h with $(h, n) = 1$.

In conclusion, the splitting field of $x^n - 1$ over $\mathbb{Q}$ is the field $\mathbb{Q}(\zeta_n)$, which is called the nth *cyclotomic extension* of $\mathbb{Q}$. For instance, for $n = 4$ we get $\mathbb{Q}(i)$ and, since $x^2 + 1$ is the minimal polynomial of i over $\mathbb{Q}$, the degree of $\mathbb{Q}(i)$ over $\mathbb{Q}$ is 2.

6.5   Finite Fields

Remember that a field is *finite* if it has a finite number of elements. As remarked previously, a finite field cannot have characteristic zero and thus always has positive characteristic, which will henceforth be denoted by p.

Let $\mathbb{F}$ be a finite field. Note that $\mathbb{Z}/p\mathbb{Z}$ is a subfield of $\mathbb{F}$; in particular, $[\mathbb{F} : \mathbb{Z}/p\mathbb{Z}] = r < \infty$ for some r and $\mathbb{F}$ has p^r elements. This implies that $\mathbb{F}^*$ is a (necessarily cyclic) group with $p^r - 1$ elements, so $a^{p^r-1} = 1$ for all $a \neq 0$, and thus $a^{p^r} - a = 0$ for all elements a in $\mathbb{F}$.

Another useful fact to remark is that the algebraic closures of $\mathbb{F}$ and $\mathbb{Z}/p\mathbb{Z}$ must coincide, because $\mathbb{F}$ is a finite extension of $\mathbb{Z}/p\mathbb{Z}$.

Now, consider the polynomial $x^{p^r} - x$; having fixed an algebraic closure Ω of $\mathbb{Z}/p\mathbb{Z}$, we find that the field $\mathbb{F}$ is the set of all roots of this polynomial in Ω. We also know that $x^{p^r} - x$ has no multiple roots by the derivative criterion, and thus Remark 4.23 yields a proof of

Theorem 6.16 *Given a prime p and a natural number r, there exists a field with p^r elements, which is the splitting field of the polynomial $x^{p^r} - x$ in some algebraic closure Ω of $\mathbb{Z}/p\mathbb{Z}$. It is the unique subfield of Ω having p^r elements.*

Having fixed an algebraic closure Ω, we denote by $\mathbb{F}_{p^r}$ its unique subfield with p^r elements. It is customary to set $q = p^r$, and from here on we will make frequent use of this convention, even without saying so explicitly.

Lest any confusion arise, remark that $\mathbb{F}_p \simeq \mathbb{Z}/p\mathbb{Z}$ but that in general $\mathbb{F}_q$ is *not* isomorphic to $\mathbb{Z}/q\mathbb{Z}$ if r is greater than 1; indeed, $\mathbb{Z}/q\mathbb{Z}$ is not even a field: the class $p + q\mathbb{Z}$ is a zero divisor.

The uniqueness of the splitting field implies

Corollary 6.17 *Any two finite fields with the same number of elements are isomorphic.*

Inside Ω we have all of the fields $\mathbb{F}_{p^r}$, where r takes all possible natural values. The next proposition states that the relation of inclusion between these subfields mimics the relation of divisibility between integers.

Proposition 6.18 *We have the inclusion $\mathbb{F}_{p^r} \subseteq \mathbb{F}_{p^s}$ if and only if r divides s.*

We shall now discuss splitting fields of polynomials over finite fields. Let $f(x)$ be an irreducible polynomial of degree r in $\mathbb{F}_q[x]$ and let a be a root of $f(x)$ in an algebraic closure Ω of $\mathbb{F}_p$. Since $f(x)$ is irreducible, $\mathbb{F}_q(a)$ has degree r over $\mathbb{F}_q$.

But there is only one subfield of Ω with q^r elements, so $\mathbb{F}_q(a)$ contains all the roots of $f(x)$. We therefore have

Proposition 6.19 *Given an irreducible polynomial $f(x) \in \mathbb{F}_q[x]$ of degree r and a root a of $f(x)$ in some algebraic closure, the field $\mathbb{F}_q(a) = \mathbb{F}_{q^r}$ is the splitting field of $f(x)$.*

As a corollary, we obtain a description of the splitting field of any polynomial.

Corollary 6.20 *Given a polynomial $f(x)$ with coefficients in $\mathbb{F}_q$ such that*

$$f(x) = f_1(x) f_2(x) \cdots f_k(x)$$

where the factors are irreducible, the splitting field of $f(x)$ is $\mathbb{F}_{q^m}$, where m is the least common multiple of the degrees of the irreducible factors $f_1(x), f_2(x), \ldots, f_k(x)$.

We shall now discuss an application of the theory developed so far: given a natural number n and a prime p, we shall study the splitting field of the polynomial $x^n - 1$ over $\mathbb{F}_p$. If $n = p^e \cdot n'$, where p does not divide n', then $x^n - 1 = (x^{n'} - 1)^{p^e}$ by Theorem 6.1, so the polynomials $x^n - 1$ and $(x^{n'} - 1)^{p^e}$ have the same splitting field. We can therefore assume without loss of generality that p does not divide n. The following simple remark is of fundamental importance:

Remark 6.21 Let n be an integer that is not divisible by p and let Ω be an algebraic closure of $\mathbb{F}_p$. The roots of the polynomial $x^n - 1$ form a cyclic subgroup of order n of Ω^*.

Consequently, the polynomial $x^n - 1$ factors in $\mathbb{F}_{p^r}$ if and only if $\mathbb{F}_{p^r}^*$ contains an element of order n, that is, if and only if n divides $p^r - 1$. Therefore, we have

Theorem 6.22 (Cyclotomic Extensions in Positive Characteristic) *Let n be an integer that is not divisible by the prime p. The splitting field of $x^n - 1$ is $\mathbb{F}_{p^r}$, where r is the order of p in the multiplicative group $(\mathbb{Z}/n\mathbb{Z})^*$.*

We now move on to another application. Studying the solvability of quadratic equations in $\mathbb{F}_p$ is already quite interesting from a mathematical standpoint; the fact that this was the first case to be tackled also makes it of some historical interest.

If $p = 2$ then the only quadratic polynomials are: x^2, $x^2 + 1 = (x + 1)^2$, $x^2 + x = x(x + 1)$ and $x^2 + x + 1$. Among these, only the last one, which has no roots, is irreducible. From now on, we will assume that $p \neq 2$.

The well-known quadratic formula still holds in $\mathbb{F}_p$ if $p \neq 2$: the solutions of the equation $ax^2 + bx + c = 0$ are

$$\frac{-b \pm \sqrt{\Delta}}{2a}$$

if $\Delta = b^2 - 4ac$ is a square in $\mathbb{F}_p$, and there are no solutions in $\mathbb{F}_p$ otherwise. The problem is thus that of determining which $a \in \mathbb{F}_p$ are squares (the *quadratic residues*) and which are not (the *quadratic non-residues*). In order to do this, we introduce the *Legendre symbol*

$$\mathbb{F}_p^* \ni a \longmapsto \left(\frac{a}{p}\right) = \begin{cases} +1 \text{ if } a \text{ is a quadratic residue,} \\ -1 \text{ if } a \text{ is a quadratic non-residue.} \end{cases}$$

It is easy to determine whether or not -1 is a quadratic residue; we have

$$\left(\frac{-1}{p}\right) = (-1)^{\frac{p-1}{2}}$$

for all odd primes p, and clearly $-1 = 1$ is a quadratic residue in $\mathbb{F}_2$. The proof of this fact can be found in Preliminary Exercise 18. A similar, slightly more complicated formula exists for $\left(\frac{2}{p}\right)$. Finally, a crucial instrument for computing Legendre symbols is the law of quadratic reciprocity, conjectured by Euler and Lagrange and proven by Gauss in 1796, which establishes a remarkable symmetry of $\left(\frac{q}{p}\right)$ when p and q are both odd primes.

7 Preliminary Exercises

This section contains several preliminary exercises, which are meant to be solved before those taken from exam papers. Their statements and the techniques used in their solutions will often serve as useful tools to solve later exercises.

Exercise 1 Let X be a nonempty finite set. Show that the number of subsets of X whose cardinality is even is the same as the number of subsets of X whose cardinality is odd.

Solution 1 We shall prove this by induction on the cardinality n of the set X. The base case is $n = 1$, where the only subset of even cardinality is the empty set and the only subset of odd cardinality is X itself. So the claim does hold for $n = 1$.

Assume that the statement holds for $|X| = n$; we shall prove it for $|X| = n + 1$. A set X of $n + 1$ elements can be written in the form $X = Y \cup \{z\}$, where Y is a set of n elements and $z \notin Y$. The subsets A of X are of two types: either $z \in A$ or $z \notin A$. In other words, any subset A of X is of the form $A = B \cup Z$, where B is a subset of Y and $Z = \varnothing$ or $Z = \{z\}$.

It is clear that the cardinality of A is even in the following two cases: if B has even cardinality and $Z = \varnothing$, or B has odd cardinality and $Z = \{z\}$. The induction hypothesis implies that each case covers 2^{n-1} subsets, so we have $2 \cdot 2^{n-1} = 2^n$ subsets of even cardinality. Since $2^n = \frac{1}{2} \cdot 2^{n+1}$ is half of the total number of subsets of X, the number of subsets of even cardinality is the same as the number of subsets of odd cardinality.

Solution 2 Let us count the subsets of X according to their cardinality and use the binomial expansion

$$(a+b)^n = \sum_{k=0}^{n} \binom{n}{k} a^k b^{n-k}.$$

If we set $a = 1$ and $b = -1$ we get

$$0 = 0^n = (1-1)^n = \sum_{k=0}^{n} (-1)^{n-k} \binom{n}{k}.$$

The sum of the binomial coefficients with even k in the expression above, which corresponds to the number of subsets of even cardinality, is the same as the sum of those with odd k, which corresponds to the number of subsets of odd cardinality.

Exercise 2 Find the number of permutations σ of $\{1, \ldots, n\}$ such that $\sigma(x) \neq x$ for all $x \in \{1, \ldots, n\}$.

Solution The best way to solve this exercise is via the inclusion-exclusion principle. The total number of permutations of $\{1, \ldots, n\}$ is $n!$. We shall count permutations σ that do not satisfy the requirement, that is, permutations such that there is x for which $\sigma(x) = x$. The correct answer will be the difference between $n!$ and the number obtained.

For $i = 1, 2, \ldots, n$, let P_i be the set of permutations σ of $\{1, \ldots, n\}$ such that $\sigma(i) = i$. What we wish to find is the number of elements in $P_1 \cup \cdots \cup P_n$, that is, permutations for which some x is such that $\sigma(x) = x$.

We compute the cardinality $|P_1 \cup \cdots \cup P_n|$ using the inclusion-exclusion principle.

For $i = 1, \ldots, n$ (that is, for each of n possible values) the cardinality of P_i is $(n-1)!$, because the permutations σ in P_i are such that $\sigma(i) = i$ and all elements $j \neq i$ can be permuted with no constraints.

For each pair $\{i, j\} \subseteq \{1, \ldots n\}$ (that is, in each of $\binom{n}{2}$ cases) the cardinality of $P_i \cap P_j$ is $(n-2)!$, because every $\sigma \in P_i \cap P_j$ satisfies $\sigma(i) = i$ and $\sigma(j) = j$, but the other $n - 2$ elements can be permuted with no constraints.

Applying this argument to all possible intersections of the P_i's, one gets the formula

$$|P_1 \cup \cdots \cup P_n| = n(n-1)! - \binom{n}{2}(n-2)! + \binom{n}{3}(n-3)! + \cdots$$

$$\cdots + (-1)^{n-2}\binom{n}{n-1}1! + (-1)^{n-1}\binom{n}{n}0!$$

$$= n!\left(\frac{1}{1!} - \frac{1}{2!} + \frac{1}{3!} + \cdots + (-1)^{n-1}\frac{1}{n!}\right).$$

The cardinality we want is therefore given by the difference

$$n!\left(\frac{1}{2!} - \frac{1}{3!} + \cdots + (-1)^n \frac{1}{n!}\right) = n!\left(\frac{1}{0!} - \frac{1}{1!} + \frac{1}{2!} - \frac{1}{3!} + \cdots + (-1)^n \frac{1}{n!}\right).$$

⟦This exercise has a fascinating alternative formulation, known as *the matching problem*. Suppose someone gives a secretary the task of mailing n letters to n different addresses, without telling them which letter has to go to which address. What is the probability that no letter is mailed to the correct address?

It is known that the series

$$\sum_{k=0}^{\infty} (-1)^k \frac{1}{k!}$$

converges to the value $1/e$, where e is Euler's number. Therefore, the probability we are considering tends to the limit $1/e$ for large n. ⟧

Exercise 3 Let X be a set of m elements and let Y be a set of n elements. Find the number of surjective maps $X \longrightarrow Y$.

Solution Again, we will use the inclusion-exclusion principle. The total number of maps from X to Y is n^m. We shall find the number of maps that are *not* surjective and then take the difference to obtain the number of surjective maps.

Let $Y = \{y_1, \ldots, y_n\}$. A map $f \colon X \to Y$ is not surjective if there is at least one element $y_i \in Y$ that does not belong to its image. In other words, the set of non-surjective maps $f \colon X \to Y$ is the union of the sets $F_1, \ldots, F_n$, where F_i is the set of maps whose image does not contain y_i.

The cardinality of each set F_i (there are n such sets, one for each value of i) is the number of maps from X to the set $Y \setminus \{y_i\}$ and therefore equal to $(n-1)^m$. The cardinality of each of the $\binom{n}{2}$ sets $F_i \cap F_j$ is the number of maps from X to the set $Y \setminus \{y_i, y_j\}$, that is, $(n-2)^m$, and so on. In conclusion, we have

$$|F_1 \cup \cdots \cup F_n| = n(n-1)^m - \binom{n}{2}(n-2)^m + \cdots + (-1)^{n-1}\binom{n}{n-1}1^m$$

and therefore the number of surjective maps is given by the difference

$$n^m - n(n-1)^m + \binom{n}{2}(n-2)^m + \cdots + (-1)^n \binom{n}{n-1}1^m.$$

⟦It is interesting and nontrivial to note that for $m < n$ the formula above gives a result of zero, that is, it confirms that there are no surjective maps $f \colon X \to Y$. ⟧

Exercise 4 Let n and k be positive integers. Find the number of solutions of the equation

$$x_1 + \cdots + x_k = n,$$

where all x_i's are positive integers.

Solution There is a bijective correspondence between solutions $(x_1, \ldots, x_k)$ of the equation above and k-tuples $(y_1, y_2, \ldots, y_k)$ constructed as

$$y_1 = x_1$$
$$y_2 = x_1 + x_2$$
$$\vdots$$
$$y_{k-1} = x_1 + x_2 + \cdots + x_{k-1}$$
$$y_k = x_1 + x_2 + \cdots + x_k = n.$$

Since $y_1 < y_2 < \cdots < y_k = n$ and y_k is fixed, the desired k-tuples are as many as the subsets $\{y_1, \ldots, y_{k-1}\}$ of the set $\{1, \ldots, n-1\}$, that is, $\binom{n-1}{k-1}$. The answer is therefore $\binom{n-1}{k-1}$.

⟦There are many possible variants of this exercise and all of them can be solved by similar methods. For example, one can consider the equation

$$x_1 + \cdots + x_k \leq n \,.$$

In this case, the k-tuples $\{y_1, \ldots, y_k\}$ no longer need to satisfy the requirement that $y_k = n$, so they correspond to the subsets of k elements of a set of n elements, and thus there are $\binom{n}{k}$ of them. One can also consider the solutions of the equation

$$x_1 + \cdots + x_k = n \,,$$

where the x_i's are *non-negative* integers. Setting $y_i = x_i + 1$, we find that $y_i > 0$ and that, since each y_i is increased by 1 with respect to the previous construction, the sum $y_1 + \ldots + y_k$ is increased by k, that is,

$$y_1 + \cdots + y_k = n + k \,.$$

Therefore, the number of solutions in this case is $\binom{n+k-1}{k-1}$. ⟧

Exercise 5 Find the number of positive divisors of the positive integer n in terms of its prime factorisation.

Solution Let $n = p_1^{a_1} \cdots p_k^{a_k}$ be the factorisation of n as a product of powers of distinct primes $p_1, \ldots, p_k$. Any divisor d of n will factor as $d = p_1^{b_1} \cdots p_k^{b_k}$ where the integers b_i satisfy the inequality $0 \leq b_i \leq a_i$. So there are $a_i + 1$ possible choices for each exponent b_i, and thus the number of positive divisors of n is

$$\prod_{i=1}^{k} (a_1 + 1) \,.$$

Exercise 6 Let n be a positive integer, let p be a prime and, given a real number x, let $\lfloor x \rfloor$ be the integer part of x, that is, the largest integer m such that $m \leq x$.

Show that

$$\sum_{h=0}^{\infty} \left\lfloor \frac{n}{p^h} \right\rfloor$$

is the power of p that exactly divides $n!$.

Solution First of all, remark that the sum above has a finite number of nonzero terms, as clearly we have $\frac{n}{p^h} < 1$ and thus $\lfloor \frac{n}{p^h} \rfloor = 0$ for h large enough.

Instead of counting the number of times that p divides each integer k with $1 \leq k \leq n$, for each h we count the number of integers between 1 and n that are divisible by p^h. Summing the quantities obtained in this way, the integer k contributes 1 for each integer h such that $p^h \mid k$, that is, it is counted as many times as the exponent of the largest power of p that divides it.

For each h, the number of integers k with $1 \leq k \leq n$ such that $p^h \mid k$ is equal to

$$\left\lfloor \frac{n}{p^h} \right\rfloor$$

so, by the trick described, the desired formula follows.

Exercise 7 Let a, b, c be integers, with a, b not both zero. Describe the set of solutions $(x, y) \in \mathbb{Z}^2$ of the equation $ax + by = c$.

Solution Let us first determine for which integers c the equation has at least one solution. From Bézout's identity, we know that there exists a pair (x_0, y_0) of integers such that $ax_0 + by_0 = m$, where $m = (a, b)$ is the greatest common divisor of a and b. If c is a multiple of m, say $c = km$, then multiplying this identity by k yields $a(kx_0) + b(ky_0) = km = c$, so the equation has a solution.

Conversely, assume that the equation has a solution (x_1, y_1). Then $ax_1 + by_1 = c$ and, since $m \mid a$ and $m \mid b$, we have that m must also divide c. In conclusion, the equation has at least a solution if and only if $(a, b) \mid c$.

Now assume that $(a, b) \mid c$ and write $a = ma_1$, $b = mb_1$, $c = mc_1$, so that $(a_1, b_1) = 1$. By simplifying, we get the equation $a_1 x + b_1 x = c_1$. We can produce a solution (x_0, y_0) of the equation $a_1 x + b_1 y = 1$ using Euclid's algorithm; we then have a particular solution of $ax_1 + by_1 = c_1$: for example, $(x, y) = (c_1 x_0, c_1 y_0)$. Letting (x', y') be any solution of $a_1 x + b_1 y = c_1$, by subtracting the two equations

$$a_1 x' + b_1 y' = c_1$$
$$a_1 x_0 + b_1 y_0 = c_1$$

we get

$$a_1(x' - x_0) = b_1(y_0 - y').$$

Since $(a_1, b_1) = 1$, this implies $a_1 \mid y_0 - y'$ and $b_1 \mid x' - x_0$. So we can write $x' - x_0 = kb_1$ for some integer k, and necessarily $y' - y_0 = -ka_1$. On the other hand, substituting the values

$$x' = x_0 + kb_1, \quad y' = y_0 - ka_1$$

into the original equation, it is easy to see that it is satisfied.

The general solution is therefore given by

$$\begin{cases} x = x_0 + kb_1 \\ y = y_0 - ka_1. \end{cases}$$

Exercise 8 Let m/n be a rational number with $(m, n) = 1$ and with $n = 2^e 5^f n'$ for some integer n' such that $(10, n') = 1$. Show that the decimal expansion of m/n is eventually periodic with a transient of length $\max\{e, f\}$.

Solution Since $(n', 2^e 5^f) = 1$, the linear Diophantine equation

$$2^e 5^f x + n' y = m$$

has a solution. Let (a, b) be one such solution; since $(m, n) = 1$, we necessarily have $(a, n') = (b, 2^e 5^f) = 1$. Dividing by n, we get

$$\frac{a}{n'} + \frac{b}{2^e 5^f} = \frac{m}{n}.$$

We perform the Euclidean division of a by n' and obtain $a = sn' + r$. Note that $(r, n') = 1$. The digits after the decimal point of a/n' are exactly the same as those of r/n'. Remember that, with the usual process of division, the $(i + 1)$th digit after the decimal point in r/n' is completely determined by the remainder of the division of $10^i r$ by n'.

Since $(10, n') = 1$, the sequence $(10^i)_i$ is eventually periodic modulo n'; in fact, it is periodic, that is, it has no preperiodic part, and its period is the multiplicative order h of 10 modulo n'. Multiplying by r, we find that the sequence $(10^i r)_i$ is also periodic modulo n', with a period h' that divides h. If we set r' to be the inverse of r modulo n', the sequence $(10^i rr')_i$ is periodic modulo n', so the period h divides h'. It follows that $h' = h$.

Now, set $k = \max\{e, f\}$; the number $\beta = b/2^e 5^f$ can be written as $b'/10^k$, with $b' \in \mathbb{Z}$, and has exactly k digits after the decimal point, because $10^k \beta \in \mathbb{Z}$, whereas $10^{k-1} \beta \notin \mathbb{Z}$.

Sum a/n' and β: the former has a periodic decimal expansion and the latter has exactly k digits after the decimal point in its decimal expansion. Therefore, their sum has a transient of length k after the decimal point, and then the digits will repeat periodically.

Exercise 9 Let g be an element of order n in a group. Show that for each positive integer k the order of g^k is $n/(k, n)$.

Solution Given a positive integer m, we have $(g^k)^m = e$ if and only if $km \equiv 0 \pmod{n}$, that is, if and only if $m \equiv 0 \pmod{n/(k, n)}$. Therefore, the smallest positive integer m such that $(g^k)^m = e$ is $n/(k, n)$.

Exercise 10 Show that if an Abelian group contains an element of order m and an element of order n then it contains an element of order $[m, n]$.

Solution Let G be an Abelian group as in the statement, let $g \in G$ be an element of order m and let $h \in G$ be an element of order n. First, assume that $(m, n) = 1$, which implies $[m, n] = mn$; we show that $z = gh$ has order mn. Indeed, we have $z^{mn} = g^{mn} h^{mn} = e \cdot e = e$, so $\text{ord}(z) \mid mn$. Moreover, the orders of the subgroups of G generated by g and by h are relatively prime, so their intersection consists of the identity only. If $z^k = e$, that is, if $g^k h^k = e$, then $g^k = h^{-k}$ and thus $g^k = h^{-k} = e$. It follows that $m \mid k$ and $n \mid k$, hence $mn \mid k$, that is, $mn \mid \text{ord}(z)$.

Now we deal with the general case: assume that the prime factorisations of m and n are

$$m = \prod_p p^{\mu_p}, \qquad n = \prod_p p^{\nu_p},$$

hence $[m, n] = \prod_p p^{\gamma_p}$, where $\gamma_p = \max\{\mu_p, \nu_p\}$. For each prime p, we have that p^{γ_p} divides the order of g (if $\mu_p \geq \nu_p$) or the order of h (if $\mu_p < \nu_p$), so there must be an element z_p of order p^{γ_p} in the subgroup generated by g or in the subgroup generated by h. The element $z = \prod_p z_p$ has order $[m, n]$.

Exercise 11 Let p be a prime and k be a positive integer. For each a with $0 \leq a \leq k$, find the number of subgroups of order p^a of the additive group $(\mathbb{Z}/p\mathbb{Z})^k$.

Solution The group $G = (\mathbb{Z}/p\mathbb{Z})^k$ has a natural vector space structure over $\mathbb{F}_p$, as the multiplication by scalars can be defined in term of the group addition: we set $\lambda \cdot x = x + \cdots + x$, with λ copies of the summand. Analogously, every subgroup of G has a natural vector space structure. Listing the subgroups of G of order p^a is therefore equivalent to listing the vector subspaces of $(\mathbb{Z}/p\mathbb{Z})^k$ of dimension a.

Each a-tuple $(v_1, \ldots, v_a)$ of linearly independent vectors generates a subspace of dimension a. The number of ordered a-tuples of linearly independent vectors is $(p^k - 1)(p^k - p) \cdots (p^k - p^{a-1})$: we have $p^k - 1$ choices for the vector v_1 (all nonzero vectors), $p^k - p$ choices for v_2 (all vectors but the p multiples of v_1), $p^k - p^2$ choices for v_3 (all vectors but the p^2 linear combinations of v_1 and v_2), and so on.

On the other hand, every subspace of dimension a can be generated by an ordered a-tuple of linearly independent vectors $(v_1, \ldots, v_a)$ in $(p^a - 1)(p^a - p) \cdots (p^a - p^{a-1})$ different ways. Indeed, there are $p^a - 1$ choices for v_1 (all nonzero vectors in the subspace), $p^a - p$ choices for v_2 (all vectors in the subspace but the p multiples of v_1), and so on.

It follows that the number of subspaces of dimension a of $(\mathbb{Z}/p\mathbb{Z})^k$, and thus the number of subgroups of order p^a, is

$$\frac{(p^k - 1)(p^k - p) \cdots (p^k - p^{a-1})}{(p^a - 1)(p^a - p) \cdots (p^a - p^{a-1})}.$$

⟦Notice that the argument of the solution implies that the fraction above is always an integer. Moreover, if we replace a with $k - a$, it is not difficult to show that the formula gives the same result. An interpretation of this symmetry is that there is a bijective correspondence pairing each subspace with its orthogonal subspace, and if a subspace has dimension a, then its orthogonal has dimension $k - a$. ⟧

Exercise 12 Find all possible orders of elements in S_3 and describe its subgroups.

Solution The group S_3 has six elements, so their orders must divide 6. We know that S_3 is not Abelian, so it does not have elements of order 6: if it did, it would not just be Abelian, but cyclic.

Clearly the neutral element, that is, the identity permutation, has order 1. The three transpositions (12), (13) and (23) have order 2 and the 2 three-cycles (123) and (132) have order 3. We have listed six different permutations, so we have exhausted the elements of S_3 and computed the order of each of them.

Now, let G be a subgroup of S_3. If G has order 1 or 6 then it is the trivial subgroup or S_3, respectively. Since by Lagrange's theorem the order of G must be a divisor of 6, there are only two other possibilities: either G has order 2 or G has order 3.

If G has order 2 then it contains the neutral element and one transposition. Therefore, there are three subgroups of order 2: $\{e, (12)\}$, $\{e, (13)\}$ e $\{e, (23)\}$.

If G has order 3 then it contains the neutral element and two elements of order 3: it must therefore be $G = \{e, (123), (132)\}$. This completes the description of all subgroups of S_3.

Exercise 13 Given a cyclic group G, describe the homomorphisms from G to itself and the group of automorphisms of G.

Solution We use the additive notation for G and let g be some fixed generator of G. First of all, we show that the choice of g induces a bijective correspondence between elements of G and homomorphisms from G to itself.

Indeed, given an element h of G, let φ_h be the map given by $\varphi_h(ng) = nh$. Remark that, since g is a generator, we must have $h = kg$ for some $k \in \mathbb{Z}$; so if $ng = 0$ then we also have $nh = nkg = kng = k \cdot 0 = 0$. This shows that φ_h is well defined. Moreover, we have $\varphi_h(ng + mg) = \varphi_h((n + m)g) = (n + m)h = nh + mh = \varphi_h(ng) + \varphi_h(mg)$ and so φ_h is a homomorphism.

On the other hand, given an homomorphism φ from G to itself, it is clear that $\varphi = \varphi_h$ for $h = \varphi(g)$. This concludes the proof that the set of homomorphisms from G to itself is in bijection with G.

Let us now describe the automorphisms of G. Since $\text{Im}(\varphi_h) = \langle h \rangle$, the homomorphism φ_h is surjective if and only if h is a generator of G.

If G is infinite then by the structure theorem for cyclic groups we have $G \simeq \mathbb{Z}$, so the only generators of G are g and $-g$. In particular, for $h = g$ and $h = -g$ we have $\varphi_h^2 = \mathrm{Id}_G$; this shows that φ_g and φ_{-g} are automorphisms and the map $\mathbb{Z}/2\mathbb{Z} \simeq \{\pm 1\} \ni k \longmapsto \varphi_{kg} \in \mathrm{Aut}(G)$ is a group isomorphism.

Finally, assume that G is finite, in which case any surjective homomorphism from G to itself is also injective. In particular, since the generators of $\mathbb{Z}/n\mathbb{Z}$ are the classes k such that $(k, n) = 1$, we find that the map $(\mathbb{Z}/n\mathbb{Z})^* \ni k \longmapsto \varphi_{kg} \in \mathrm{Aut}(G)$ is an isomorphism.

Exercise 14 Describe the nilpotent elements in $\mathbb{Z}/n\mathbb{Z}$ in terms of the prime factorisation of n.

Solution Let $n = p_1^{e_1} \cdots p_k^{e_k}$ be the prime factorisation of n, where $p_1, \ldots, p_k$ are distinct and $e_i > 0$ for $i = 1, \ldots, k$. A nilpotent element $\bar{x} \in \mathbb{Z}/n\mathbb{Z}$ is the residue class of an integer x for which there is a positive integer m such that $x^m \equiv 0 \pmod{n}$, that is, $n \mid x^m$.

Since $p_i \mid n$ for $i = 1, \ldots k$, we must have $p_i \mid x^m$ and thus (by definition of a prime) $p_i \mid x$. Conversely, if $p_1 \mid x, \ldots, p_k \mid x$, which implies $p_1 \cdots p_k \mid x$, then we have $n \mid x^m$ for $m \geq \max\{e_1, \ldots, e_k\}$.

In conclusion, the nilpotent elements of $\mathbb{Z}/n\mathbb{Z}$ are the classes represented by multiples of $p_1 \cdots p_k$.

Exercise 15 Let $\mathbb{K}$ be a field of characteristic different from 2. Show that

(i) every extension of $\mathbb{K}$ of degree 2 is of the form $\mathbb{K}(\sqrt{a})$ for some $a \in \mathbb{K}$;
(ii) if $[\mathbb{K}(\sqrt{a}) : \mathbb{K}] = [\mathbb{K}(\sqrt{b}) : \mathbb{K}] = 2$, then $\mathbb{K}(\sqrt{a}) = \mathbb{K}(\sqrt{b})$ if and only if ab is the square of an element of $\mathbb{K}$.

Solution

(i) An extension $\mathbb{F}$ of $\mathbb{K}$ of degree 2 is of the form $\mathbb{F} = \mathbb{K}(\alpha)$ for some algebraic element α whose minimal polynomial over $\mathbb{K}$ has degree 2. Suppose that the minimal polynomial of α over $\mathbb{K}$ is $\mu_\alpha(x) = x^2 + rx + s$. Since the characteristic of $\mathbb{K}$ is not 2, the solutions of the equation $\mu_\alpha(x) = 0$ can be computed using the quadratic formula

$$x_1, x_2 = \frac{-r \pm \sqrt{r^2 - 4s}}{2},$$

because 2 is invertible in $\mathbb{K}$. It is now clear that, setting $a = r^2 - 4s$, the solutions of the equation are contained in $\mathbb{K}(\sqrt{a})$ and, conversely, a is contained in $\mathbb{K}(\alpha)$.
(ii) We first show that if $\mathbb{K}(\sqrt{a}) = \mathbb{K}(\sqrt{b})$ then ab is the square of an element of $\mathbb{K}$. Since in particular we have $\mathbb{K}(\sqrt{a}) \subseteq \mathbb{K}(\sqrt{b})$, there exist elements $c, d \in \mathbb{K}$ such that $\sqrt{a} = c + d\sqrt{b}$. By squaring we obtain

$$a = c^2 + d^2 b + 2cd\sqrt{b}.$$

By the linear independence over $\mathbb{K}$ of the elements 1 and $\sqrt{b}$ and because the characteristic of $\mathbb{K}$ is different from 2, we must have $cd = 0$, so $c = 0$ or $d = 0$. If $c = 0$ then $a = d^2b$, hence $ab = d^2b^2$ as wanted. If $d = 0$ then $a = c^2$, which contradicts the assumption that the degree $[\mathbb{K}(\sqrt{a}) : \mathbb{K}]$ is 2. Conversely, suppose ab is the square of an element of $\mathbb{K}$, that is, $ab = c^2$ for some $c \in \mathbb{K}$. By hypothesis, a and b are nonzero, so $\sqrt{a} = \pm\frac{c}{\sqrt{b}} \in \mathbb{K}(\sqrt{b})$ and $\sqrt{b} = \pm\frac{c}{\sqrt{a}} \in \mathbb{K}(\sqrt{a})$.

Exercise 16 For each positive integer k, let $f_k(x) = x^k - 1 \in \mathbb{Q}[x]$. Show that for all $m, n > 0$ the greatest common divisor of $f_m(x)$ and $f_n(x)$ is $f_d(x)$, where d is the greatest common divisor of m and n.

Solution Let $d = (n, m)$; assume $m = da$ and $n = db$. Clearly,

$$x^d \equiv 1 \pmod{x^d - 1},$$

and raising both sides of the congruence to the ath power and to the bth power we get $x^m \equiv (x^d)^a \equiv 1 \pmod{x^d - 1}$ and $x^n \equiv (x^d)^b \equiv 1 \pmod{x^d - 1}$, so $x^d - 1 \mid (x^m - 1, x^n - 1)$.

Conversely, let $f(x) = (x^m - 1, x^n - 1)$. Each root $\alpha \in \mathbb{C}$ of $f(x)$ is both a root of $x^m - 1$ and a root of $x^n - 1$, so $\alpha^m = \alpha^n = 1$. This implies that the multiplicative order of α divides both m and n, so it divides their greatest common divisor d. In other word, every root of $f(x)$ is also a root of $x^d - 1$. By the derivative criterion for multiple roots, all roots of $x^m - 1$ are simple, so the same is true for $f(x)$; we thus obtain that $f(x) \mid x^d - 1$. Therefore, $f(x)$ and $x^d - 1$ divide each other, which, since they are both monic, implies that they coincide.

Exercise 17 Let $f(x) = x^2 + a$ be a polynomial with rational coefficients and let $\overline{f}(x)$ be its class in the ring $\mathbb{Q}[x]/(x^3 - x^2)$. Find the values of a for which $\overline{f}(x)$ is invertible and compute its inverse.

Solution . We know that the class of $f(x)$ is invertible in $A = \mathbb{Q}[x]/(x^3 - x^2)$ if and only if the greatest common divisor of $f(x)$ and $x^3 - x^2$ is 1. Since $x^3 - x^2 = x^2(x - 1)$, the two polynomials are coprime if and only if 0 and 1 are not roots of $f(x)$. We therefore impose the conditions $f(0) = a \neq 0$ and $f(1) = a + 1 \neq 0$: the class $\overline{f}(x)$ is invertible in A if and only if $a \neq 0, -1$, whereas if $a = 0$ or $a = -1$ then it is a zero divisor.

Suppose $a \neq 0, -1$; we use Euclid's algorithm to compute the inverse of $\overline{f}(x)$. We have

$$x^3 - x^2 = (x^2 + a)(x - 1) - a(x - 1)$$

$$x^2 + a = a(x - 1)(\frac{1}{a}x + \frac{1}{a}) + a + 1$$

from which we obtain

$$a+1 = (x^2+a)-a(x-1)(-\frac{1}{a}x+\frac{1}{a}) = (x^2+a)(1-\frac{1}{a}(x^2-1))+\frac{1}{a}(x+1)(x^3-x^2)$$

and thus

$$(x^2+a)\left(-\frac{1}{a(a+1)}x^2+\frac{1}{a}\right) \equiv 1 \pmod{x^3-x^2}.$$

We conclude that the inverse of $\overline{f}(x)$ in A is the class of

$$-\frac{1}{a(a+1)}x^2+\frac{1}{a}.$$

Exercise 18 Let p be a prime different from 2. Show that

(i) the elements $a \in \mathbb{F}_p^*$ that are squares form a subgroup of $\mathbb{F}_p^*$ of order $(p-1)/2$;
(ii) the Legendre symbol

$$\mathbb{F}_p^* \ni a \longmapsto \left(\frac{a}{p}\right) \in \{\pm 1\}$$

is a surjective group homomorphism, and in particular the product of two non-squares is a square;

(iii) -1 is the square of an element in $\mathbb{F}_p^*$ if and only if $p \equiv 1 \pmod 4$, that is, we have

$$\left(\frac{-1}{p}\right) = (-1)^{\frac{p-1}{2}}.$$

Solution

(i) The map $\mathbb{F}_p^* \ni a \xrightarrow{f} a^2 \in \mathbb{F}_p^*$ is clearly a homomorphism, because $f(ab) = (ab)^2 = a^2b^2 = f(a)f(b)$. The kernel of f is the set $\{a \in \mathbb{F}_p^* \mid a^2 = 1\}$. Since $\mathbb{F}_p$ is a field, the only solutions of $x^2 = 1$ are ± 1, so $\mathrm{Ker}(f) = \{\pm 1\}$. By the fundamental homomorphism theorem, the image of f is thus a subgroup of $\mathbb{F}_p^*$ of order $(p-1)/2$. Since $\mathrm{Im}(f)$ is the set of squares $\mathbb{F}_p^*$, we have shown the required statement.

(ii) We saw above that the subset Q of all squares in $\mathbb{F}_p^*$ is a subgroup of order $(p-1)/2$. The quotient $\mathbb{F}_p^*/Q$ has therefore two elements, and the composite map $\mathbb{F}_p^* \longrightarrow \mathbb{F}_p^*/Q \longrightarrow \{\pm 1\}$ is a group homomorphism. The result of the composition is clearly

$$\mathbb{F}_p^* \ni a \longmapsto \left(\frac{a}{p}\right) \in \{\pm 1\}.$$

We thus find that the Legendre symbol is multiplicative and, in particular, that the product of two non-squares is a square.

(iii) Assume that there exists $a \in \mathbb{F}_p^*$ such that $a^2 = -1$. We have $a^4 = (-1)^2 = 1$, so the order of a must be a divisor of 4; on the other hand, it cannot be less than 4 because $a^2 = -1 \neq 1$. It follows that 4 must divide the order of the group, that is, $4 \mid p - 1$.

Conversely, suppose $4 \mid p - 1$. Since $\mathbb{F}_p^*$ is a cyclic group of order $p - 1$, $\mathbb{F}_p^*$ has a subgroup of order d for each divisor d of $p - 1$. In particular, $\mathbb{F}_p^*$ has a cyclic subgroup of order 4 and thus an element a of order 4. But, if a has order 4, then $b = a^2 \neq 1$ and $b^2 = a^4 = 1$, which implies $b = -1$ because -1 is the only element of order 2 in $\mathbb{F}_p^*$.

Exercise 19 Factor $x^8 - 1$ in $\mathbb{K}[x]$ for $\mathbb{K} = \mathbb{C}, \mathbb{R}, \mathbb{Q}, \mathbb{F}_{17}$ and $\mathbb{F}_{43}$.

Solution Remark that, if a field $\mathbb{K}$ is contained in a field $\mathbb{F}$, then any factorisation of a polynomial in $\mathbb{K}[x]$ is also valid in $\mathbb{F}[x]$; in particular, the factorisation of the polynomial in $\mathbb{F}[x]$ refines its factorisation in $\mathbb{K}[x]$. In other words, the factorisation in $\mathbb{K}[x]$ is obtained by possibly grouping together some of the factors in $\mathbb{F}[x]$.

Moreover, remark that every factorisation of a polynomial in $\mathbb{Z}[x]$ is also valid in $(\mathbb{Z}/m\mathbb{Z})[x]$ for every positive integer m.

By the fundamental theorem of algebra, the polynomial $x^8 - 1$ factors in $\mathbb{C}[x]$ as a product of eight linear factors, which correspond to eighth roots of unity. If we denote by ζ a primitive eighth root of unity (that is, one of order exactly 8, such as for example $\zeta = (1 + i)/\sqrt{2}$), the factorisation of $x^8 - 1$ in $\mathbb{C}[x]$ is

$$x^8 - 1 = \prod_{h=0}^{7} (x - \zeta^h).$$

Every polynomial with real coefficient having a complex root also has its complex conjugate as a root. Therefore, from the factorisation in $\mathbb{C}[x]$ we can obtain the factorisation in $\mathbb{R}[x]$ by preserving the factors $x - 1$ e $x + 1$ and pairing together the factors that correspond to complex conjugate roots. We obtain

$$(x - \zeta)(x - \zeta^{-1}) = x^2 - \sqrt{2}x + 1,$$
$$(x - \zeta^2)(x - \zeta^{-2}) = x^2 + 1,$$
$$(x - \zeta^3)(x - \zeta^{-3}) = x^2 + \sqrt{2}x + 1.$$

Therefore, the factorisation of $x^8 - 1$ in $\mathbb{R}[x]$ is

$$x^8 - 1 = (x - 1)(x + 1)(x^2 - \sqrt{2}x + 1)(x^2 + 1)(x^2 + \sqrt{2}x + 1).$$

The factors $x - 1$, $x + 1$ and $x^2 + 1$ have rational coefficients and are irreducible in $\mathbb{R}[x]$, so they are irreducible in $\mathbb{Q}[x]$. The factors $x^2 - \sqrt{2}x + 1$ and $x^2 + \sqrt{2}x + 1$, which do not have rational coefficients, must be grouped together: $(x^2 - \sqrt{2}x$

$+ 1)(x^2 + \sqrt{2}x + 1) = x^4 + 1$. Therefore, the factorisation of $x^8 - 1$ in $\mathbb{Q}[x]$ is

$$x^8 - 1 = (x - 1)(x + 1)(x^2 + 1)(x^4 + 1).$$

By Gauss's lemma, a primitive polynomial is irreducible in $\mathbb{Z}[x]$ if and only if it is irreducible in $\mathbb{Q}[x]$, so the factorisation of $x^8 - 1$ in $\mathbb{Z}[x]$ coincides with that in $\mathbb{Q}[x]$.

Now consider the field $\mathbb{K} = \mathbb{F}_{17}$. The multiplicative group of $\mathbb{K}$ is cyclic of order 16, so it has a unique cyclic subgroup of order 8. An element $\alpha \in \mathbb{K}^*$ belongs to this subgroup if and only if $\alpha^8 = 1$; so the polynomial $x^8 - 1$ has eight roots in $\mathbb{F}_{17}$, namely, the elements of this cyclic subgroup. Simple calculations show that a generator of this cyclic group is given by the class of 2 modulo 17. This yields the factorisation

$$x^8 - 1 = (x - 2)(x - 4)(x - 8)(x + 1)(x + 2)(x + 4)(x + 8)(x - 1).$$

Finally, consider the case of $\mathbb{K} = \mathbb{F}_{43}$. Given the factorisation in $\mathbb{Z}[x]$, we certainly can write $x^8 - 1 = (x-1)(x+1)(x^2+1)(x^4+1)$ in $\mathbb{F}_{43}[x]$. The polynomial $x^2 + 1$ is still irreducible over $\mathbb{K}$: it has no roots in $\mathbb{K}$ because any root would have order 4 in $\mathbb{K}^*$, but 4 does not divide 42, which is the order of $\mathbb{K}^*$.

The polynomial $x^4 + 1$ also has no roots, since a root would have order 8. Remark, however, that the order of $\mathbb{F}^*_{43^2}$ is divisible by 8, so it contains all roots of $x^8 - 1$. The degree of the splitting field of a polynomial with coefficients in a finite field is the least common multiple of the degrees of its irreducible factors, so the irreducible factors of $x^4 + 1$ must have degree 2.

In order to explicitly compute the factorisation, we write

$$x^4 + 1 = (x^2 + ax + b)(x^2 + cx + d)$$

which needs to be satisfied for some $a, b, c, d \in \mathbb{K}$. Equating the corresponding coefficients on the two sides we obtain

$$\begin{cases} a + c = 0 \\ b + ac + d = 0 \\ ad + bc = 0 \\ bd = 1 \end{cases}$$

and, substituting $c = -a$,

$$\begin{cases} b + d - a^2 = 0 \\ a(d - b) = 0 \\ bd = 1. \end{cases}$$

The second equation of this system implies that $a = 0$ or $d = b$. If $a = 0$ then $b + d = 0$ and $bd = 1$, that is, $(x - b)(x - d) = x^2 + 1$; but, as seen before, the equation $x^2 + 1$ has no roots in $\mathbb{K}$, so this case yields no solutions. If $d = b$ then we get $2b - a^2 = 0$ and $b^2 = 1$, that is, $b = \pm 1$ and $a^2 = \pm 2$.

Note that $2^7 = 128 \equiv -1 \pmod{43}$, hence $(-2)^8 \equiv -2 \pmod{43}$ and so $a = \pm 16$ is a solution of the system and we have $d = b = 1$ and $c = \mp 16$. These values yield the factorisation $x^4 + 1 = (x^2 + 16x - 1)(x^2 - 16x - 1)$. It is now clear that there is no solution such that $b = -1$, which would imply $a^2 = 2$, since there cannot be four factors of degree 2 of the polynomial $x^4 + 1$ by unique factorisation.

In conclusion, the factorisation of $x^8 - 1$ in $\mathbb{F}_{43}[x]$ is

$$x^8 - 1 = (x - 1)(x + 1)(x^2 + 1)(x^2 + 16x - 1)(x^2 - 16x - 1).$$

⟦We can also show that $a^2 = 2$ has no solutions in $\mathbb{F}_{43}$ in the following way. From the fact that $2^7 \equiv -1 \pmod{43}$ we find that the order of 2 in $\mathbb{F}_{43}^*$ is 14. But then 2 is not a square in $\mathbb{F}_{43}$, because the nonzero squares form the image of $\mathbb{F}_{43}^* \ni x \longmapsto x^2 \in \mathbb{F}_{43}^*$, which has order $(43 - 1)/2 = 21$, not divisible by 14.⟧

Exercise 20 Let $\mathbb{K}$ be a field and let $f(x) = a_n x^n + \cdots + a_0 \in \mathbb{K}[x]$ be a polynomial of degree n; we call the *reciprocal polynomial* of $f(x)$ the polynomial $\hat{f}(x) = a_0 x^n + \cdots + a_n$. Show that if $f(0) \neq 0$ then $f(x)$ is irreducible in $\mathbb{K}[x]$ if and only if $\hat{f}(x)$ is irreducible in $\mathbb{K}[x]$.

Solution For this solution, we shall assume that all polynomials mentioned have a nonzero constant coefficient.

First of all, remark that $\hat{f}(x) = x^n f(1/x)$ with $n = \deg f$; it follows immediately that the reciprocal polynomial of $f(x)g(x)$ is $\hat{f}(x)\hat{g}(x)$. Moreover, $\deg(\hat{f}(x)) = \deg(f)$ and $\hat{\hat{f}}(x) = f(x)$.

The statement we need to prove is equivalent to the following: $f(x)$ is reducible in $\mathbb{K}[x]$ if and only if $\hat{f}(x)$ is reducible in $\mathbb{K}[x]$. Suppose that $f(x)$ is reducible and factors as $f(x) = g(x)h(x)$; we then have that $\hat{f}(x) = \hat{g}(x)\hat{h}(x)$ is also reducible.

The opposite implication is a consequence of $\hat{\hat{f}}(x) = f(x)$.

Exercise 21 Find all irreducible polynomials of degree up to 5 in $\mathbb{F}_2[x]$.

Solution The polynomials of degree 1 are clearly irreducible; they are x and $x + 1$.

Remark that the only roots a polynomial can have in $\mathbb{F}_2 = \mathbb{Z}/2\mathbb{Z}$ are 0 and 1. Now, a polynomial has 0 as a root if and only if its constant coefficient is zero, and it has 1 as a root in $\mathbb{F}_2$ if and only if it is the sum of an even number of monomials. Since polynomials of degree 2 or 3 are irreducible if and only if they have no roots, it follows from our remark that the irreducible polynomials of degree 2 or 3 are those with constant coefficient 1 and with an odd number of monomials, namely, $x^2 + x + 1$, $x^3 + x + 1$ and $x^3 + x^2 + 1$.

A polynomial of degree 4 or 5 is irreducible if it has no roots and no irreducible factors of degree 2. Because of our argument above, the polynomials of degree 4 that have no roots are $x^4 + x^3 + x^2 + x + 1$, $x^4 + x^3 + 1$, $x^4 + x + 1$ and $x^4 + x^2 + 1$.

In order to find which of them are irreducible we need to exclude those that factor as a product of irreducible polynomials of degree 2. Since there is only one such polynomial, the only polynomial we need to exclude is $(x^2+x+1)^2 = x^4+x^2+1$. Therefore, the irreducible polynomials of degree 4 are $x^4+x^3+x^2+x+1, x^4+x^3+1$ and $x^4 + x + 1$.

We can argue in a similar way for polynomials of degree 5: those with no roots have constant coefficient 1 and an odd number of monomials. In order to obtain the irreducible ones, we need to exclude those that factor as a product of x^2+x+1 and an irreducible polynomial of degree 3: these are $(x^2+x+1)(x^3+x+1) = x^5+x^4+1$ and $(x^2+x+1)(x^3+x^2+1) = x^5+x+1$. The irreducible polynomials of degree 5 are therefore $x^5+x^3+x^2+x+1, x^5+x^4+x^2+x+1, x^5+x^4+x^3+x+1, x^5+x^4+x^3+x^2+1, x^5+x^3+1$ and x^5+x^2+1.

Exercise 22 Find the number of irreducible polynomials in $\mathbb{F}_2[x]$ of degree up to 6.

Solution Denote by $\overline{\mathbb{F}_2}$ a fixed algebraic closure of $\mathbb{F}_2$. The roots in $\overline{\mathbb{F}_2}$ of irreducible polynomials of degree d in $\mathbb{F}_2[x]$ are precisely the elements of degree d in $\overline{\mathbb{F}_2}$, that is, the elements of the field $\mathbb{F}_{2^d}$ that do not belong to any of its proper subfields. Since an irreducible polynomial of degree d in $\mathbb{F}_2[x]$ has d different roots in $\overline{\mathbb{F}_2}$, we can compute the number of irreducible polynomials of degree d using the inclusion-exclusion principle. In particular, letting n_d be the number of irreducible polynomials of degree d, we find that

$$n_1 = |\mathbb{F}_2| = 2,$$

$$n_2 = \frac{1}{2}|\mathbb{F}_{2^2} \setminus \mathbb{F}_2| = \frac{2^2 - 2}{2} = 1,$$

$$n_3 = \frac{1}{3}|\mathbb{F}_{2^3} \setminus \mathbb{F}_2| = \frac{2^3 - 2}{3} = 2,$$

$$n_4 = \frac{1}{4}|\mathbb{F}_{2^4} \setminus \mathbb{F}_{2^2}| = \frac{2^4 - 2^2}{4} = 3,$$

$$n_5 = \frac{1}{5}|\mathbb{F}_{2^5} \setminus \mathbb{F}_2| = \frac{2^5 - 2}{5} = 6,$$

$$n_6 = \frac{1}{6}|\mathbb{F}_{2^6} \setminus (\mathbb{F}_{2^3} \cup \mathbb{F}_{2^2})| = \frac{(2^6 - (2^3 + 2^2 - 2))}{6} = 9.$$

Exercise 23 Factor the following polynomials in $\mathbb{Q}[x]$

(i) $4x^3 + 11x^2 + 19x + 21$;
(ii) $x^4 + 8x^2 - 5$.

Solution

(i) The polynomial $f(x) = 4x^3 + 11x^2 + 19x + 21$ has degree 3, so it is reducible if and only if it has a root. Any rational root of $f(x)$ must be of the form a/b, where a is a divisor of 21 and b is a divisor of 4. The possible rational roots are thus

$$\pm 1, \ \pm\frac{1}{2}, \ \pm\frac{1}{4}, \ \pm 3, \ \pm\frac{3}{2}, \ \pm\frac{3}{4}, \ \pm 7, \ \pm\frac{7}{2}, \ \pm\frac{7}{4}, \ \pm 21, \ \pm\frac{21}{2}, \ \pm\frac{21}{4}.$$

By substituting these values into the polynomial, we find that $f(-7/4) = 0$. By Ruffini's theorem, $f(x)$ is thus divisible by $x + 7/4$ in $\mathbb{Q}[x]$. Moreover, since $f(x)$ has integer coefficients, by Gauss' lemma it is divisible by the polynomial obtained from $x + 7/4$ by eliminating denominators, namely, $4x + 7$. If we perform the division, we find that

$$4x^3 + 11x^2 + 19x + 21 = (4x + 7)(x^2 + x + 3).$$

Since the polynomial $x^2 + x + 3$ has no rational roots, the expression above is indeed the factorisation of $f(x)$ in $\mathbb{Q}[x]$.

(ii) First of all, we look for rational roots. By the same argument as above, the only potential rational roots are $\pm 1, \pm 5$. Substituting these values into $g(x) = x^4 + 8x^2 - 5$ we never get zero, so $g(x)$ has no rational roots; equivalently, by Ruffini's theorem, $g(x)$ has no irreducible linear factors.

As for irreducible factors of degree 2, remark that the polynomial gives a biquadratic equation. Setting $y = x^2$, we find that the roots of the polynomial $y^2 + 8y - 5$ are $-4 \pm \sqrt{21}$, so the roots of $g(x)$ are

$$\pm\sqrt{-4 + \sqrt{21}}, \quad \pm\sqrt{-4 - \sqrt{21}} = \pm i\sqrt{4 + \sqrt{21}}.$$

The first two roots are real, the other two are imaginary, and they are complex conjugates. Therefore, if $g(x)$ has irreducible factors of degree 2 in $\mathbb{Q}[x]$ (which must in particular be in $\mathbb{R}[x]$), one must have both complex conjugate roots, so it must be

$$(x - i\sqrt{4 + \sqrt{21}})(x + i\sqrt{4 + \sqrt{21}}) = x^2 + 4 + \sqrt{21}.$$

But this polynomial does not have rational coefficients, so it is not a factor of $g(x)$ in $\mathbb{Q}[x]$. It follows that $g(x)$ is irreducible in $\mathbb{Q}[x]$.

Exercise 24 Factor the polynomial $x^4 + x^3 + x^2 + 1$ in $\mathbb{Q}[x]$.

Solution By Gauss's lemma, $f(x) = x^4 + x^3 + x^2 + 1$ is irreducible in $\mathbb{Q}[x]$ if and only if it is irreducible in $\mathbb{Z}[x]$. Any factorisation of $f(x)$ in $\mathbb{Z}[x]$ yields a factorisation modulo p (whose factors are not necessarily irreducible) for each prime p. The class of $f(x)$ modulo 2 factors as $(x + 1)(x^3 + x + 1)$, with both factors being irreducible in $(\mathbb{Z}/2\mathbb{Z})[x]$. Consequently, either $f(x)$ is irreducible in $\mathbb{Z}[x]$, or it has a root in $\mathbb{Z}$. The only possible roots are ± 1, but $f(\pm 1) \neq 0$, so $f(x)$ is irreducible.

Exercise 25 Find the minimal polynomial of α^2 over a field $\mathbb{K}$ knowing the minimal polynomial of α.

Solution Let n be the degree of α over $\mathbb{K}$. Degrees are multiplicative in towers, hence $[\mathbb{K}(\alpha) : K] = [\mathbb{K}(\alpha) : \mathbb{K}(\alpha^2)][\mathbb{K}(\alpha^2) : \mathbb{K}]$. Since α satisfies the equation $x^2 - \alpha^2 = 0$ over $\mathbb{K}(\alpha^2)$, the first of the two factors is either 1 or 2, and the second is either n or $n/2$.

Let $\mu_\alpha(x) = x^n + a_{n-1}x^{n-1} + \cdots + a_1 x + a_0$ be the minimal polynomial of α over $\mathbb{K}$. By separating terms of even and odd degree, we can write $\mu_\alpha(x) = p(x^2) + xd(x^2)$, where $p(x^2)$ is the sum of the even degree monomials in $\mu_\alpha(x)$ and $xd(x^2)$ is the sum of the odd degree monomials.

We have two cases.

① The polynomial $d(x)$ is zero, so $\mu_\alpha(x) = p(x^2)$. We have $p(\alpha^2) = 0$ and $\deg p = n/2$, so $p(x)$ is the minimal polynomial of α^2 over $\mathbb{K}$.

② The polynomial $d(x)$ is nonzero. In this case, $\alpha = -p(\alpha^2)/d(\alpha^2)$, so α^2 has degree n over $\mathbb{K}$ because $\mathbb{K}(\alpha) \subseteq \mathbb{K}(\alpha^2) \subseteq \mathbb{K}(\alpha)$.

Remark that the polynomial

$$g(x^2) = (p(x^2) - xd(x^2))(p(x^2) + xd(x^2)) = p(x^2)^2 - x^2d(x^2)^2$$

is zero when evaluated at α and has degree $2n$. Therefore, $g(x)$ is zero when evaluated at α^2 and has degree n; moreover, its leading coefficient is $(-1)^n$. It follows that the minimal polynomial of α^2 over $\mathbb{K}$ is $(-1)^n g(x)$.

Chapter 2
Exercises

1 Sequences

1 Let $a_0, a_1, a_2, \ldots$ be the sequence defined by recurrence as

$$\begin{cases} a_0 = 2, \ a_1 = 3; \\ a_{n+1} = \dfrac{a_n + a_{n-1}}{6} & \text{for } n \geq 1. \end{cases}$$

(i) Show that for all $n \geq 2$ we have $a_n = b_n/6^{n-1}$ with $b_n \equiv -1 \pmod 6$.
(ii) For each $n \geq 0$, set $c_n = 5a_n + (-1)^n 4/3^{n-1}$. Show that for all $n \geq 0$ we have $c_n = 22 \cdot 2^{-n}$.

2 Let $a_0, a_1, a_2, \ldots$ be the sequence defined by recurrence as

$$\begin{cases} a_0 = 0, \ a_1 = 1; \\ a_{n+1} = 5a_n - 6a_{n-1} & \text{for } n \geq 1. \end{cases}$$

Show that

(i) $(a_n, 6) = 1$ for all $n > 0$;
(ii) $5 \mid a_n$ if and only if n is even.

3 Let $a_1, a_2, a_3, \ldots$ be the sequence defined by recurrence as

$$\begin{cases} a_1 = 1, \ a_2 = 2; \\ a_{n+1} = \frac{1}{2}a_n + a_{n-1} & \text{for } n \geq 2. \end{cases}$$

(i) Show that $a_{n+1} \geq a_n$ for all $n \geq 1$.
(ii) Show that $a_{2n+2} = 9a_{2n}/4 - a_{2n-2}$ for all $n \geq 2$.

© Springer Nature Switzerland AG 2020
R. Chirivì et al., *Selected Exercises in Algebra*, UNITEXT 119,
https://doi.org/10.1007/978-3-030-36156-3_2

4 Consider the sequence $a_0, a_1, a_2, \ldots$ defined by recurrence as

$$\begin{cases} a_0 = 2, \ a_1 = 1; \\ a_{n+1} = a_n + a_{n-1} \quad \text{for } n \geq 1. \end{cases}$$

Show that

(i) $a_0^2 + a_1^2 + \cdots + a_n^2 = a_n a_{n+1} + 2$ for all $n \geq 0$;
(ii) a_n is even if and only if $n \equiv 0 \pmod 3$.

5 Let $k > 0$ be a natural number. Show that there exists a unique sequence of real numbers $a_0, a_1, a_2, \ldots$ such that

$$\begin{cases} a_0 = 0, \ a_k = 1; \\ a_{n+1} = a_n + a_{n-1} \quad \text{for } n \geq 1 \end{cases}$$

and that for this sequence we have $a_1 = 1/F_k$, where F_k is the kth Fibonacci number.

6 Let $a_0, a_1, a_2, \ldots$ be the sequence defined by the recurrence relation

$$\begin{cases} a_0 = 9, \ a_1 = 12, \ a_2 = 38; \\ a_{n+2} = 7a_n - 6a_{n-1} \qquad \text{for } n \geq 1. \end{cases}$$

(i) Find all values of n for which $3 \mid a_n$.
(ii) Find all values of n for which we have $a_{n+1} > a_n$.

7 We set inductively $a_0 = 31$, $a_{n+1} = a_n^3$ for $n \geq 0$. Show that there exists a positive integer k such that for all n we have $a_{n+k} \equiv a_n \pmod{44}$ and find the minimum such k.

8 Given $k \in \mathbb{N}$, let $a_1, a_2, a_3, \ldots$ be the sequence of natural numbers given by

$$\begin{cases} a_1 = k, \\ a_{n+1} = a_n + (202, a_n) \quad \text{for } n \geq 1. \end{cases}$$

Show that there exists $n_0 \in \mathbb{N}$ such that for all $n \geq n_0$ we have $202 \mid a_n$.

9 Let a be an integer not divisible by 3 and let $a_0, a_1, a_2 \ldots$ be the sequence defined by

$$\begin{cases} a_0 = 1, \ a_1 = a; \\ a_{n+1} = 5a_n + 3a_{n-1} \quad \text{for } n \geq 1. \end{cases}$$

Show that $(a_{n+1}, a_n) = 1$ for all $n \geq 1$.

10 For each integer $n \geq 0$, set $a_n = 3^n + 5^n$.

(i) Find real numbers h, k such that $a_{n+1} = ha_n + ka_{n-1}$ for all $n \geq 1$.
(ii) Determine whether or not there exists n such that $7 \mid a_n$.

11 Let $a_0, a_1, a_2, \ldots$ be the sequence defined by

$$\begin{cases} a_0 = 2, \ a_1 = 3, \ a_2 = 5; \\ a_{n+1} = a_n - a_{n-1} + 2a_{n-2} \quad \text{for } n \geq 2. \end{cases}$$

Show that $a_n < a_{n+1}$ for all $n \geq 0$.

12 Let h, k be integers such that $(h, k) = 1$ and let $a_0, a_1, a_2 \ldots$ be the sequence defined by

$$\begin{cases} a_0 = 1, \ a_1 = 1; \\ a_{n+1} = ha_n + ka_{n-1} \quad \text{for } n \geq 1. \end{cases}$$

(i) Show that $(a_n, a_{n+1}) = 1$ for all $n \geq 0$.
(ii) Having set $h = 35$ and $k = 71$, find the greatest common divisor of all numbers in the set $\{a_n^2 - 1 \mid n = 0, 1, 2, \ldots\}$.

13 Denote by F_n, for $n \geq 0$, the nth Fibonacci number. Show the following:

(i) $\binom{n}{0} F_1 + \binom{n}{1} F_2 + \cdots + \binom{n}{n-1} F_n + \binom{n}{n} F_{n+1} = F_{2n+1}$;
(ii) $\binom{n}{1} F_1 + \binom{n}{2} F_2 + \cdots + \binom{n}{n-1} F_{n-1} + \binom{n}{n} F_n = F_{2n}$.

14 Consider the sequence $a_1, a_2, a_3, \ldots$ of natural numbers defined by

$$\begin{cases} a_1 = 1, \ a_2 = 4; \\ a_{n+1} = a_n + 3a_{n-1} \quad \text{for } n \geq 2. \end{cases}$$

(i) Show that there exist real constants α, β such that for all $n \geq 1$

$$a_n = \alpha \left(\frac{1 + \sqrt{13}}{2} \right)^n + \beta \left(\frac{1 - \sqrt{13}}{2} \right)^n.$$

(ii) Find all values of n for which a_n is even.

2 Combinatorics

15 Let $X = \{1, 2, \ldots, n\}$.

(i) How many ordered triples (A, B, C) of disjoint subsets of X are such that $A \cup B \cup C = X$?

(ii) Show that the number of ordered triples (A, B, C) of subsets of X such that $A \cup B \cup C = X$ is 7^n.

16 Let X be the set of all pairs (m, n) of relatively prime integers such that $1 \leq m, n \leq 100$. Show that $|X| + 1 = 2 \sum_{k=1}^{100} \phi(k)$.

17 Find the cardinality of the set $X = \{1 \leq n \leq 10000 \,|\, (n, 18) = 6 \text{ and } n \equiv 2 \pmod{7}\}$.

18 Find the number of positive divisors of $3^{40} \cdot 5^{25}$ that are congruent to 1 modulo 7.

19 Find all positive integers n such that $\phi(n) = 12$.

20 Find the number of triples (x, y, n) of integers such that $0 \leq x, y < 50$, $n \in \mathbb{N}$ and $x + y = n^2$.

21 Find all positive integers n such that

$$\phi(n) = \frac{2}{5}n.$$

22 For each positive integer n, let $d(n)$ be the number of its positive divisors.

(i) Show that $d(n) + \phi(n) \leq n + 1$ for all positive integers n.
(ii) Find all positive integers n such that $d(n) + \phi(n) = n$.

23 Find all natural numbers $n \leq 120$ such that $(n, \phi(n)) = 3$.

24 Find the number of ordered triples of integers (a, b, c) with the following properties: $1 \leq a, b, c \leq 60$, exactly two among a, b, c are even and exactly one among a, b, c is divisible by 3.

25 Given a positive integer m, let $\omega(m)$ be the number of distinct prime factors of m. Show that

$$\frac{\phi(m)}{m} \geq \frac{1}{\omega(m) + 1}.$$

26 Find the number of integers n satisfying all of the following properties: $1000 < n < 10000$, none of the digits in the decimal representation of n is equal to 9 and at least two digits are the same.

27 For each integer $n > 0$, let S_n be the set of permutations of $\{1, \dots, n\}$.

(i) Find the cardinality of the set

$$\{f \in S_n \,|\, f(i) \leq i + 1 \text{ for } 1 \leq i \leq n\}.$$

(ii) Show that the cardinality of the set

$$\{f \in S_n \,|\, i - 1 \leq f(i) \leq i + 1 \text{ for } 1 \leq i \leq n\}$$

is given by the $(n + 1)$th Fibonacci number.

28 Let $X = \{1, 2, \ldots, 100\}$.

(i) Find the number of all subsets of X having exactly three elements, at least two of which are congruent modulo 5.
(ii) Find the number of maps $f : X \longrightarrow X$ such that $f(n) \equiv n + 1 \pmod{5}$ for all $n \in X$.

29 Let $X = \{1, 2, \ldots, 100\}$. Find the cardinality of the following sets:

(i) $\{(x, y) \in X^2 \mid (xy, 6) = 1\}$;
(ii) $\{(x, y) \in X^2 \mid x < y + 6\}$.

30 Let $X = \{1, 2, \ldots, 100\}$. Find the cardinality of the following sets:

(i) $A = \{f : X \longrightarrow X \mid f \text{ is injective and } f^2(x) \equiv f(x) \pmod 2 \ \forall x \in X\}$;
(ii) $B = \{f : X \longrightarrow X \mid f^2(x) = 1 \ \forall x \in X\}$.

31 Find the cardinality of the following sets:

(i) $X = \{d \in \mathbb{N} \mid d \mid 144000 \text{ and } d \text{ has an even number of divisors}\}$;
(ii) $Y = \{d \in \mathbb{N} \mid d \mid 144000 \text{ and } d \text{ is a perfect square but not a perfect cube}\}$.

32 Let $X = \{1, 2, \ldots, 100\}$. Find the cardinality of the following sets:

(i) $\mathcal{A} = \left\{ A \subseteq X \mid \sum_{a \in A} a \equiv 0 \pmod 2 \right\}$;
(ii) $\mathcal{B} = \left\{ A \subseteq X \mid \prod_{a \in A} a \equiv 0 \pmod 8 \right\}$.

33 Find the number of integers n, with $2 \leq n \leq 1000$, such that $\phi(n) \mid n$.

34 A single player game consists in tossing a coin indefinitely. For each coin toss, the probability of heads is the same as that of tails. The player's initial score is 0 and, for each toss, the player is awarded two points for tails and one point for heads. For each $k \geq 1$, let x_k be the player's score after k coin tosses.

  Show that for all $n \geq 1$ the probability p_n that there exists k such that $x_k = n$ is equal to

$$\frac{2}{3} + \frac{(-1)^n}{3 \cdot 2^n}.$$

35 For each bijection $f : \{1, 2, \ldots, 10\} \longrightarrow \{1, 2, \ldots, 10\}$, set

$$S(f) = \sum_{i=1}^{10} |f(i) - i|.$$

Find the number of bijections such that

(i) $S(f) = 2$;
(ii) $S(f) = 3$;
(iii) $S(f) = 4$.

36 Let $X = \{1, 2, \ldots, 100\}$. Find the number of subsets A of X such that

(i) A has 96 elements and the sum of all elements of A is even;
(ii) A has 97 elements and the sum of all elements of A is divisible by 3.

37 Find the cardinalities of the three following sets:

$$A = \{f : \{1, \ldots, 5\} \to \{1, \ldots, 100\} \mid f(i) < f(i+1) \; \forall \, i = 1, 2, 3, 4\},$$
$$B = \{f \in A \mid \exists i \text{ with } f(i+1) > f(i) + 1\},$$
$$C = \{f \in A \mid f(i+1) > f(i) + 1 \; \forall \, i = 1, 2, 3, 4\}.$$

38

(i) Given $4n$ people, how many ways are there to form n bridge teams, each consisting of four people?
(ii) Given $4n$ people, $2n$ men and $2n$ women, how many ways are there to form n bridge teams, each consisting of two men and two women?

39 Let f be a permutation of $\{1, 2, \ldots, n\}$. Suppose that for all $x, y \in \{1, 2, \ldots, n\}$ the following holds: x divides y if and only if $f(x)$ divides $f(y)$.

(i) Is it always true that f sends the product of any k distinct primes to the product of k distinct primes?
(ii) Is it always true that f sends powers of a prime to powers of a prime?
(iii) If $n = 10$, how many possibilities are there for f? And if $n = 13$?

40 Consider a deck of 40 playing cards, 10 for each of four suits—coins, swords, clubs and cups.

(i) How many orderings of the deck are such that the cards of each suit appear in increasing order?
(ii) How many orderings of the deck are such that all coins cards precede all swords cards?

41 Given a positive integer n, how many subsets of $\{1, 2, 3, \ldots, n\}$ contain at least three numbers of the same parity?

42 A four-colouring of $\mathbb{Z}/40\mathbb{Z}$ is a map $c : \mathbb{Z}/40\mathbb{Z} \to \{0, 1, 2, 3\}$. How many four-colourings $c : \mathbb{Z}/40\mathbb{Z} \to \{0, 1, 2, 3\}$ are such that for all $x \in \mathbb{Z}$ we have $c(\overline{x}) \neq c(\overline{x + 10})$?

43 Consider an $n \times n$ table, each cell of which is coloured either black or white.

(i) How many colourings are such that no row is completely black or completely white?
(ii) How many colourings are such that each row and each column contains exactly one black cell?
(iii) Assume that n is even; how many colourings are such that every row contains the same number of white cells as of black cells?

44 Show that for all $n \geq 1$ we have

(i) $\displaystyle\sum_{k=0}^{n} k\binom{n}{k} = n2^{n-1}$;

(ii) $\displaystyle\sum_{k=0}^{n} k^2\binom{n}{k} = (n + n^2)2^{n-2}$.

45 Find the cardinality of the set

$$\{(a_1, \ldots, a_{30}) \in \{0, 1\}^{30} \mid a_1 + a_3 + \cdots + a_{29} \leq 2 \text{ and } a_2 + a_4 + \cdots + a_{30} \leq 2\}.$$

46 Let $X = \{1, 2, \ldots, 100\}$.

(i) Find the number of three-element subsets of X containing two elements whose sum is 10.

(ii) Find the number of three-element subsets of X containing at least two elements that are divisible by 5.

47 Find the number of ordered pairs $(x, y) \in \mathbb{Z}/2^{100}\mathbb{Z} \times \mathbb{Z}/2^{100}\mathbb{Z}$ such that $xy = \bar{0}$.

48 Consider the set X of all possible teams of four you can form from a set of 13 people.

(i) Having fixed two people p and q, pick a team from X at random; what is the probability that p and q do not both belong to that team?

(ii) Find the cardinality of any subset of X with the following property: given any two people, they must appear together in exactly one team belonging to the subset.

49 Find the number of all positive divisors d of $2^{100}3^{100}$ such that $d \equiv 4 \pmod 5$.

50 Let $X = \{1, 2, \ldots, 100\}$.

(i) Find the number of two-element subsets of X such that the sum of their elements is divisible by 4.

(ii) Find the number of three-element subsets of X containing no consecutive numbers.

51 Let X be the set of maps from $\{1, 2, \ldots, 10\}$ to itself and, for each $f \in X$, let

$$M_f = \max\{f(x) \mid 1 \leq x \leq 10\}, \quad m_f = \min\{f(x) \mid 1 \leq x \leq 10\}.$$

(i) Find the number of all maps $f \in X$ such that $M_f - m_f = 1$.

(ii) Find the number of all maps $f \in X$ such that $M_f = 10$ and $m_f = 1$.

52 Find all natural numbers n such that $\phi(n) = n - 8$.

53 Let $X = \{1, 2, \ldots, 100\}$.

(i) Find the number of maps $f : X \longrightarrow X$ with exactly ten fixed points, that is, ten elements $x \in X$ such that $f(x) = x$.

(ii) Find the number of maps $f : X \longrightarrow X$ such that

$$\sum_{x \in X} |f(x) - x| = 2.$$

54 Let $X = \{1, 2, \ldots, 20\}$.

(i) Find the number of all ordered pairs (A, B) of subsets of X such that $|A| = 5$ and $|A \cup B| = 12$.
(ii) Find the number of all ordered triples (A, B, C) of subsets of X such that $|(A \cup B) \cap C| = 8$.

55

(i) How many strings $(a_0, \ldots, a_9)$, with $a_i \in \{0, 1, 2, 3, 4\}$ for each i, are such that there are fewer odd a_i's than even a_i's?
(ii) Find the number of all strings $(a_0, \ldots, a_9)$, with $a_i \in \{0, 1, 2, 3, 4\}$ for each i, such that

$$\sum_{i=0}^{9} (-1)^i a_i \equiv 0 \pmod 6.$$

56 Let $N = \{1, 2, \ldots, 100\}$. Find the cardinality of the following sets:

(i) $X = \{A \subseteq N \mid \max A - \min A = 60\}$;
(ii) $Y = \{f : N \longrightarrow N \mid f(1) \cdot f(2) \cdots f(100) \not\equiv 0 \pmod{10}\}$.

57 Let W be the set of words of length 3 one can form with 26 letters. Find the number of pairs (α, β) with $\alpha, \beta \in W$ such that α, β do not have any letters in common.

58 Let $X = \{1, 2, \ldots, 100\}$. Find the number of

(i) all ordered pairs (A, B) of subsets of X with $|A \cup B| = 40$ and $|A| = 10$;
(ii) all subsets A of X with $|A| = 5$ and $\prod_{x \in A} x \equiv 0 \pmod 9$.

59 Consider a six-sided die whose faces are numbered from 1 to 6; to each die roll assign a score given by the value of the face the die lands on. Find the probability that, after rolling the same die n times, the sum of the scores obtained is a multiple of 7.

60

(i) Find the number of positive integer solutions (x, y) of the equation $2x + 3y = 100$.
(ii) Find the number of all three-element subsets A of $\{1, 2, \ldots, 100\}$ such that the sum of their elements is equal to 100.

61 Find all natural numbers a such that there exists a natural number n for which

$$\phi(n) = \frac{a}{43} n.$$

62 Let $X = \{1, 2, \ldots, 100\}$.

(i) Find the number of all two-element subsets $\{a, b\}$ of X such that $ab \equiv a + b$ (mod 3).

(ii) Find the number of all two-element subsets $\{a, b\}$ of X such that $ab(a+b) \equiv 0$ (mod 3).

63 Let $X = \{1, \ldots, 100\}$ and let $S(X)$ be the set of permutations of X.

(i) Given $\sigma \in S(X)$ and $i \in \{0, 1, 2\}$, let $X_{i,\sigma} = \{x \in X \mid \sigma(x) - x \equiv i$ (mod 3)$\}$. Show that, for all $\sigma \in S(X)$, we have $|X_{1,\sigma}| \equiv |X_{2,\sigma}|$ (mod 3).

(ii) Find the number of all permutations $\sigma \in S(X)$ such that $\sigma \circ \sigma(x) \equiv x$ (mod 2) for all $x \in X$.

64 Find the number of all ordered triples (x, y, z) of positive integers such that

(i) $xyz = 10^{100}$;
(ii) $x^2yz = 10^{100}$.

65 Let $X = \{1, 2, \ldots, 10\}$. Find the number of maps $f : X \longrightarrow X$ such that for all a, b in X, the number $f(a) \cdot f(b)$ is not prime.

66 Consider a set of n pairs of twins.

(i) How many ways are there to form a team of six people containing exactly two pairs of twins?

(ii) Let $n = 12$. How many ways are there to partition the 24 people into four teams of 6 in such a way that at least one pair of twins ends up separated into different teams?

67 Let $X = \{1, 2, \ldots, 100\}$. Find the cardinality of the following sets:

(i) $\{A \in \mathcal{P}(X) \mid \sum_{x \in A} x^{100} \equiv 0$ (mod 2)$\}$;
(ii) $\{(A, B) \in \mathcal{P}(X)^2 \mid 4$ exactly divides $\prod_{x \in A \cap B} x\}$.

3 Congruences

68 Find, as a function of the integer a, all solutions of the following system of congruences:

$$\begin{cases} 2^{ax} \equiv 13 & \text{(mod 17)} \\ (x - a)(x - 2) \equiv 0 & \text{(mod 4)}. \end{cases}$$

69 Find all integer values of a for which the following system has a solution:

$$\begin{cases} 2^x \equiv a & \text{(mod 9)} \\ x \equiv a^2 & \text{(mod 3)}. \end{cases}$$

70 Find all integer values of a for which the following system has a solution:

$$\begin{cases} 2^x \equiv 3^a \pmod{7} \\ 4x^2 \equiv a^2 \pmod{24}. \end{cases}$$

71 Find all values of $a \in \mathbb{Z}$ such that the following system has a solution, then solve it:

$$\begin{cases} 3^{x^2-1} \equiv 2^a \pmod{13} \\ x - 1 \equiv 0 \pmod{3}. \end{cases}$$

72 Find all solutions of the following system of congruences:

$$\begin{cases} 2^x \equiv x \pmod{7} \\ x^2 \equiv 1 \pmod{15}. \end{cases}$$

73 Find all natural values of n such that the following system is satisfied:

$$\begin{cases} \binom{n}{3} \equiv 0 \pmod{2} \\ \binom{n}{4} \equiv 0 \pmod{2}. \end{cases}$$

74 Solve the following system of congruences:

$$\begin{cases} x^2 \equiv 4 \pmod{14} \\ x \equiv 3 \pmod{5}. \end{cases}$$

75 Solve the following system of congruences:

$$\begin{cases} x^{660} \equiv 1 \pmod{847} \\ x \equiv 11 \pmod{13}. \end{cases}$$

76 Find all $a \in \mathbb{Z}$ for which the congruence $x^3 - a^3 \equiv 0 \pmod{85}$ has solutions other than $x \equiv a \pmod{85}$.

77 Solve the congruence $2^x \equiv 5 \pmod{3^3}$. Then solve the systems

$$\begin{cases} 2^x \equiv 5 \pmod{3^3} \\ x \equiv 2 \pmod{15}, \end{cases} \qquad \begin{cases} 2^x \equiv 5 \pmod{3^4} \\ x \equiv 3 \pmod{15}. \end{cases}$$

78

(i) For which integers b does the congruence $81^x \equiv b \pmod{125}$ have a solution?
(ii) Assume that $81^x \equiv b_0 \pmod{125}$ has a solution x_0 and describe the set of all solutions.

79 Solve the congruence $2^x \equiv 3 \pmod{125}$. How could one solve the congruence $2^x \equiv 3 \pmod{625}$?

80 Solve the system of congruences

$$\begin{cases} 5^x \equiv 3 & \pmod{11} \\ x^2 \equiv -3 & \pmod{21}. \end{cases}$$

81 For each $a \in \mathbb{Z}$, find all integer values of x such that

$$\frac{1}{3}x^3 - \frac{8}{21}ax^2 + \frac{3}{7}x + \frac{1}{7}a$$

is an integer.

82 Find the number of solutions modulo 77 of the congruence $x^{15} \equiv x^{27} \pmod{77}$.

83 Find, for each $k \in \mathbb{N}$, all solutions of the following system of congruences:

$$\begin{cases} x^k \equiv x & \pmod{7} \\ x^3 \not\equiv x & \pmod{7}. \end{cases}$$

84 Find all integer values of a for which the following system has a solution

$$\begin{cases} ax \equiv 4 & \pmod{25} \\ x^2 + a \equiv 0 & \pmod{15} \end{cases}$$

and find all solutions for $a = -1$.

85 For each $a \in \mathbb{Z}$, solve the following system of congruences:

$$\begin{cases} ax \equiv 1 & \pmod{9} \\ a^x \equiv 1 & \pmod{9}. \end{cases}$$

86 For each integer a, find all solutions of the system

$$\begin{cases} (6a - 1)x \equiv 1 & \pmod{21} \\ x \equiv a & \pmod{35}. \end{cases}$$

87 Find all integer values of a for which the following system of congruences has a solution

$$\begin{cases} 9^{ax} \equiv 1 & \pmod{34} \\ x^2 - 9ax \equiv 6 & \pmod{15} \end{cases}$$

and solve it for $a = 4$.

88 For each integer a, determine whether the following system of congruences has any solutions and, if so, find them:

$$\begin{cases} 3x \equiv a \pmod{42} \\ 6x \equiv 1 \pmod{35}. \end{cases}$$

89 Solve the following system of congruences:

$$\begin{cases} 5^x \equiv 9 \pmod{2^4} \\ x^2 + 2x + 8 \equiv 0 \pmod{176}. \end{cases}$$

90 Find all integer values of a such that the following system has a solution:

$$\begin{cases} x^2 \equiv 5a \pmod{120} \\ 6x \equiv a \pmod{21}. \end{cases}$$

Find all solutions of the system for $a = 45$.

91 Find the number of solutions of the congruence $x^{100} \equiv a \pmod{77}$ for each integer a.

92

(i) Find, for each integer a, the number of solutions of the congruence $x^a \equiv 1 \pmod{92}$.
(ii) For each integer a, solve the following system of congruences:

$$\begin{cases} x^a \equiv 1 \pmod{92} \\ 6x \equiv 8 \pmod{23}. \end{cases}$$

93 For each integer a, find the number of solutions of the following system of congruences:

$$\begin{cases} 2x \equiv a \pmod{22} \\ x^2 \equiv 7a \pmod{84}. \end{cases}$$

94 For each integer a, solve the following system of congruences:

$$\begin{cases} a^x \equiv 3 \pmod{8} \\ x^{2a} \equiv 4 \pmod{9}. \end{cases}$$

95 Find the number of all ordered pairs $(x, y) \in \mathbb{Z}/100\mathbb{Z} \times \mathbb{Z}/100\mathbb{Z}$ such that $xy = \bar{0}$.

96 For each integer a, find the number of solutions of the following system of congruences:

$$\begin{cases} 6x \equiv 4a \pmod{72} \\ 5x \equiv 2 \pmod{39}. \end{cases}$$

97 Solve the following system of congruences:

$$\begin{cases} 8^{x^2-1} \equiv -1 \pmod{27} \\ x^{22} + 2x \equiv 8 \pmod{44}. \end{cases}$$

98 For each $a \in \mathbb{Z}$, find the number of solutions of the following system of congruences modulo an appropriate integer:

$$\begin{cases} 3^x \equiv 2^a \pmod{5} \\ x^3 \equiv a + 2 \pmod{24}. \end{cases}$$

99 For each $a \in \mathbb{Z}$, find all solutions of the following system of congruences:

$$\begin{cases} a^x \equiv 3 \pmod{7} \\ x^2 \equiv a \pmod{8}. \end{cases}$$

100 For each $a \in \mathbb{Z}$, find all solutions of the following system of congruences:

$$\begin{cases} a^x \equiv 1 \pmod{5} \\ ax \equiv 2 \pmod{8}. \end{cases}$$

101 For each $a \in \mathbb{Z}$, find all solutions of the following system of congruences:

$$\begin{cases} 5^{x^2-1} \equiv 2^a \pmod{13} \\ x^3 \equiv 0 \pmod{64}. \end{cases}$$

102 Find all integer values of a for which the following system has a solution:

$$\begin{cases} a^x \equiv 11 \pmod{14} \\ x^a \equiv 1 \pmod{9}. \end{cases}$$

103

(i) Find all $x \in \mathbb{Z}$ such that $3^x \equiv 7 \pmod{10}$.
(ii) Find all $x \in \mathbb{Z}$ such that $3^x \equiv 4 + x \pmod{10}$.

104 Find all solutions of the following system of congruences:

$$\begin{cases} x^{2x+1} \equiv 1 \pmod{7} \\ \quad\ 4x \equiv 7 \pmod{15}. \end{cases}$$

105

(i) Let k be a natural number. Find the number of $x \in \mathbb{Z}$ with $0 \le x \le k$ such that $x \equiv 1 \pmod{n}$ for all n with $1 \le n \le 10$.

(ii) How many integers x are such that $x \equiv -1 \pmod{n}$ for all positive integers n?

(iii) How many integers x are such that $x \equiv n \pmod{2n}$ for all positive integers n?

106 Find all pairs of positive integers (x, n) that satisfy the congruence $x^n \equiv 39 \pmod{10x}$.

107 For each $n \in \mathbb{Z}$, find the number of solutions of the congruence

$$x^{5n} \equiv 1 \pmod{55}.$$

108

(i) Solve the congruence $x^2 - x + 43 \equiv 0 \pmod{55}$.

(ii) For each integer a, solve the following system of congruences:

$$\begin{cases} x^2 - x + 43 \equiv 0 \pmod{55} \\ \quad\quad\ x^{11^4} \equiv x^a \pmod{5}. \end{cases}$$

109 Solve the following two congruences:

(i) $x^2 + 2x + 5 \equiv 0 \pmod{65}$;

(ii) $3^{2x} + 2 \cdot 3^x + 5 \equiv 0 \pmod{65}$.

110 Solve the following system of congruences:

$$\begin{cases} x^2 + 2x + 2 \equiv 0 \pmod{10} \\ \quad\quad\ 7x \equiv 20 \pmod{22}. \end{cases}$$

111 For each integer a, find the number of solutions modulo 180 of the following system of congruences:

$$\begin{cases} ax \equiv 2 \pmod{12} \\ 9x \equiv a^2 + 2a - 3 \pmod{81}. \end{cases}$$

112 Solve the following system of congruences:

$$\begin{cases} x^{131} \equiv x \pmod{55} \\ x^6 + x \equiv 0 \pmod{125}. \end{cases}$$

113 For each integer a, determine whether the following system has any solutions and, if so, find them:

$$\begin{cases} ax \equiv 12 \pmod{77} \\ 13x \equiv 25 \pmod{133}. \end{cases}$$

114 Find all possible integer values of a for which the following system has a solution

$$\begin{cases} x^{80} \equiv 2 \pmod 7 \\ 80^x \equiv 2 \pmod 7 \\ 7x \equiv a \pmod{10} \end{cases}$$

and find its solutions as a function of a.

115 Find all solutions of the following system of congruences:

$$\begin{cases} x^{41} \equiv x \pmod{700} \\ 45x \equiv 25 \pmod{700}. \end{cases}$$

116 Find all solutions of the congruence $x^{x+1} \equiv 1 \pmod{27}$.

117 Find all integer values of a for which the system of congruences

$$\begin{cases} ax \equiv 12 \pmod{27} \\ a^3x^2 \equiv 9 \pmod{39} \end{cases}$$

has a solution.

118 Let a be an integer; consider the system of congruences

$$\begin{cases} x^2 - 7a \equiv 0 \pmod 5 \\ a^x \equiv 3 \pmod{35}. \end{cases}$$

(i) Find all values of a for which the system has a solution.
(ii) Find, for each value of a, the number of solutions of the system expressed as congruence classes for an appropriate modulus.

119 Find all integer values of a for which the system of congruences

$$\begin{cases} x^2 + x + 1 \equiv 0 \pmod{13} \\ ax \equiv 27 \pmod{78} \end{cases}$$

has a solution, then solve it.

120 Solve the following system of congruences:

$$\begin{cases} x^2 - 4x + 3 \equiv 0 \quad \text{(mod 15)} \\ \qquad\quad 30x \equiv -6 \quad \text{(mod 81)}. \end{cases}$$

121 Find the number of solutions modulo 1001 of the congruence $x^{101} \equiv x$ (mod 1001).

122 Find the number of solutions modulo 2^{10} of the congruence $x^5 - 16x \equiv 0$ (mod 2^{10}).

123 Find all integer values of a for which the following system of congruences has a solution:

$$\begin{cases} 2^x \equiv 3^{x+a^2} \quad \text{(mod 17)} \\ 3x \equiv a^{23} \qquad \text{(mod 24)}. \end{cases}$$

124 For each integer a, find the number of solutions modulo 90 of the following system of congruences:

$$\begin{cases} \qquad\qquad 3x \equiv a+1 \quad \text{(mod 9)} \\ (x-1)(x-a) \equiv 0 \qquad \text{(mod 15)}. \end{cases}$$

125 Find all integer values of a for which the following system of congruences has a solution, then solve it:

$$\begin{cases} x^{27} \equiv x^2 \quad \text{(mod 144)} \\ 10x \equiv a \quad \text{(mod 25)} \\ 2^{x-1} \equiv 4 \quad \text{(mod 11)}. \end{cases}$$

126 Find all integer pairs (x, y) such that

$$\begin{cases} \qquad\qquad\qquad 2^{2y^2-5y+4} \equiv 2 \quad \text{(mod 36)} \\ (2x^2 + 17)(2x^2 + xy + 4x + 2y)^{-1} \equiv 1 \quad \text{(mod 592)} \\ \qquad\qquad\qquad\quad x^{23} + 1 \equiv 0 \quad \text{(mod 100)}. \end{cases}$$

127 For each $a \in \mathbb{Z}$, find all solutions of the following system of congruences:

$$\begin{cases} a^x \equiv 1 \quad \text{(mod 77)} \\ ax \equiv 1 \quad \text{(mod 10)}. \end{cases}$$

128 For each $a \in \mathbb{Z}$, find all integer solutions of the system

$$\begin{cases} 7^x \equiv a \pmod{8} \\ (x+a)^4 \equiv 0 \pmod{200}. \end{cases}$$

129 For each $a \in \mathbb{Z}$, find all integer solutions of the system

$$\begin{cases} 7ax \equiv a \pmod{49} \\ x^a \equiv 1 \pmod{3}. \end{cases}$$

130 Find the number of integer solutions of the system

$$\begin{cases} x^3 \equiv 8 \pmod{1000} \\ x \equiv 2 \pmod{3} \end{cases}$$

with $0 \le x < 3001$.

131

(i) Find all solutions of

$$x^{36} \equiv x \pmod{9}.$$

(ii) Solve the following system of congruences:

$$\begin{cases} x^{36} \equiv x \pmod{9} \\ x^2 - x \equiv 0 \pmod{64}. \end{cases}$$

132 For each $a \in \mathbb{N}$, find the number of solutions modulo 584 of the congruence

$$x^{a+5} - x^a - x^5 + 1 \equiv 0 \pmod{584}.$$

133 Solve the following congruence and find the number of its solutions modulo 10^{10}:

$$x^5 - 4x + 400 \equiv 0 \pmod{10^{10}}.$$

4 Groups

134 Let $(G, +)$, $(G', +)$ be Abelian groups. Let H be a proper nontrivial subgroup of G and let H' be a proper nontrivial subgroup of G'. Moreover, let

$$\text{Hom}(G, G') = \{f : G \to G' \mid f \text{ is a group homomorphism}\}$$

be the group of all homomorphisms from G to G', with the operation $+$ given by $(f + g)(x) = f(x) + g(x)$ for all $x \in G$. Determine whether the following subsets are subgroups of $\mathrm{Hom}(G, G')$:

$$A = \{f \in \mathrm{Hom}(G, G') \mid \mathrm{Ker}(f) \subseteq H\};$$
$$B = \{f \in \mathrm{Hom}(G, G') \mid \mathrm{Ker}(f) \supseteq H\};$$
$$C = \{f \in \mathrm{Hom}(G, G') \mid f(G) \subseteq H'\};$$
$$D = \{f \in \mathrm{Hom}(G, G') \mid f(G) \supseteq H'\}.$$

135 Let $G = \mathbb{Z}/6\mathbb{Z} \times \mathbb{Z}/20\mathbb{Z}$.

(i) How many elements of G have order 60?
(ii) How many cyclic subgroups of order 30 does G have?
(iii) How many injective homomorphisms $f : \mathbb{Z}/12\mathbb{Z} \longrightarrow G$ are there?

136 Let G be the group of all bijective maps $f : \mathbb{Z}/72\mathbb{Z} \longrightarrow \mathbb{Z}/72\mathbb{Z}$ of the form $f(x) = ax$ with $(a, 72) = 1$ and let $H = \{f \in G \mid f(\overline{12}) = \overline{12}\}$.

(i) Show that H is a subgroup of G and find the order of H.
(ii) Is the subgroup H cyclic?

137 Let $G = \mathbb{Z}/12\mathbb{Z}$ and let a, b be elements of G. Moreover, let $f : G \longrightarrow G \times G$, $g : G \times G \longrightarrow G$ be the homomorphisms given by $f(x) = (ax, bx)$ and $g(y, z) = y + z$.

(i) How many pairs (a, b) correspond to an injective homomorphism f?
(ii) How many pairs (a, b) are such that $g \circ f$ is injective?

138 Given any prime p, show that the group $(\mathbb{Z}/p^2\mathbb{Z})^*$ contains both elements of order p and elements of order $p - 1$.

139 Let G be a group and let H, K be normal subgroups of G. Show that $HK = \{hk \mid h \in H, k \in K\}$ is a normal subgroup of G.

140 Let G be a group and let $f : G \longrightarrow G$ be a homomorphism such that $f \circ f = f$. Show that

(i) $\mathrm{Ker}(f) \cap \mathrm{Im}(f) = \{e\}$;
(ii) $G = \mathrm{Ker}(f) \cdot \mathrm{Im}(f)$.

141

(i) Find the number of elements of order 2 and the number of elements of order 3 in the group $(\mathbb{Z}/49\mathbb{Z})^*$.
(ii) How many group homomorphisms are there from $\mathbb{Z}/6\mathbb{Z}$ to $(\mathbb{Z}/49\mathbb{Z})^*$?

142 Let G be a group, let H be a subgroup of G and let $Z(H) = \{g \in G \mid gh = hg \ \forall h \in H\}$ be the centraliser of H in G. Show that

(i) $Z(H)$ is a subgroup of G;
(ii) if H is normal in G then $Z(H)$ is as well;

(iii) for all group homomorphisms $f : G \longrightarrow G'$ we have $f(Z(H)) \subseteq Z(f(H))$.
(iv) Find an example of a group homomorphism $f : G \longrightarrow G'$ and a subgroup H of G such that $Z(H) = G$ and $Z(f(H)) \neq G'$.

143 Let G be an Abelian group and let H be the subset of elements of G whose order is finite.

(i) Show that H is a subgroup of G and find an example where H is infinite.
(ii) Show that every element of G/H other than the neutral element has infinite order.
(iii) Show that G/H is isomorphic to G if and only if H is trivial.
(iv) Show that the kernel of every homomorphism $G \longrightarrow \mathbb{Z}$ contains H.

144 Let G_1 and G_2 be groups and let $G_1 \times G_2$ be their direct product. Let $\pi_1 : G_1 \times G_2 \longrightarrow G_1$ and $\pi_2 : G_1 \times G_2 \longrightarrow G_2$ be the two projection maps. Show the following:

(i) if H_1 is a normal subgroup of G_1 and H_2 is a normal subgroup of G_2 then $H_1 \times H_2$ is a normal subgroup of $G_1 \times G_2$;
(ii) if $\mathcal{H}$ is a subgroup of $G_1 \times G_2$ then $\mathcal{H} \subseteq \pi_1(\mathcal{H}) \times \pi_2(\mathcal{H})$;
(iii) if $|G_1| = m$, $|G_2| = n$ and $(m, n) = 1$ then for all subgroups $\mathcal{H}$ of $G_1 \times G_2$ we have $\mathcal{H} = \pi_1(\mathcal{H}) \times \pi_2(\mathcal{H})$.

145 Given two finite groups G_1 and G_2, let $f : G_1 \longrightarrow G_2$ be a group homomorphism and let H be a subgroup of G_1 that contains the kernel of f.

(i) Show that $[G_1 : H] = [f(G_1) : f(H)]$.
(ii) Does the above equality hold even if we drop the requirement that $\mathrm{Ker}(f) \subseteq H$?
(iii) Does the above equality hold if $G_1 = \mathbb{Z}$ and G_2 is a finite group?

146 Given a group G, a subgroup M of G is *maximal* if $M \neq G$ and, for all subgroups H of G such that $M \subsetneq H \subseteq G$, we have $H = G$. We shall denote the intersection of all maximal subgroups of G by N.

(i) Show that N is a normal subgroup of G.
(ii) Let $G = \mathbb{Z}/n\mathbb{Z}$. Show that $N = \{\overline{0}\}$ if and only if n is squarefree.
(iii) Find N for $G = \mathbb{Z}/100\mathbb{Z}$.

147 Given a group G, let N be a normal subgroup of G. Let f be an automorphism of G such that $f(N) = N$. For each $g \in G$, set $\varphi(gN) = f(g)N$: is φ an automorphism of G/N?

148 Let $G = (\mathbb{Z}/35\mathbb{Z})^*$.

(i) For each positive integer n, find the number of elements of order n in G.
(ii) Find the number of subgroups of G whose order is 6.

149 Given a group G, let $f, g : G \longrightarrow \mathbb{Z}/12\mathbb{Z}$ be two group homomorphisms.

(i) Show that the set $\{x \in G \mid f(x) = g(x)\}$ is a normal subgroup of G.

(ii) Assume $G = S_3 \times \mathbb{Z}/2\mathbb{Z}$ and $H = \langle ((123), \bar{0}) \rangle$; find all homomorphisms $f : G \longrightarrow \mathbb{Z}/12\mathbb{Z}$ such that $f(h) = \bar{0}$ for all $h \in H$.

150 Given two primes p, q with $p < q$ let $G = (\mathbb{Z}/pq\mathbb{Z})^*$, $G^{(2)} = \{x^2 \mid x \in G\}$ and $G^{(3)} = \{x^3 \mid x \in G\}$.

(i) Find the order of each of the subgroups $G^{(2)}$ and $G^{(3)}$.
(ii) Find all p, q for which the subgroup $G^{(2)}$ is cyclic.
(iii) Find all p, q for which the subgroup $G^{(3)}$ is cyclic.

151 Let $(G, +)$ be an Abelian group and let H, K be subgroups of G such that $|G/H| = m$ and $|G/K| = n$, where $(m, n) = 1$. Show that

(i) $G = H + K$;
(ii) $G/(H \cap K) \simeq G/H \times G/K$.

152 Let $G = \mathbb{Z}/20\mathbb{Z} \times \mathbb{Z}/8\mathbb{Z}$. Find the number of homomorphisms $f : G \longrightarrow G$. Moreover, for each $n \in \mathbb{N}$, let $f_n : G \to G$ be the homomorphism given by $f_n(x) = nx$ for all $x \in G$.

(i) For which values of n is the kernel of f_n a cyclic group?
(ii) For which values of n is the image of f_n a cyclic group?

153 Let G be a finite Abelian group and let H be a cyclic subgroup of G such that G/H is also cyclic. Let $m = \text{ord}(H), n = \text{ord}(G/H)$.

(i) Show that if $(m, n) = 1$ then G is cyclic.
(ii) Give an example where $(m, n) > 1$ and G is not cyclic.

154 Given a positive integer n, let G be a group of order n and let $f_k : G \longrightarrow G$ be the map defined by $f_k(x) = x^k$ for all $x \in G$.

(i) Show that if f_{n-1} is a homomorphism then G is Abelian.
(ii) Show that if $n = 62$ and f_8 is a homomorphism then G is Abelian.
(iii) Give an example of a finite group and an integer k such that f_k is not a homomorphism.

155 Let G be a finite Abelian group of order n. For each prime divisor p of n, let G_p be the set of elements of G whose order is a power of p.

(i) Show that G_p is a subgroup of G of order a power of p.
(ii) Show that the order of every element of G/G_p is coprime to p.
(iii) Show that, if $n = p^a q^b$, where p, q are distinct primes and $a, b \in \mathbb{N}$, then G/G_p is isomorphic to G_q.

156 Let L be the additive subgroup of rational numbers consisting of elements that can be written in the form $m/10^n$ for some $m \in \mathbb{Z}, n \in \mathbb{N}$. For each positive integer k, find

(i) the number of elements of order k in $\mathbb{Q}/L$;
(ii) the number of solutions of the equation $kx = 0$ in $\mathbb{Q}/L$.

157 Let G be the group of rigid transformations of three dimensional space that send a cube to itself.

(i) Determine whether or not G has any subgroups of order 3.
(ii) Cut each face of the cube with a line segment as in the figure. Compute the index of the subgroup H of G that sends the figure to itself.
(iii) Is H normal in G?

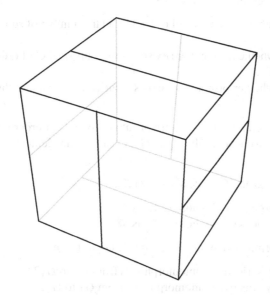

158 Let G be the group $\mathbb{Z}/6\mathbb{Z} \times \mathbb{Z}/6\mathbb{Z}$.

(i) Find the number of cyclic subgroups of G.
(ii) Find the number of elements $x \in G$ such that $G/\langle x \rangle$ is a cyclic group.

159 Let G be the set of bijections $f : \mathbb{Z}/60\mathbb{Z} \longrightarrow \mathbb{Z}/60\mathbb{Z}$ such that for all $x \in \mathbb{Z}$ we have $f(\overline{x}) \equiv \overline{x} \pmod{20}$.

(i) Show that G, endowed with the composition operation, is a group; compute its cardinality.
(ii) For each $m \in \{6, 8, 10, 12\}$, determine whether G has a subgroup of order m and whether it has a cyclic subgroup of order m.

160 Let $(G, +)$ be a finite Abelian group such that $3G = \{3x \mid x \in G\}$ is a cyclic subgroup of G. Show that every subgroup of G whose order is coprime to 3 is cyclic.

161 For each natural number n, find the number of homomorphisms and the number of injective homomorphisms from $\mathbb{Z}/n\mathbb{Z}$ to $\mathbb{Z}/10 \times \mathbb{Z}/20\mathbb{Z}$.

162 Denote by G the group $\mathbb{Z}/5\mathbb{Z} \times \mathbb{Z}/10\mathbb{Z} \times \mathbb{Z}/36\mathbb{Z}$ and let $f : G \longrightarrow G$ be the homomorphism given by $f(g) = 78g$ for all $g \in G$.

(i) Find the cardinalities of the kernel and of the image of f.
(ii) Find the largest possible order of an element in the image of f.

163 Let $(G, +)$ be an Abelian group and let $Q = \{2g \mid g \in G\}$.

(i) Show that Q is a subgroup of G.
(ii) For which values of $m \in \{1, 2, 3, 4\}$ is it possible that $|G/Q| = m$? For each $m \in \{1, 2, 3, 4\}$, either give an example where this equality holds or show that it cannot.

164 Let G be the group $\mathbb{Z}/99\mathbb{Z} \times \mathbb{Z}/33\mathbb{Z}$.

(i) Find the number of elements of order 11 and the number of subgroups of order 11 in G.
(ii) Determine whether or not there exists a subgroup H of G of order 11 such that G/H is cyclic.
(iii) Determine whether or not there exists a surjective homomorphism from G to $\mathbb{Z}/121\mathbb{Z}$.

165 Let G be a group, let H be a normal subgroup of G of order n and let m be a positive integer coprime to n. Show that G/H contains an element of order m if and only if G does.

166 Consider the group $G = \mathbb{Z}/8\mathbb{Z} \times Z/12\mathbb{Z}$.

(i) Find all subgroups of G whose order is 4.
(ii) Find all subgroups of G whose order is 48.

167 Let p be a prime and let $G_1 = (Z/p\mathbb{Z})^*$, $G_2 = (\mathbb{Z}/p^2\mathbb{Z})^*$.

(i) Are there any surjective homomorphisms from G_2 to G_1?
(ii) Are there any injective homomorphisms from G_1 to G_2?

168 Let $(G, +)$ be a finite Abelian group, let p be a prime and let a be a natural number such that p^a exactly divides $|G|$.

(i) Show that $H = \{x \in G \mid p^a x = 0\}$ is a subgroup of G.
(ii) Show that G/H does not contain any elements of order p.
(iii) Show that $|H| = p^a$.

169 Given positive integers m, n, let G be the group $\mathbb{Z}/m\mathbb{Z} \times \mathbb{Z}/n\mathbb{Z}$ and let $\mathbb{C}^*$ be the multiplicative group of nonzero complex numbers.

(i) Denote by $\mathrm{Hom}(G, \mathbb{C}^*)$ the set of homomorphisms from G to $\mathbb{C}^*$. Show that for all f and g in $\mathrm{Hom}(G, \mathbb{C}^*)$ the map $G \ni x \longmapsto (fg)(x) \doteq f(x) \cdot g(x) \in \mathbb{C}^*$ is a homomorphism and that the operation $(f, g) \longmapsto fg$ makes the set $\mathrm{Hom}(G, \mathbb{C}^*)$ into a group.
(ii) Find all pairs (m, n) for which the set $\mathrm{Hom}(G, \mathbb{C}^*)$ contains an injective homomorphism.

170 Let $(G, +)$ be a finite Abelian group of order n and, for each prime p, let pG be the subgroup of G given by $pG = \{px \mid x \in G\}$. Show that

(i) if p and q are different primes, then $G = pG + qG$;
(ii) if p and q are different primes, then $G = pG \cup qG$ if and only if $pq \nmid n$;

(iii) if p, q, r are different odd primes, then $G = pG \cup qG \cup rG$ if and only if $pqr \nmid n$.

171 Find the minimum and maximum possible number of elements whose order is prime in a group G as G varies among all Abelian groups of order 200.

172 Let G be an Abelian group of order n and let p, q be two distinct prime divisors of n. Let h_p, h_q and h_{pq} be the number of subgroups of G of order p, q and pq, respectively. Similarly, let m_p, m_q and m_{pq} be the number of elements of G of order p, q and pq, respectively.

Show that $h_{pq} = h_p h_q$ and $m_{pq} = m_p m_q$.

173 Let G be a group, let p be a prime and let H and K be two distinct normal subgroups of G of index p such that $H \cap K = \{e\}$.

(i) Show that G is isomorphic to $\mathbb{Z}/p\mathbb{Z} \times \mathbb{Z}/p\mathbb{Z}$.
(ii) Find the number of subgroups of order p in G.

174 Let G be an Abelian group, let k be a positive integer, and set $G^k = \{g^k \mid g \in G\}$.

(i)   Show that G^k is a subgroup of G and that all elements of G/G^k have finite order.
(ii) Assuming G is a cyclic group of order n, find the cardinality of G/G^k.
(iii) Give an example of a group G such that $G/G^{10} \simeq \mathbb{Z}/2\mathbb{Z} \times \mathbb{Z}/10\mathbb{Z}$.

175 Given positive integers m and n, denote by $\mathrm{Hom}(\mathbb{Z}/m\mathbb{Z}, \mathbb{Z}/n\mathbb{Z})$ the group of homomorphisms from $\mathbb{Z}/m\mathbb{Z}$ to $\mathbb{Z}/n\mathbb{Z}$, endowed with the addition operation.

(i) Show that $\mathrm{Hom}(\mathbb{Z}/m\mathbb{Z}, \mathbb{Z}/n\mathbb{Z}) \simeq \mathbb{Z}/d\mathbb{Z}$, where d is the greatest common divisor of m and n.
(ii) Describe the subgroup of order 12 of $\mathrm{Hom}(\mathbb{Z}/360\mathbb{Z}, \mathbb{Z}/420\mathbb{Z})$.

176 Let G be a group and let $\Delta = \{(x, x) \mid x \in G\}$.

(i)   Show that Δ is a subgroup of $G \times G$.
(ii) Show that Δ is normal in $G \times G$ if and only if G is Abelian.
(iii) Show that if G is Abelian then $(G \times G)/\Delta$ is isomorphic to G.

177 Let $(G, +)$ be a finite Abelian group of order n. Show that

(i) for all primes p that divide n, the set

$$G_p = \{x \in G \mid \exists k \in \mathbb{N} \text{ such that } p^k x = 0\}$$

is a subgroup of G;
(ii) given $x, y \in G$ of order a and b, respectively, with $(a, b) = 1$, we have $\mathrm{ord}(x + y) = ab$;
(iii) G is cyclic if and only if for all prime divisors p of n the set G_p is a cyclic subgroup of G.

178 Let $G = \mathbb{Z}/18\mathbb{Z} \times \mathbb{Z}/12\mathbb{Z}$, $G' = \mathbb{Z}/36\mathbb{Z}$.

(i) Show that, given homomorphisms $f, g \in \mathrm{Hom}(G, G')$, the map $f + g$ given by $(f + g)(x) = f(x) + g(x)$ for all $x \in G$ is a homomorphisms from G to G', and the operation

$$\mathrm{Hom}(G, G') \times \mathrm{Hom}(G, G') \ni (f, g) \longmapsto f + g \in \mathrm{Hom}(G, G')$$

makes the set $\mathrm{Hom}(G, G')$ into a group. Find the cardinality of $\mathrm{Hom}(G, G')$.

(ii) Find the number of surjective homomorphism from G to G'.

(iii) Show that for all $(a, b) \in G$ the map $\varphi_{(a,b)} : \mathrm{Hom}(G, G') \longrightarrow G'$ given by $\varphi_{(a,b)}(f) = f(a, b)$ is a homomorphism. Find the cardinalities of the kernel and of the image of $\varphi_{(\bar{1},\bar{1})}$.

179 Let $(G, +)$ be an Abelian group and let H, K be subgroups of G such that $[G : H] = m$, $[G : K] = n$. Set $d = [G : H \cap K]$. Show that

(i) $d \leq mn$;

(ii) $d \mid mn$;

(iii) $d = mn$ if and only if $H + K = G$.

180 Let G be the subgroup of the additive group $\mathbb{Q}$ of rational numbers given by

$$G = \{\frac{a}{b} \mid a \in \mathbb{Z}, \ (b, 10) = 1\}.$$

(i) Show that G does not contain any subgroups isomorphic to $\mathbb{Z} \times \mathbb{Z}$.

(ii) Show that G has an infinite number of cyclic quotients.

(iii) Show that G does not have any cyclic quotients of order 3.

181 Let G and H be nontrivial cyclic groups and let $f : G \longrightarrow H$ be an injective homomorphism. Show that the following conditions are necessary and sufficient for the existence of a homomorphism $g : H \longrightarrow G$ such that $g \circ f$ is an isomorphism.

(i) If G is finite then H is finite and, letting $|G| = a$, $|H| = b$, we have $a \mid b$ and $(a, b/a) = 1$.

(ii) If G is infinite then $H \simeq \mathbb{Z}$ and f is surjective.

182 A subgroup M of a group G is *maximal* if $M \neq G$ and for all subgroups L of G such that $M \subsetneq L \subseteq G$ we have $L = G$.

(i) Let K, M be subgroups of a group G such that $K \trianglelefteq G$ and $K \subseteq M$. Show that M is a maximal subgroup of G if and only if M/K is a maximal subgroup of G/K.

(ii) Show that in a finite Abelian group G every subgroup $H \neq G$ is contained in a maximal subgroup of G.

(iii) Show that in a finite nontrivial Abelian group a subgroup is maximal if and only if its index is prime.

183

(i) Find the number of homomorphisms and the number of injective homomorphisms from $\mathbb{Z}/12\mathbb{Z}$ to the group $\mathbb{Z}/4\mathbb{Z} \times S_3$.

(ii) Describe all homomorphisms $\varphi : \mathbb{Z}/12\mathbb{Z} \longrightarrow \mathbb{Z}/4\mathbb{Z} \times S_3$ such that $\varphi(\overline{10})$ has order 3.

184 Let $G = (\mathbb{Z}/1000\mathbb{Z})^*$.

(i) Show that G is not a cyclic group.

(ii) Let $H = \{g \in G \mid \mathrm{ord}(g) \text{ is a power of 2}\}$; show that H is a subgroup of G and compute its order.

(iii) Show that G contains an element of order 25 and deduce that G/H is cyclic.

185 Let G be an Abelian group and, for each $k \in \mathbb{N}$, set $G^k = \{g^k \mid g \in G\}$.

(i) Show that for all k the set G^k is a subgroup of G.

(ii) Assuming that G is finite of order n, find all integers k such that $G^k = G$.

(iii) Give an example of a group G such that $G^k \neq G$ for all $k > 1$.

(iv) Give an example of a nontrivial group G such that $G^k = G$ for all $k \geq 1$.

186 Let $G = \mathbb{Z}/3\mathbb{Z} \times S_3$, with the direct product group structure.

(i) Find all possible orders of subgroups of G.

(ii) Find the number of cyclic subgroups of G.

5 Rings and Fields

187 Let $f(x) = x^4 + x^3 - 3 \in \mathbb{F}_7[x]$. Find the number of zero divisors and the inverse of $x + 1$ in $\mathbb{F}_7[x]/(f(x))$.

188 For each integer m, find the degree of the splitting field of the polynomial $(x^2 - m)(x^4 - 25)$ over $\mathbb{Q}$.

189 Let $f(x) = x^3 + 3x - 1$, $g(x) = x^2 - 2$.

(i) Letting α be a complex root of $f(x)$, find the minimal polynomial of $1/(\alpha + 2)$ over $\mathbb{Q}$.

(ii) Find all primes p such that the polynomials $f(x), g(x)$, considered as elements of $\mathbb{F}_p[x]$, have a common root.

190 Let $f(x) = x^6 + 4x^3 + 2$.

(i) Letting α be a complex root of $f(x)$, find the minimal polynomial of $1/\alpha^2$ over $\mathbb{Q}$.

(ii) Find the splitting field of $f(x)$ over $\mathbb{F}_7$.

191 Find the degree of the splitting field of $x^6 - 4$ over $\mathbb{Q}$ and over $\mathbb{F}_{11}$.

192 Find all primes p for which the polynomial $x^6 + 1 \in \mathbb{F}_p[x]$ has at least one root in $\mathbb{F}_p$.

193 Let α be a complex root of $x^4 - 2x^3 + x - 1$.

(i) Find a polynomial $g(x) \in \mathbb{Q}[x]$ such that $\alpha^2 g(\alpha) = 1$.
(ii) For each integer k, compute the degree $[\mathbb{Q}(\alpha^2 + k\alpha) : \mathbb{Q}]$.

194 Let $\alpha \in \mathbb{C}$ be a root of the polynomial $x^4 + 2x^2 + 2$. Compute the minimal polynomial of $\alpha^2 + 1$ and of $1/(\alpha + 2)$ over $\mathbb{Q}$.

195 Compute the degree of the splitting field of $(x^3 - 2)(x^4 - 3)$ over $\mathbb{Q}$, over $\mathbb{F}_3$ and over $\mathbb{F}_{11}$.

196 Let $\alpha = \sqrt{5} + i \in \mathbb{C}$.

(i) Find the minimal polynomial $f(x)$ of α over $\mathbb{Q}$.
(ii) Compute the degree of the splitting field of the polynomial $f(x)$ over $\mathbb{Q}$ and over $\mathbb{F}_7$.

197 Let $\alpha \in \mathbb{C}$ be a root of the polynomial $x^4 - x - 1$. Find the minimal polynomial of $2\alpha - 1$ and of α^2 over $\mathbb{Q}$.

198 Find the degree of the splitting field of $x^4 - 6x^2 - 3$ over $\mathbb{Q}$ and over $\mathbb{F}_{13}$.

199 Let $\alpha = \sqrt{2 + \sqrt{7}} \in \mathbb{C}$.

(i) Find the degree of $\mathbb{Q}(\alpha)$ over $\mathbb{Q}$.
(ii) Find the degree of the splitting field of the minimal polynomial of α over $\mathbb{Q}$.

200 Let $\alpha = \sqrt{2 + i\sqrt{2}}$.

(i) Find the minimal polynomial of α and the minimal polynomial of $\alpha^2 + 1$ over $\mathbb{Q}$.
(ii) Find a polynomial $f(x) \in \mathbb{Q}[x]$ such that $f(\alpha) = (\alpha^2 + 2\alpha)^{-1}$.

201 Find the degree of the splitting field of $x^8 - 4$ over $\mathbb{Q}$ and over $\mathbb{F}_3$.

202 Find the degree of the splitting field of $x^4 + 26$ over $\mathbb{Q}$, over $\mathbb{F}_5$ and over $\mathbb{F}_7$.

203 Find the degree of the splitting field of $x^6 - 12x^3 + 27$ over $\mathbb{Q}$ and over $\mathbb{F}_5$.

204 Find the minimal polynomial of $\sqrt{2 + \sqrt{3}}$ over $\mathbb{Q}$ and the degree of its splitting field.

205 Let $f(x)$ be the polynomial $(x^3 - 7)(x^2 + 3)$.

(i) Find the degree of the splitting field of $f(x)$ over $\mathbb{Q}$.
(ii) Find the number of zero divisors in $\mathbb{F}_5[x]/(f(x))$.

206 Find the degree of the splitting field of $2x^4 + 6x^2 - 5$ over $\mathbb{Q}$ and over $\mathbb{F}_{19}$.

207 Let $\mathbb{K} \subseteq \mathbb{F}$ be fields, let $f(x) = x^5 + 3x + 3 \in \mathbb{K}[x]$ and let α be a root of $f(x)$ in $\mathbb{F}$. Compute all possible values of $[\mathbb{K}(\alpha) : \mathbb{K}]$ and $[\mathbb{K}(\alpha^7) : \mathbb{K}]$ for $\mathbb{K} = \mathbb{Q}$ and $\mathbb{K} = \mathbb{F}_2$.

208 Let $\mathbb{K}$ be a field and $f(x) \in \mathbb{K}[x]$ a polynomial of positive degree, and set. Show that every zero divisor in $\mathbb{K}[x]/(f(x))$ is nilpotent if and only if $f(x)$ is a power of a single irreducible polynomial.

209 Find the number of zero divisors in the ring $\mathbb{F}_5[x]/(x^3 - 2x + 1)$ that are not nilpotent.

210 Let $x^4 - a \in \mathbb{Z}[x]$ be a *reducible* polynomial. Show that

(i) if $a > 0$ then there is $b \in \mathbb{N}$ such that $a = b^2$;
(ii) if $a < 0$ then there is $c \in \mathbb{N}$ such that $a = -c^2$; moreover, there is $d \in \mathbb{N}$ such that $c = 2d^2$.

211 Compute the degree of the splitting field of $x^4 + 5x^2 + 5$ over $\mathbb{Q}$ and over $\mathbb{F}_{11}$.

212 Let $\alpha \in \mathbb{C}$ be a root of the polynomial $x^3 - x^2 - 2x - 1$ and set $\beta = \alpha^4 - 3\alpha^2$. Find

(i) the minimal polynomial of β over $\mathbb{Q}$;
(ii) a polynomial $g(x)$ with rational coefficients such that $\beta g(\alpha) = 1$.

213 Find the degree of the splitting field of the polynomial $(x^2 + 3)(x^3 - 5)$ over $\mathbb{Q}$.

214 Let $f(x)$ be the polynomial $x^{15} - 1$.

(i) Find the irreducible factors of $f(x)$ and the degree of its splitting field, first over $\mathbb{F}_3$ and then over $\mathbb{F}_5$.
(ii) What are the possible degrees of the splitting field $\mathbb{K}$ of $f(x)$ over $\mathbb{F}_p$, where p is a prime other than 3 or 5?
(iii) For each degree d from the previous question, give an explicit example of a prime p other than 3 or 5 for which $[\mathbb{K} : \mathbb{F}_p] = d$.

215 Find the degree of the splitting field of the polynomial $(x^2 + 2)(x^4 - 2)$ over $\mathbb{Q}$ and over $\mathbb{F}_7$.

216 Let $\mathbb{K}$ be a field and let α, β be roots in an algebraic closure of $\mathbb{K}$ of the polynomials $x^2 - 5$ and $x^2 + 5$, respectively.

(i) Compute the degree of the minimal polynomial of $\alpha + \beta$ over $\mathbb{K}$, for $\mathbb{K} = \mathbb{Q}$.
(ii) Assume $\mathbb{K} = \mathbb{F}_p$; what are the possible degrees of the minimal polynomial of $\alpha + \beta$ over $\mathbb{F}_p$, as p varies among all primes? For each possible degree, give an example.
(iii) Compute the degree of the minimal polynomial of $\alpha + \beta$ over $\mathbb{F}_{2011}$.

217 Let $\mathbb{K}$ be a field and let α be a root of $f(x) = x^4 - 3$ in an algebraic closure of $\mathbb{K}$.

(i) Compute the degree of $\mathbb{K}(\alpha)$ over $\mathbb{K}$, first for $\mathbb{K} = \mathbb{Q}$ and then for $\mathbb{K} = \mathbb{Q}(\sqrt{-3})$.
(ii) Compute the degree of the splitting field of $f(x)$ over $\mathbb{K}$ for $\mathbb{K} = \mathbb{Q}$ and for $\mathbb{K} = \mathbb{Q}(\sqrt{-3})$.

218 Let $f(x) = x^5 + x^2 - x + 4$.

(i) Determine the splitting field of $f(x)$ over $\mathbb{F}_2$ and over $\mathbb{F}_3$.
(ii) For each positive integer k, find the degrees of the irreducible factors of $f(x)$ as a polynomial in $\mathbb{F}_{3^k}[x]$.

219 Let α be a root of the polynomial $x^4 + 2x^3 + 2x^2 + x + 3$. Find the minimal polynomial of $\alpha + 1$ and of $\alpha^2 + \alpha$ over $\mathbb{Q}$.

220 Let $f(x) = x^4 + 3x^2 + 1 \in \mathbb{Q}[x]$ and let $\alpha \in \mathbb{C}$ be a complex root of $f(x)$. Find

(i) the degree $[\mathbb{Q}(\alpha) : \mathbb{Q}]$;
(ii) the degree of the splitting field of $f(x)$ over $\mathbb{Q}$;
(iii) the minimal polynomial of $1/(\alpha + 1)$ over $\mathbb{Q}$.

221 Find the degree of the splitting field of $x^4 - 4x^2 + 2$ over $\mathbb{Q}$, over $\mathbb{Q}(i)$ and over $\mathbb{F}_7$.

222 Let $\mathbb{K}, \mathbb{E}, \mathbb{F}$ be the splitting fields over $\mathbb{Q}$ of $x^{24} - 1$, of $x^8 - 1$ and of $x^3 - 1$, respectively.

(i) Show that $\mathbb{K} = \mathbb{E}\mathbb{F}$.
(ii) Find a basis of $\mathbb{K} \cap \mathbb{R}$ as a vector space over $\mathbb{Q}$.

223

(i) Find the values of $n \in \mathbb{N}$ for which the polynomial $x^{2n} + x^n + 1$ is divisible by $x^2 + x + 1$ in $\mathbb{Q}[x]$.
(ii) Compute the degree of the splitting field $x^8 + x^4 + 1$ over $\mathbb{Q}$ and over $\mathbb{F}_7$.

224 Let α be a complex root of $x^4 - x^3 + x^2 - x + 1$; for each $c \in \mathbb{Q}$, compute the degree $[\mathbb{Q}(\alpha + c\alpha^{-1}) : \mathbb{Q}]$.

225 Let $f(x) = x^3 + 3x + 1$.

(i) Find the degree of the splitting field of $f(x)$ over $\mathbb{Q}$.
(ii) Find all primes p for which $f(x)$, considered as a polynomial in $\mathbb{F}_p[x]$, has a multiple root in $\mathbb{F}_p$.

226 Let $f(x) = x^9 - 1$.

(i) Show that $f(x)$ has an irreducible factor of degree 6 over $\mathbb{F}_{11}$.
(ii) Find the degree of the splitting field of $f(x)$ over $\mathbb{Q}$ and over $\mathbb{Q}(\zeta)$, where $\zeta \in \mathbb{C}$ is a primitive third root of unity.

227 Let $\alpha \in \mathbb{C}$ be a root of the polynomial $f(x) = x^4 + x + 1$.

(i) Find the minimal polynomial of $1/(\alpha + 1)$ and the minimal polynomial of α^2 over $\mathbb{Q}$.
(ii) Find the splitting field of the polynomial $f(x)$ over $\mathbb{F}_5$.

228 Given a prime p, consider $a \in \mathbb{F}_p^*$ and let $f(x) = (x^4 - a)(x^4 + a) \in \mathbb{F}_p[x]$.

(i) Show that if $p \equiv 3 \pmod 4$ then the splitting field of $f(x)$ over $\mathbb{F}_p$ has degree 2.

(ii) Show that one can choose a and p with $p \equiv 1 \pmod 4$ so that the splitting field of $f(x)$ over $\mathbb{F}_p$ has degree 1, 2 or 4.

229 Let p be an odd prime and let $f(x) = x^6 + ax^3 + b \in \mathbb{F}_p[x]$.

(i) Show that the degree of the splitting field of $f(x)$ over $\mathbb{F}_{p^2}$ can only be 1 or 3.

(ii) Show that the degree of the splitting field of $f(x)$ over $\mathbb{F}_p$ can neither be 4 nor 5.

(iii) Show that if $p \equiv 2 \pmod 3$ then the degree of the splitting field of $f(x)$ over $\mathbb{F}_p$ cannot be 3.

230 Let $f(x) = (x^{15} - 1)(x^{12} - 1)$.

(i) Find the degree of the splitting field of $f(x)$ over $\mathbb{F}_2$ and over $\mathbb{F}_7$.

(ii) Find all possible degrees of the splitting field of $f(x)$ over $\mathbb{F}_p$ as p varies among all primes.

231 Find the number of solutions of $2x^4 - 41x^3 + 201x^2 - 71x - 91 = 0$ in $\mathbb{Z}/1635\mathbb{Z}$ and list at least six distinct solutions.

232 Let $\alpha \in \mathbb{C}$ be a root of $x^3 - x - 1$.

(i) Write $1/(\alpha + 2)$ as a polynomial in α with rational coefficients.

(ii) Compute the degrees $[\mathbb{Q}(\alpha^2) : \mathbb{Q}]$ and $[\mathbb{Q}(\alpha^3) : \mathbb{Q}]$.

233 Let $\alpha \in \mathbb{C}$ be a root of the polynomial $f(x) = x^4 - 3x - 5$.

(i) Show that $f(x)$ is irreducible in $\mathbb{Q}[x]$.

(ii) Find the minimal polynomial of $2\alpha - 3$ over $\mathbb{Q}$.

(iii) Find the minimal polynomial of α^2 over $\mathbb{Q}$.

234 Show that the ring $\mathbb{Z}[x]/(2x^2 + 17, x^2 + 6)$ is a vector space of dimension 2 over $\mathbb{F}_5$.

235 Let $\mathbb{K}$ be the field $\mathbb{Q}(\sqrt[3]{2}, i)$.

(i) Compute the degree $[\mathbb{K} : \mathbb{Q}]$.

(ii) Is it true that $\mathbb{K} = \mathbb{Q}(\sqrt[3]{2} + i)$?

(iii) Find the minimal polynomial of $\sqrt[3]{2} + i$ over $\mathbb{Q}$.

236

(i) Compute the degrees $[\mathbb{Q}(\sqrt{3}, \sqrt{5}) : \mathbb{Q}]$ and $[\mathbb{Q}(\sqrt{3} - \sqrt{5}) : \mathbb{Q}]$.

(ii) Find the minimal polynomials of $\sqrt{3} - \sqrt{5}$ and $\sqrt{\sqrt{3} - \sqrt{5}} - 1$ over $\mathbb{Q}$.

237 Factor the polynomial $x^7 + x^6 + x^5 + x^4 + x^3 + x^2 + x + 1$ as a product of irreducible factors in $\mathbb{C}[x]$, $\mathbb{Z}[x]$, $\mathbb{F}_5[x]$ and $\mathbb{F}_{17}[x]$.

238 Let $f(x) = x^4 + 5x^3 + 5x^2 - x + 4$.

(i) Factor $f(x)$ in $\mathbb{F}_7[x]$.

(ii) Find the number of zero divisors and the number of invertible elements in the ring $\mathbb{F}_7[x]/(f(x))$.

239 Let $\alpha = 2 + \sqrt{5 + \sqrt{-5}} \in \mathbb{C}$. Compute $[\mathbb{Q}(\alpha) : \mathbb{Q}]$ and $[\mathbb{Q}(\alpha^2) : \mathbb{Q}]$.

240 Let $f(x) = x^4 + 3x^3 + x + 1$.

(i) For each positive integer k, find the degree of the splitting field of $f(x)$ over $\mathbb{F}_{2^k}$ and over $\mathbb{F}_{3^k}$.

(ii) Letting $\alpha \in \mathbb{C}$ be a root of $f(x)$, compute $[\mathbb{Q}(\alpha) : \mathbb{Q}]$.

241 Find the degree of the splitting field of $x^4 - 2$ over $\mathbb{Q}$, over $\mathbb{F}_3$ and over $\mathbb{F}_{17}$.

Chapter 3
Solutions

1 Sequences

1

(i) We proceed by induction. Since the inductive step relies on the previous two cases, we shall first prove the statement for both $n = 2$ and $n = 3$. If $n = 2$, then we have $a_2 = 5/6$ and therefore the statement holds with $b_2 = 5 \equiv -1$ (mod 6). For $n = 3$ we have $a_3 = 23/36$, so the statement holds with $b_3 = 23 \equiv -1$ (mod 6).

Let us now assume that $a_m = b_m/6^{m-1}$, where $b_m \equiv -1$ (mod 6), for all $m \leq n$; we shall show that a_{n+1} can be written in the same form. From the definition we have

$$a_{n+1} = \frac{a_n + a_{n-1}}{6} = \frac{1}{6}\left(\frac{b_n}{6^{n-1}} + \frac{b_{n-1}}{6^{n-2}}\right) = \frac{b_n + 6b_{n-1}}{6^n}.$$

We can thus simply set $b_{n+1} = b_n + 6b_{n-1}$ and we also have $b_{n+1} \equiv b_n \equiv -1$ (mod 6).

(ii) We proceed by induction and again begin by checking the statement in the first two cases. For $n = 0$ we have $c_0 = 5a_0 + (-1)^0 4/3^{-1} = 22$ and for $n = 1$ we have $c_1 = 5a_1 + (-1)^1 4/3^0 = 15 - 4 = 11$; the statement is therefore true for the first two possible values of n. Let us now suppose that $c_m = 22 \cdot 2^{-m}$ for

© Springer Nature Switzerland AG 2020
R. Chirivì et al., *Selected Exercises in Algebra*, UNITEXT 119,
https://doi.org/10.1007/978-3-030-36156-3_3

all $m \leq n$. Then we have

$$c_{n+1} = 5a_{n+1} + (-1)^{n+1}\frac{4}{3^n}$$

$$= \frac{1}{6}(5a_n + 5a_{n-1}) - (-1)^n\frac{4}{3^n}$$

$$= \frac{1}{6}\left(c_n - (-1)^n\frac{4}{3^{n-1}} + c_{n-1} - (-1)^{n-1}\frac{4}{3^{n-2}}\right) - (-1)^n\frac{4}{3^n}$$

$$= \frac{22}{6}(2^{-n} + 2^{-(n-1)}) + (-1)^n\left(-\frac{2}{3^n} + \frac{2}{3^{n-1}} - \frac{4}{3^n}\right)$$

$$= \frac{22}{2^{n+1}}$$

as required.

2

(i) Proceed by induction on n. If $n = 1$, then $a_1 = 1$ and therefore the statement holds. Assume that the statement holds for n; then we have

$$(a_{n+1}, 6) = (5a_n - 6a_{n-1}, 6) = (5a_n, 6) = (a_n, 6) = 1.$$

(ii) By induction on n: for $n = 0, 1$ the statement holds by definition of a_0, a_1. Now assume that the statement holds for all a_i with $i \leq n$. From the definition

$$a_{n+1} = 5a_n - 6a_{n-1}$$

we get that $5 \mid a_{n+1}$ if and only if $5 \mid a_{n-1}$. This concludes the proof.

3

(i) Both the required statement and the additional inequality $a_n \geq \frac{1}{2}a_{n+1}$ will be shown by induction on n.

For $n = 1$, both inequalities immediately follow from the fact that $a_1 = 1$ and $a_2 = 2$.

Suppose both inequalities hold for $n - 1$. Since $a_{n+1} = a_n/2 + a_{n-1}$, from the induction hypothesis $a_{n-1} \geq a_n/2$ it follows that $a_{n+1} \geq a_n$. Furthermore, we can use the hypothesis that $a_{n-1} \leq a_n$ to obtain that $a_{n+1} \leq 3a_n/2$; the latter is less than $2a_n$ because, since $a_1 \leq a_2 \leq \cdots \leq a_n$, the number a_n is strictly positive.

⟦As an alternative, we can show the required inequality directly, using the induction hypothesis for two consecutive values of n. We have to check the statement for *two* initial values: $a_2 = 2 \geq 1 = a_1$ and $a_3 = 2 \geq 2 = a_2$. Assuming the statement holds for $n - 1$ and n, we have $a_{n+1} = a_n/2 + a_{n-1} \geq a_{n-1}/2 + a_{n-2} = a_n$.⟧

(ii) In order to show the required statement we shall rely on the three recurrence relations

$$a_{2n} = \frac{1}{2}a_{2n-1} + a_{2n-2},$$

$$a_{2n+1} = \frac{1}{2}a_{2n} + a_{2n-1},$$

$$a_{2n+2} = \frac{1}{2}a_{2n+1} + a_{2n},$$

which are all satisfied for $n \geq 2$. Isolating a_{2n-1} in the first equation and substituting its value into the second one we obtain that $a_{2n+1} = 5a_{2n}/2 - 2a_{2n-2}$. Substituting this expression for a_{2n+1} into the third equation yields the desired equality.

4

(i) Proceed by induction on n. For $n = 0$ and $n = 1$ the required equality is easily checked. Assume it holds for n; we then have $a_0^2 + a_1^2 + \cdots + a_n^2 + a_{n+1}^2 = a_n a_{n+1} + 2 + a_{n+1}^2 = a_{n+1}(a_n + a_{n+1}) + 2 = a_{n+1}a_{n+2} + 2$, that is, the desired equality for $n + 1$.

(ii) Proceed by induction on n. For $n = 0, 1, 2$ the statement can be checked directly. Assume it holds for all a_i with $i \leq n$. By employing the recurrence relation twice we obtain that $a_{n+3} = 2a_{n+1} + a_n$ for all $n \geq 0$, hence $2 \mid a_{n+3}$ if and only if $2 \mid a_n$, that is, if and only if $n \equiv 0 \pmod{3}$.

5 The sequence $a_n = F_n/F_k$ does satisfy the requirements, since we have $a_0 = F_0/F_k = 0$, $a_k = F_k/F_k = 1$ and

$$a_{n+1} = \frac{F_{n+1}}{F_k} = \frac{F_n + F_{n-1}}{F_k} = a_n + a_{n-1} \text{ for } n \geq 1.$$

A sequence satisfying the requirements of the problem does therefore exist.

We shall now show that any such sequence must also satisfy $a_1 = 1/F_k$. Indeed, assume $a_0, a_1, \ldots$ is a sequence satisfying the required conditions; we shall show by induction that this sequence coincides with the sequence $b_n = a_1 F_n$.

The above equality holds for $n = 0$ and $n = 1$ by construction. Assume it does for all natural numbers up to n; we have $a_{n+1} = a_n + a_{n-1} = a_1 F_n + a_1 F_{n-1} = a_1 F_{n+1}$. It follows that we necessarily have $1 = a_k = b_k = a_1 F_k$, that is, $a_1 = 1/F_k$.

Finally, we can show uniqueness: since any sequence satisfying the requirements is such that $a_0 = 0$ and $a_1 = 1/F_k$, the induction above shows that $a_n = F_n/F_k$ for all n.

6

(i) Because of the initial conditions, $3 \mid a_0$, $3 \mid a_1$, $3 \nmid a_2$. Assume now that $n \geq 1$ and notice that $3 \mid 6a_{n-1}$ for all n. If $3 \mid a_n$ then $3 \mid 7a_n$ and therefore $3 \mid 7a_n -$

$6a_{n-1} = a_{n+2}$. Conversely, if $3 \mid a_{n+2}$ then $3 \mid a_{n+2} + 6a_{n-1} = 7a_n$; since 3 is prime and $3 \nmid 7$ this entails that $3 \mid a_n$. Consequently, for $n \geq 1$, $3 \mid a_n$ if and only if $3 \mid a_{n+2}$, so the values of n we are looking for are zero and all odd natural numbers.

(ii) For $n = 0$ we have $a_1 > a_0$. We shall show by induction that, for all $n \geq 1$, we have $a_{n+1} > a_n$ if n is odd and $a_{n+1} < a_n$ if n is even.

Consider the first three cases: the recursion yields $a_3 = 30$ and $a_4 = 194$, so we have $a_2 > a_1$, $a_3 < a_2$ and $a_4 > a_3$. Now suppose $n \geq 2$; from the recursion we obtain

$$a_{n+2} - a_{n+1} = 7a_n - 6a_{n-1} - 7a_{n-1} + 6a_{n-2} = 7(a_n - a_{n-1}) - 6(a_{n-1} - a_{n-2}).$$

If $n + 1$ is odd, $n - 1$ is also odd and the induction hypothesis implies $a_n - a_{n-1} > 0$, while $n - 2$ is even and, again by the induction hypothesis, we have $a_{n-1} - a_{n-2} < 0$. It follows that $a_{n+2} - a_{n+1}$ is the sum of two positive numbers and is therefore positive.

Conversely, if $n + 1$ is even then the induction hypothesis yields $a_n - a_{n-1} < 0$ and $a_{n-1} - a_{n-2} > 0$, so $a_{n+2} - a_{n+1}$ is the sum of two negative numbers and therefore negative.

7 First of all we shall show by induction on n that $a_n = 31^{3^n}$ for all n. For $n = 0$ one can check that $31^{3^0} = 31^1 = 31 = a_0$. If we assume the statement for n, then we have that $a_{n+1} = a_n^3 = (31^{3^n})^3 = 31^{3^{n+1}}$.

Since $(31, 44) = 1$, powers of 31 are periodic modulo 44, with period given by the multiplicative order of 31 in the group $(\mathbb{Z}/44\mathbb{Z})^*$, which is a divisor of $\phi(44) = 20$. The order in question is the least common multiple of the multiplicative order of 31 modulo 4 and modulo 11.

Now, $31 \equiv -1 \pmod{4}$, so the former is 2. We also have $31 \equiv -2 \pmod{11}$ and, since $-2 \not\equiv 1 \pmod{11}$ while $(-2)^5 = -32 \equiv 1 \pmod{11}$, the latter is 5. Consequently, the order of 31 in $(\mathbb{Z}/44\mathbb{Z})^*$ is 10.

Therefore, we have $31^{3^{n+k}} \equiv 31^{3^n} \pmod{44}$ if and only if $3^{n+k} \equiv 3^n \pmod{10}$. Since the order of 3 in $(\mathbb{Z}/10\mathbb{Z})^*$ is 4, the sequence $a_n = 31^{3^n}$ is periodic modulo 44 with period 4.

⟦Since the order of 3 in $(\mathbb{Z}/20\mathbb{Z})^*$ is 4 as well, just observing that the order of 31 in $(\mathbb{Z}/44\mathbb{Z})^*$ divides 20 is enough to conclude that k is at most 4, and one could simply exclude $k = 1$ and $k = 2$ via a direct check to obtain the desired result.⟧

8 Remark that $202 = [2, 101] \mid a_n$ if and only if $2 \mid a_n$ and $101 \mid a_n$.

Moreover, for $p = 2$ and for $p = 101$, if $p \mid a_{n_0}$ then $p \mid a_n$ for all $n \geq n_0$. We can show this by induction on n: it holds by assumption for $n = n_0$ and, if $p \mid a_n$, then we have $p \mid (202, a_n)$ and therefore $p \mid a_{n+1} = a_n + (202, a_n)$.

It is thus enough to show the existence of m and n such that $2 \mid a_m$ and $101 \mid a_n$.

Consider the prime 2: if k is even then $(202, k)$ is also even, and if k is odd then $(202, k)$ is also odd; therefore, $a_2 = k + (202, k)$ is always even.

Let us now find n such that $101 \mid a_n$. Given $n \geq 2$, if $101 \nmid a_n$, which as explained before implies $101 \nmid a_m$ for all $m \leq n$, one easily obtains that $a_{n+1} = a_2 + 2(n-1)$. Since the equation $2(n-1) + a_2 \equiv 0 \pmod{101}$ admits $n \equiv -51a_2 + 1 \pmod{101}$ as a solution, any $n \geq 0$ for which the congruence holds is such that $101 \mid a_{n+1}$.

9 First of all, remark that $3 \nmid a_n$ for all $n \geq 0$: for $n = 0$ and $n = 1$ this is true by hypothesis; for $n > 1$ it holds by induction, since, if we had that $3 \mid a_{n+1}$, we would also have that $3 \mid 5a_n$ and thus (because $(3, 5) = 1$) $3 \mid a_n$, which would be a contradiction.

Let us now prove the problem statement, again by induction on n. If $n = 1$ then we have $a_2 = 5a + 3$, which implies $(a_2, a_1) = (5a + 3, a) = (3, a) = 1$. Let us now assume $(a_n, a_{n-1}) = 1$ for some $n \geq 2$, and let d be a common divisor of a_{n+1} and a_n. The integer d must also be a divisor of $a_{n+1} - 5a_n = 3a_{n-1}$. But, since $d \mid a_n$, we have $(d, 3) = 1$ and thus $d \mid a_{n-1}$. Finally, this yields that $d \mid (a_n, a_{n-1}) = 1$.

10

(i) Via direct calculations, one finds $a_0 = 2$, $a_1 = 8$, $a_2 = 34$, $a_3 = 152$. Any real numbers h and k satisfying the requirements must therefore also satisfy

$$\begin{cases} 34 = 8h + 2k \\ 152 = 34h + 8k. \end{cases}$$

The only solution to the system above is $h = 8$, $k = -15$. And indeed, those values for h and k satisfy the original requirement for all $n \geq 1$, because

$$8(3^n + 5^n) - 15(3^{n-1} + 5^{n-1}) = (3+5)(3^n + 5^n) - 5 \cdot 3^n - 3 \cdot 5^n = 3^{n+1} + 5^{n+1}.$$

(ii) Notice that 5 is the multiplicative inverse of 3 modulo 7, so that $7 \mid a_n$ is equivalent to $3^n + 3^{-n} \equiv 0 \pmod 7$. In other words, since 3 is invertible modulo 7, the divisibility condition is the same as $3^{2n} + 1 \equiv 0 \pmod 7$. When examining remainders of the powers of 3, one finds that $3^k + 1 \equiv 0 \pmod 7$ if and only if $k \equiv 3 \pmod 6$. In particular, all admissible values for k are odd, while we must have $k = 2n$ in order for 7 to divide a_n. Consequently, no integer possesses the required property.
⟦Alternatively, one could observe that $(\mathbb{Z}/7\mathbb{Z})^*$ has six elements and, by evaluating $3^n + 3^{-n}$ $\pmod 7$ for $n = 0, 1, 2, 3, 4, 5$, conclude that there is no n such that $7 \mid a_n$.⟧

11 The recurrence relation tells us that $a_n < a_{n+1}$ if and only if $-a_{n-1} + 2a_{n-2} > 0$, that is, $a_{n-1} < 2a_{n-2}$. We shall show by induction that both $a_n < a_{n+1}$ and $a_{n+1} < 2a_n$ hold for all $n \geq 0$. The initial assigned values immediately yield both inequalities for $n = 0, 1$. Now assume both inequalities hold for all indices strictly smaller than n; we shall show them for n by means of the recurrence relation.

By the induction hypothesis we have $a_{n-1} < 2a_{n-2}$, which implies $a_{n+1} = a_n - a_{n-1} + 2a_{n-2} > a_n - 2a_{n-2} + 2a_{n-2} = a_n$ and therefore $a_n < a_{n+1}$.

Also by the induction hypothesis, $a_{n-2} < a_{n-1} < a_n$, hence $a_{n+1} = a_n - a_{n-1} + 2a_{n-2} < a_n - a_{n-1} + 2a_{n-1} = a_n + a_{n-1} < 2a_n$, which yields $a_{n+1} < 2a_n$.

12

(i) We shall show the statement by induction on n. For $n = 0$ we have $(a_0, a_1) = (1, 1) = 1$, so the statement holds. Now, given $n > 0$, assume the statement holds for all natural numbers strictly less than n. We have

$$(a_n, a_{n+1}) = (a_n, ha_n + ka_{n-1}) = (a_n, ka_{n-1}) = (a_n, k),$$

where the last equality is implied by the induction hypothesis. It is therefore enough to show that $(a_n, k) = 1$ for all $n \geq 0$. Again, we can do this by induction on n.

If $n = 0, 1$, then $(a_0, k) = (a_1, k) = (1, k) = 1$. Let us assume the result up to n and show it is true for $n + 1$. We have

$$(a_{n+1}, k) = (ha_n + ka_{n-1}, k) = (ha_n, k) = (h, k) = 1$$

where the second to last equality holds by induction hypothesis. This completes the proof.

(ii) By computing the fist values of $b_n = a_n^2 - 1$ we get $b_0 = 0$, $b_1 = 0$, $b_2 = 106^2 - 1 = 105 \cdot 107$, so the greatest common divisor we are looking for divides $105 \cdot 107$. Moreover, we have $a_2 \equiv 1 \pmod{105}$, while $a_2 \equiv -1 \pmod{107}$. When considering the congruence modulo 105 one immediately obtains by induction on n that $a_{n+1} = 35a_n + 71a_{n-1} \equiv 35 + 71 \equiv 1 \pmod{105}$, hence $b_n \equiv 1^2 - 1 \equiv 0 \pmod{105}$ for all $n \geq 0$.

The congruence modulo 107 yields

$$a_3 \equiv -35 + 71 \equiv 36 \pmod{107}$$

so that $b_3 \equiv 36^2 - 1 \equiv 35 \cdot 37 \not\equiv 0 \pmod{107}$ and, since 107 is prime, $(b_3, 107) = 1$.

The greatest common divisor we were after is therefore 105.

13   We shall show both statements by induction. If $n = 0$, then (i) and (ii) correspond to $F_1 = F_1$ and $0 = F_0$ respectively, so they both trivially hold.

Now assume *both* (i) and (ii) hold for $0, 1, \ldots, n$, and let us show each for $n + 1$. In order to show (i), we want to prove that $\sum_{i=0}^{n+1} \binom{n+1}{i} F_{i+1} = F_{2n+3}$. We have:

$$\sum_{i=0}^{n+1} \binom{n+1}{i} F_{i+1} = \sum_{i=0}^{n+1} \left(\binom{n}{i} + \binom{n}{i-1} \right) F_{i+1}$$

$$= \sum_{i=0}^{n+1} \binom{n}{i} F_{i+1} + \sum_{i=0}^{n+1} \binom{n}{i-1} F_{i+1}.$$

Since $\binom{n}{n+1} = 0$, the fist sum above is $\sum_{i=0}^{n} \binom{n}{i} F_{i+1}$, that is, F_{2n+1}. As for the second sum, setting $j = i - 1$ and remarking that $\binom{n}{-1} = 0$, we find

$$\sum_{i=0}^{n+1} \binom{n}{i-1} F_{i+1} = \sum_{j=0}^{n} \binom{n}{j} F_{j+2} = \sum_{j=0}^{n} \binom{n}{j} (F_j + F_{j+1})$$

$$= \sum_{j=0}^{n} \binom{n}{j} F_j + \sum_{j=0}^{n} \binom{n}{j} F_{j+1}$$

$$= F_{2n} + F_{2n+1} .$$

Putting all of this together, we obtain

$$\sum_{i=0}^{n+1} \binom{n+1}{i} F_{i+1} = F_{2n+1} + F_{2n} + F_{2n+1} = F_{2n+2} + F_{2n+1} = F_{2n+3} .$$

For (ii), we have

$$\sum_{i=1}^{n+1} \binom{n+1}{i} F_i = \sum_{i=1}^{n} \binom{n+1}{i} F_i + \binom{n+1}{n+1} F_{n+1}$$

$$= \sum_{i=1}^{n} \left(\binom{n}{i} + \binom{n}{i-1} \right) F_i + \binom{n}{n} F_{n+1}$$

$$= \sum_{i=1}^{n} \binom{n}{i} F_i + \sum_{j=0}^{n} \binom{n}{j} F_{j+1} = F_{2n} + F_{2n+1} = F_{2n+2}.$$

14

(i) Fist of all, notice that the system

$$\begin{cases} \alpha + \beta = a_1 = 1 \\ \dfrac{1 + \sqrt{13}}{2}\alpha + \dfrac{1 - \sqrt{13}}{2}\beta = a_2 = 4 \end{cases}$$

obtained for $n = 1$ and $n = 2$ does have a solution, since the two equations are independent.

$[\![$Carrying out the computations explicitly one immediately finds $\alpha = 1 + \sqrt{13}/26$ and $\beta = 1 - \sqrt{13}/26.]\!]$

Let us now show the values α and β that solve this system also satisfy the requirement of the problem for all $n \geq 1$. The base cases $n = 1$ and $n = 2$

obviously work. For the inductive step, let $n \geq 2$ and let us assume the statement for all positive integers up to n. We then have

$$a_n + 3a_{n-1} = \alpha(\frac{1 + \sqrt{13}}{2})^{n-1}(\frac{1 + \sqrt{13}}{2} + 3) +$$

$$\beta(\frac{1 - \sqrt{13}}{2})^{n-1}(\frac{1 - \sqrt{13}}{2} + 3)$$

$$= \alpha(\frac{1 + \sqrt{13}}{2})^{n-1}(\frac{7 + \sqrt{13}}{2}) + \beta(\frac{1 - \sqrt{13}}{2})^{n-1}(\frac{7 - \sqrt{13}}{2})$$

$$= \alpha(\frac{1 + \sqrt{13}}{2})^{n+1} + \beta(\frac{1 - \sqrt{13}}{2})^{n+1}$$

so the statement holds for $n + 1$.

(ii) We show that a_n is even if and only if $n \equiv 2 \pmod 3$. Indeed, using the recurrence relation yields that $a_n = 4a_{n-2} + 3a_{n-3}$ for all $n \geq 4$. Therefore, a_n is even if and only if a_{n-3} is even; in other words, the parity of a_n only depends on the remainder of n modulo 3. In conclusion, the statement is immediately proven by remarking that, if a_1 is odd, then a_2 is even and $a_3 = a_2 + 3a_1 = 7$ is, again, odd.

2 Combinatorics

15

(i) Each element $x \in X$ must belong either to A, or to B, or to C; the three options, since A, B and C have pairwise empty intersections, are mutually exclusive. The number of possible triples is therefore 3^n.

(ii) We can argue as above: each element $x \in X$ must belong to one of the sets

$$A \setminus (B \cup C), \quad B \setminus (A \cup C), \quad C \setminus (A \cup B), \quad (A \cap B) \setminus C,$$

$$(A \cap C) \setminus B, \quad (B \cap C) \setminus A, \quad A \cap B \cap C.$$

The fact that these seven option are mutually exclusive is obvious. The number of triples is therefore 7^n, as claimed.

16 First of all, remark that the only pair in X consisting of two copies of the same integer is $(1, 1)$. We can therefore partition X into the following disjoint subsets: $\{(1, 1)\}$, $X_1 = \{(m, n) \mid 1 \leq m < n \leq 100, (m, n) = 1\}$ and $X_2 = \{(m, n) \mid 1 \leq n < m \leq 100, (m, n) = 1\}$.

It is also clear that X_1 and X_2 contain the same number of elements, since $X_1 \ni (m, n) \longmapsto (n, m) \in X_2$ is a bijection between the two sets.

Moreover, by grouping together all elements whose second component is n for $n = 1, 2, \ldots, 100$, we obtain that $|X_1| = \sum_{n=2}^{100} \phi(n)$.

We can therefore conclude that, indeed, $|X| = 1 + |X_1| + |X_2| = 1 + 2|X_1| = 1 + 2\sum_{n=2}^{100} \phi(n) = 2\sum_{n=1}^{100} \phi(n) - 1$.

17 First, let us consider the set of all n satisfying the conditions $(n, 18) = 6$ and $n \equiv 2 \pmod 7$; that is, let us set aside for the moment the further restriction given by the inequalities in the statement.

The first condition is equivalent to 6 dividing n and 9 not dividing n, that is, $n = 6h$ and $h \not\equiv 0 \pmod 3$ for some integer h. The second condition now becomes $6h \equiv 2 \pmod 7$, that is, $h \equiv -2 \pmod 7$, so we must have $h = 7k - 2$ for some integer k. Using what we just proved, we can now rewrite $h \not\equiv 0 \pmod 3$ as $7k - 2 \not\equiv 0 \pmod 3$, that is, $k \not\equiv -1 \pmod 3$, so we must have $k = 3u$ or $k = 3u + 1$ for some integer u. Substituting into previous expressions we obtain $n = 6h = 6(7k - 2) = 6(7 \cdot 3u - 2) = 126u - 12$ or $n = 6h = 6(7k - 2) = 6(7 \cdot (3u + 1) - 2) = 126u + 30$.

Let us now impose the condition $1 \le n \le 10000$: in the first case, this gives $1 \le 126u - 12 \le 10000$ and so $1 \le u \le 79$, which yields 79 solutions in total; in the second case, $1 \le 126u + 30 \le 10000$ so $0 \le u \le 79$, hence another 80 solutions.

In conclusion, the set in the problem statement contains $79 + 80 = 159$ elements.

18 The positive divisors of $3^{40} \cdot 5^{25}$ are the integers of the form $3^n 5^m$ with $0 \le n \le 40, 0 \le m \le 25$.

The required congruence then becomes $3^n 5^m \equiv 1 \pmod 7$. Powers of 3 modulo 7 are congruent to $1, 3, 2, -1, -3$ and -2, in this order, and then cycle back to 1. In particular, $5 \equiv -2 \equiv 3^5 \pmod 7$. We may therefore rewrite the requirement for n and m as $3^{n+5m} \equiv 1 \pmod 7$.

We just saw that 3 has multiplicative order 6 modulo 7, hence $n + 5m \equiv 0 \pmod 6$, that is, $n \equiv m \pmod 6$.

In order to obtain the required cardinality we must find the number of pairs (n, m) such that $0 \le n \le 40, 0 \le m \le 25$ and $n \equiv m \pmod 6$.

Since $40 = 6 \cdot 6 + 4$, we have $7 = 6 + 1$ solutions for each m congruent to $0, \ldots, 4$ modulo 6 and, since $25 = 4 \cdot 6 + 1$, there are 22 such m, which amounts to 154 solutions.

We must now count solutions such that $m \equiv 5 \pmod 6$: each such m gives six possibilities for n and, since m can only be 5, 11, 17 or 23, this yields another 24 solutions.

In conclusion, the required cardinality is $154 + 24 = 178$.

19

Solution 1 We shall proceed in several steps.

If n were a power of 2, $n = 2^a$ say, then $\phi(n) = 2^{a-1}$ would be as well. There must therefore be an odd prime factor of n.

Given a prime p that divides n, the number $p - 1$ divides 12, so $p - 1$ can only be 1, 2, 3, 4, 6 or 12. But, since p is prime, the only options are $p = 2, 3, 5, 7, 13$.

Given a prime q such that q^2 divides n, q must also divide $\phi(n)$: such a prime can only be 2 or 3.

If five divides n, then 5^2 does not divide n thanks to the remark above, so we must have $n = 5m$ with $(5, m) = 1$. But then $12 = \phi(n) = 4\phi(m)$ and so $\phi(m) = 3$, which is impossible since the only odd value that Euler's totient function takes is 1.

If seven divides n, then 7^2 does not divide n, so $n = 7m$ with $(7, m) = 1$. Then $12 = \phi(n) = 6\phi(m)$ and so $\phi(m) = 2$.

Let now q be a prime that divides m; then $q - 1$ divides $\phi(m) = 2$, so the only possibilities are $q = 2$ and $q = 3$. Moreover, 9 cannot divide m or 3 would divide $\phi(m) = 2$. Similarly, 8 cannot divide m or 4 would divide $\phi(m)$. We thus have $m = 2^a 3^b$ with $a = 0, 1, 2$ and $b = 0, 1$. Checking these six values directly yields $\phi(m) = 2$ for $m = 3, 4, 6$.

We have ended up with the three possibilities $n = 21$, $n = 28$ and $n = 42$.

If 13 divides n, then 13^2 does not divide n, so $n = 13m$ with $(13, m) = 1$. Then $12 = \phi(n) = 12\phi(m)$ and so $\phi(m) = 1$, which gives $m = 1$ or $m = 2$. This yields the solutions $n = 13$ and $n = 26$.

If none of the above holds then $n = 2^a 3^b$ for natural numbers a and b with $b > 0$. In this case, $12 = \phi(n) = 2^{a-1} \cdot 2 \cdot 3^{b-1} = 2^a 3^{b-1}$, so $a = 2$ and $b = 2$, that is, $n = 36$.

Finally, we can conclude that $\phi(n) = 12$ if and only if n belongs to the set $\{13, 21, 26, 28, 36, 42\}$.

Solution 2 Let $n = p_1^{\alpha_1} \cdots p_k^{\alpha_k}$ be the prime factorisation of n. Since Euler's totient function is multiplicative, we have $\phi(n) = \prod_{i=1}^{k} \phi(p_i^{\alpha_i})$. Unless $p_i^{\alpha_i} = 2$, $\phi(p_i^{\alpha_i}) = (p_i - 1)p_i^{\alpha_i}$ is even, so we can only decompose 12 as $1 \cdot 2 \cdot 6$, $2 \cdot 6$, $1 \cdot 12$ or 12, where the factor 1 only appears if $2|n$ and $4 \nmid n$.

Consider the case where $\phi(p_i^{\alpha_i}) = 2$: if $p_i = 2$ then $p_i^{\alpha_i} = 4$; if not, then $p_i^{\alpha_i} = 3$.

If $\phi(p_i^{\alpha_i}) = 6$, then if $p_i = 3$ we must have $p_i^{\alpha_i} = 9$; the only other possibility is $p_i^{\alpha_i} = 7$.

Finally, consider the case where $\phi(p_i^{\alpha_i}) = 12$: one can immediately check that $p_i \neq 2, 3$, so we must have $p_i - 1 = 12$, that is, $p_i^{\alpha_i} = 13$.

Taking into account the fact that factors $p_i^{\alpha_i}$ must be pairwise relatively prime, we are left with only the following possibilities for n: $4 \cdot 9 = 36$, $4 \cdot 7 = 28$, $3 \cdot 7 = 21$, $2 \cdot 3 \cdot 7 = 42$, 13, $2 \cdot 13 = 26$. These all satisfy the required condition.

20   Since $n^2 = x + y \leq 49 + 49 = 98$ we have $n \leq 9$. We shall first deal with the case $n \leq 7$.

In this case, all triples $(h, n^2 - h, n)$ with $h = 0, \ldots, n^2$ satisfy the conditions. We therefore have $n^2 + 1$ triples for each $n \leq 7$.

If $n = 8$ then we get the triples $(h, 64 - h, 8)$ with $h = 15, \ldots, 49$, where the values of h are those for which $0 \leq x, y < 50$; this yields $49 - 15 + 1 = 35$ triples.

Similarly, for $n = 9$ we get the triples $(h, 81 - h, 9)$ with $h = 32, \ldots, 49$, which are $49 - 32 + 1 = 18$.

The total number of triples is therefore $1+2+5+10+17+26+37+50+35+18 = 201$.

21 We can write the expression for Euler's totient function ϕ as

$$\frac{\phi(n)}{n} = \prod_{p|n} \frac{p-1}{p}.$$

Suppose q is the largest prime divisor of n; q is then also the largest prime divisor of the denominator of $\phi(n)/n$: this is because q divides the denominator of $(q-1)/q$ and cannot cancel out with any factor of q dividing the numerator of $(p-1)/p$, since we have $p-1 < q$.

Consequently, if $\phi(n)/n = 2/5$, then the largest prime factor of n must be 5. Consider all integers n for which this is true. If n is divisible by 2 and by 5 but not by 3, then

$$\frac{\phi(n)}{n} = \frac{1}{2} \cdot \frac{4}{5} = \frac{2}{5}$$

and the condition is satisfied. In all other cases, that is, when n is not divisible by 2 or is divisible by 3, it is easily checked that the desired equality does not hold.

22 Set $D = \{1 \le d \le n \mid d \text{ divides } n\}$ and $\Phi = \{1 \le k \le n \mid (k, n) = 1\}$. If $x \in D \cap \Phi$ then $x = (x, n) = 1$, hence $|D \cap \Phi| = 1$.

Statement (i) is a simple consequence of the inclusion-exclusion principle: we have

$$|D| + |\Phi| = |D \cup \Phi| + |D \cap \Phi| = |D \cup \Phi| + 1 \le n + 1.$$

As for (ii), remark that the required equality holds if and only if $|D \cup \Phi| = n - 1$, that is, if and only if there is *exactly one* integer k, with $1 \le k \le n$, which is not a divisor of n nor is it coprime to n.

If n is prime, then $d(n) = 2$ and $\phi(n) = n - 1$, thus $d(n) + \phi(n) = n + 1$. It is therefore enough to consider composite numbers. If $n = ab$ with $a > 1$ and $b > 4$, then there are at least two numbers, that is, $a(b-1)$ and $a(b-2)$, that are not divisors of n nor coprime to n, so no such n satisfies the required equality.

Assume that $n = p_1 \ldots p_k$ is the prime factorisation of n, with $p_1 \le p_2 \le \cdots \le p_k$ not necessarily distinct. If we set $a = p_1$ and $b = p_2 \ldots p_k$, we can use what we showed above to obtain that we can only have $p_2 \ldots p_k = 2, 3, 4$, so $n = 4, 6, 9, 8$. Direct verification shows that the case $n = 4$ must be excluded, because $d(4) + \phi(4) = 3 + 2 = 5$, while all other values are indeed solutions:

$$d(6) + \phi(6) = 4 + 2 = 6, \quad d(9) + \phi(9) = 3 + 6 = 9, \quad d(8) + \phi(8) = 4 + 4 = 8.$$

23 Remark that all natural numbers with the required property must be multiples of 3, so can be written in the form $n = 3^a m$, where $(m, 3) = 1$. We must also have

$a \leq 2$, otherwise both n and $\phi(n) = 2 \cdot 3^{a-1}\phi(m)$ would be multiples of 9. Now, n must be odd, because from $3 \mid n$ we obtain that $2 \mid \phi(n)$ and therefore if n were even 2 would divide $(n, \phi(n))$. Finally, all primes $p > 3$ that divide n must appear in the prime factorisation of n with an exponent of 1, because if that were not the case we would have $p \mid (n, \phi(n))$. Two separate cases present themselves:

① The integer n is a multiple of 9, that is, $n = 9m$ with $(m, 3) = 1$ and m odd and squarefree. Then $n \leq 120$ implies that $m \leq 13$. If $p \mid m$ then $3 \nmid p - 1$, or we would have $9 \mid (n, \phi(n))$. The only admissible values for m are therefore $m = 1, 5, 11$, which yield the solutions $n = 9, 45, 99$.

② Conversely, suppose that n is not a multiple of 9, that is, $n = 3m$ with $(m, 3) = 1$ and m odd and squarefree. If $n = 3p$ for some prime p, then $\phi(n) = 2(p - 1)$ and the required condition is equivalent to $3 \mid p - 1$. Since we ask that $n \leq 120$, we must have $p \leq 40$: the only primes satisfying the requirements are $7, 13, 19, 31, 37$, and from those we get the solutions $n = 21, 39, 57, 93, 111$.

Finally, if $n = 3m$ where m is the product of at least two distinct primes, each of which must be strictly greater 3, the fact that $m \leq 40$ implies that the only possibility for m is $m = 5 \cdot 7$, that is, $n = 105$.

24   There are three possible configurations of remainders modulo 2 for a triple (a, b, c) satisfying the requirements: $(0, 0, 1)$, $(0, 1, 0)$ and $(0, 0, 1)$. There are 12 possibilities for the triple's remainders modulo 3: $(0, x, y)$, $(x, 0, y)$ or $(x, y, 0)$, where x and y are nonzero remainders.

There are therefore 36 possible configurations modulo 6 for the triples under consideration. For each configuration modulo 6, there are $10^3 = 1000$ triples pertaining to it, since we have $60/6 = 10$. Hence, the total number of triples satisfying the requirements is $36 \cdot 1000 = 36000$.

25   We shall show the statement by induction on $k = \omega(m)$. We have $k = 0$ if and only if $m = 1$, in which case

$$1 = \frac{\phi(1)}{1} \geq \frac{1}{0 + 1} = 1.$$

Now assume the statement for $\omega(m) = k$; we shall show it for $\omega(m) = k + 1$. Suppose $m = p_1^{\alpha_1} \cdots p_{k+1}^{\alpha_{k+1}}$ with $p_1 < p_2 < \cdots < p_{k+1}$ and $\alpha_i > 0$ for all i. Set $n = p_1^{\alpha_1} \cdots p_k^{\alpha_k}$.

By the induction hypothesis, $\phi(n)/n \geq 1/(k + 1)$; moreover, it is clear that $p_i \geq i + 1$ for all i, so in particular

$$1 - \frac{1}{p_{k+1}} \geq 1 - \frac{1}{k + 2} = \frac{k + 1}{k + 2}.$$

Therefore, we have

$$\frac{\phi(m)}{m} = \frac{\phi(n)}{n} \cdot \left(1 - \frac{1}{p_{k+1}}\right) \geq \frac{1}{k + 1} \cdot \frac{k + 1}{k + 2} = \frac{1}{k + 2}.$$

26 There are $8 \cdot 9 \cdot 9 \cdot 9 - 1$ integers satisfying the first two requirements: indeed, there are eight possibilities for the first digit and nine for each of the other digits, except that the number 1000 must be excluded. Among these integers, those that *do not* satisfy the last requirement are exactly those whose digits are all different: there are eight possibilities for the first digit; eight possibilities for the second (9 minus the one value already chosen for the first digit); seven for the third (9 minus the values used for the first two digits) and six for the fourth (9 minus the values used for the first three digits).

Numbers of this type are therefore $8 \cdot 8 \cdot 7 \cdot 6$. Consequently, there are $8 \cdot 9 \cdot 9 \cdot 9 - 1 - 8 \cdot 8 \cdot 7 \cdot 6 = 3143$ integers satisfying the original requirements.

27

(i) There are two possibilities for $f(1)$: $f(1) = 1$ and $f(1) = 2$. Suppose $2 \le i \le n$; given the values of $f(1), \ldots, f(i-1)$, which we know must belong to $\{1, \ldots i\}$ and be distinct, there are two possibilities for $f(i)$, given by the two elements of $\{1, \ldots, i+1\} \setminus \{f(1), \ldots, f(i-1)\}$. Finally, given the values of $f(1), \ldots, f(n-1)$, there is but one possible value for $f(n)$, that is, the element of $\{1, \ldots, n\}$ different from $f(1), \ldots, f(n-1)$. The cardinality we are looking for is therefore 2^{n-1}.

(ii) Let x_n be the cardinality of the set in the statement, which we shall denote by X_n. For $n = 1, 2$ all permutations satisfy the requirement, so $x_1 = 1 = F_2$ and $x_2 = 2 = F_3$.
In order to show the claim it is therefore enough to show that $x_{n+1} = x_n + x_{n-1}$ for $n \ge 2$. Since we necessarily have $f(n+1) = n+1$ or $f(n+1) = n$, let us partition the set X_{n+1} into two subsets: $Y_{n+1} = \{f \in X_{n+1} \mid f(n+1) = n+1\}$ and $Z_{n+1} = \{f \in X_{n+1} \mid f(n+1) = n\}$.
Elements of Y_{n+1} can be made to bijectively correspond to elements of X_n, so $|Y_{n+1}| = x_n$. On the other hand, if $f \in Z_{n+1}$ then the fact that $f(i) \le n$ for $i = 1, \ldots, n-1$ implies that necessarily $f(n) = n+1$. It follows that Z_{n+1} is in bijection with X_{n-1} and therefore has x_{n-1} elements.

28

(i) The number of all subsets satisfying the requirements can be obtained as the sum of the number N of all subsets of X consisting of three elements that are all congruent modulo 5, and the number M of all subsets of X consisting of three elements, exactly two of which are congruent modulo 5. Since X has 20 elements in each congruence class modulo 5, the number N is given by the number of possible classes modulo five times the number of three-element subsets of the chosen class, that is, $N = 5 \cdot \binom{20}{3}$. Similarly, M is given by the number of possible choices of a class modulo five times the number of possible two-element subsets of the chosen class, times the number of ways we can choose a third element outside that class, so $M = 5 \cdot \binom{20}{2} \cdot 80$. Taking the sum we obtain $N + M = 81700$.

⟦The same result could be obtained by subtracting the number of subsets of X consisting of three elements from distinct congruence classes, that is, $\binom{5}{3}20^3$, from the number $\binom{100}{3}$ of all possible subsets of X with three elements.⟧

(ii) The set X contains 20 elements from each congruence class modulo 5. Maps from X to X satisfying the requirement send each of the 100 elements $n \in X$ to one of the 20 elements of X in the congruence class of $n + 1$ modulo 5. The number we want is therefore 20^{100}.

29

(i) The condition $(xy, 6) = 1$ is equivalent to $(x, 6) = 1$ and $(y, 6) = 1$, so we have to find the number of ordered pairs of elements of X such that both components are coprime to 6. Now, set $X_i = \{x \in X \mid x \equiv i \pmod 6\}$ for $i \in \{1, 2, 3, 4, 5, 6\}$; we have $X = \sqcup_{i=1}^{6} X_i$ and for each $x \in X_i$ we have $(x, 6) = (i, 6)$. It follows that $x \in X$ is coprime to 6 if and only if $x \in X_1$ or $x \in X_5$. Since $100 = 16 \cdot 6 + 4$, we get $|X_1| = 17$ and $|X_5| = 16$. Therefore, X has 33 elements that are coprime to 6.

The number of ordered pairs of elements in X such that both components are coprime to 6 is thus 33^2.

(ii) Let $A = \{(x, y) \in X^2 \mid x < y + 6\}$. For $y = 1, 2, \ldots, 95$, set $X_y = \{1, \ldots, y + 5\}$, and, for $96 \leq y \leq 100$, set $X_y = X$. Clearly, we have

$$A = \bigsqcup_{y=1}^{100} X_y \times \{y\}$$

and so, this being a disjoint union,

$$|A| = \sum_{y=1}^{100} |X_y|$$
$$= \sum_{y=1}^{95} (y + 5) + 5 \cdot 100$$
$$= 500 + \sum_{y=6}^{100} y$$
$$= 500 + 100 \cdot 101/2 - (1 + 2 + 3 + 4 + 5)$$
$$= 5535.$$

30

(i) Maps in the set A are bijective, so for all $y \in X$, there is $x \in X$ such that $y = f(x)$. The condition $f^2(x) \equiv f(x) \pmod 2$ implies that $f(y) \equiv y \pmod 2$ for all $y \in X$, so the elements of A are exactly those bijections from X to X sending even numbers to even numbers and odd numbers to odd numbers. Consequently, $|A| = 50!50!$.

(ii) Given $f \in B$, set $Y = f^{-1}(1)$; the set Y cannot be empty, because $1 \in f(X)$. Moreover, $f(Y) = \{1\}$ and $1 \notin f(X \setminus Y)$. It is easy to check that the condition $f^2(X) = \{1\}$ is equivalent to $\{1\} = f(Y) \subseteq Y$ and $f(X \setminus Y) \subseteq Y$, which, because of the argument above, implies $f(X \setminus Y) \subseteq Y \setminus \{1\}$.

Maps in B having $f^{-1}(1) = Y$ for some fixed subset Y of X such that $1 \in Y$ and $|Y| = k+1$ are therefore as many as all possible maps from $X \setminus Y$ to $Y \setminus \{1\}$, that is, k^{99-k}. For each $k \geq 0$ there are $\binom{99}{k}$ subsets Y of X that have cardinality $k+1$ and contain 1, so

$$|B| = \sum_{k=0}^{99} \binom{99}{k} k^{99-k}.$$

31

(i) Clearly, $d \mid 144000 = 2^7 3^2 5^3$ if and only if $d = 2^a 3^b 5^c$ with $0 \leq a \leq 7$, $0 \leq b \leq 2$, and $0 \leq c \leq 3$. The number of divisors of an integer d as above is $(a+1)(b+1)(c+1)$, so it is even if and only if at least one among a, b, c is odd. The number of values for d such that this holds can be obtained as a sum by considering the following cases: a is odd and b, c have no further constraints, $4 \cdot 3 \cdot 4 = 48$ possibilities; a is even, b is odd and c has no further constraints, $4 \cdot 1 \cdot 4 = 16$ possibilities; finally, a and b are both even and c is odd, $4 \cdot 2 \cdot 2 = 16$ possibilities.
So X has 80 elements.
⟦The same result could be obtained by subtracting from the number of all divisors of 144000, that is, $8 \cdot 3 \cdot 4 = 96$, the number of those having an odd number of divisors, so those given by a, b, c all even, that is, $4 \cdot 2 \cdot 2 = 16$.⟧

(ii) As for the set Y, remark that a number is both a perfect square and a perfect cube if and only if it is a 6th power. It follows that $d = 2^a 3^b 5^c \in Y$ if and only if a, b, c are all even, but not all divisible by 6. Therefore, we have $|Y| = 4 \cdot 2 \cdot 2 - 2 \cdot 1 \cdot 1 = 14$.

32

(i) Consider $X_0 = \{x \in X \mid x \equiv 0 \pmod 2\}$ and $X_1 = \{x \in X \mid x \equiv 1 \pmod 2\}$. For all $A \subseteq X$, setting $A_i = A \cap X_i$ for $i = 0, 1$, we have $A = A_0 \cup A_1$ and $\sum_{a \in A} a \equiv |A_1| \pmod 2$.
Therefore, A belongs to $\mathcal{A}$ if and only if the cardinality of A_1 is even. So, sets $A \in \mathcal{A}$ are obtained by choosing a subset A_0 of X_0, which can be done in 2^{50} ways, and a subset A_1 of X_1 with an even number of elements in one of $\sum_{k=0}^{25} \binom{50}{2k} = 2^{49}$ ways. It follows that $|\mathcal{A}| = 2^{99}$.

(ii) Using the notation we introduced in order to answer the previous question, first remark that the condition $A \in \mathcal{B}$ does not impose any constraints on A_1, which can be any subset of X_1: we have 2^{50} possibilities for A_1. The subset A_0 can be of one of the following types.
It can have at least three elements: $|A_0| \geq 3$, which gives $2^{50} - \binom{50}{0} - \binom{50}{1} - \binom{50}{2}$ possibilities. Or we have $|A_0| = 2$, in which case at least one of its two elements must be divisible by 4; equivalently, we must choose two elements of X_0 that do not both lie in the subset containing the 25 elements divisible by 2 but not by 4, which gives $\binom{50}{2} - \binom{25}{2}$ possibilities. Or, finally, $|A_0| = 1$ and the element

it contains must be one of the 12 elements of X_0 that are divisible by 8, which gives 12 possibilities.

In conclusion, $|\mathcal{B}| = 2^{50}(2^{50} - \binom{50}{0} - \binom{50}{1} - \binom{25}{2} + 12) = 2^{50}(2^{50} - 339)$.

33 Let $X = \{2, \ldots, 1000\}$ and consider $n \in X$. We shall write $n = 2^a p_1^{e_1} \cdots p_r^{e_r}$ with $p_1, \ldots p_r$ distinct odd primes, $e_i \geq 1$ and $a \geq 0$. Moreover, since $\phi(n)$ is even for all $n > 2$, the required divisibility implies that n must be even, that is, $a \geq 1$. It follows that

$$\phi(n) = 2^{a-1} \prod_{i=1}^{r} p_i^{e_i-1}(p_i - 1) \mid n \iff \prod_{i=1}^{r}(p_i - 1) \mid 2 \prod_{i=1}^{r} p_i.$$

Since the p_i's are odd, we have $2^r \mid \prod_{i=1}^{r}(p_i - 1)$: but then, we must have $r = 0$ or $r = 1$, because $2 \prod_{i=1}^{r} p_i$ cannot be a multiple of 4.

If $r = 0$, then $n = 2^a$ with $a \geq 1$: there are exactly nine integers of this form in the set X and all of them satisfy the condition $\phi(n) \mid n$.

On the other hand, if $r = 1$ then we have $n = 2^a p^e$ for some odd prime p and some $a, e \geq 1$: the condition $\phi(n) \mid n$ is equivalent to $p - 1 \mid 2p$, so to $p - 1 \mid 2$, since $(p - 1, p) = 1$. The only possibility is that $p = 3$.

Integers of the form $n = 2^a 3^e$, with $a, e \geq 1$, in the set X can be enumerated as follows: we have $2^a 3^e \in X$ if and only if $2^a \leq 1000/3^e$, which immediately yields eight possible values of a for $e = 1$, plus 6 values for $e = 2$, plus 5 for $e = 3$, plus 3 for $e = 4$, plus 2 for $e = 5$, while there are none for $e \geq 6$.

In conclusion, the number of integers n satisfying the required conditions is $9 + 8 + 6 + 5 + 3 + 2 = 33$.

34 We shall check the formula by induction on n. For $n = 1$ we clearly have $p_1 = 1/2$, that is the probability that the outcome of the first coin toss is heads, and indeed $1/2 = 2/3 + (-1)/(3 \cdot 2)$. For $n = 2$ we have $p_2 = 1/2 \cdot 1/2 + 1/2 = 3/4$, that is, x_2 is given by the probability of the first two outcomes being heads, plus the probability that the first outcome is tails; and indeed, $3/4 = 2/3 + 1/(3 \cdot 2^2)$.

Now assume the formula from the statement is valid for all $m < n$; we shall show it for n. Notice that there are two mutually exclusive ways to obtain a score $x_k = n$ for some k: either there is k such that $x_{k-1} = n - 1$ and that the outcome of the kth coin toss is heads, or there is k such that $x_{k-1} = n - 2$ and that the outcome of the kth coin toss is tails.

By the induction hypothesis, the probability of the first event is

$$\frac{1}{2}\left(\frac{2}{3} + \frac{(-1)^{n-1}}{3 \cdot 2^{n-1}}\right)$$

and the probability of the second event is

$$\frac{1}{2}\left(\frac{2}{3} + \frac{(-1)^{n-2}}{3 \cdot 2^{n-2}}\right).$$

By summing the two, we find that the probability of having $x_k = n$ for some k is equal to

$$\frac{2}{3} + \frac{1}{2} \cdot \frac{(-1)^n}{3 \cdot 2^{n-1}}(-1 + 2) = \frac{2}{3} + \frac{(-1)^n}{3 \cdot 2^n}$$

as required.

35 We shall write $S(f) = S_+(f) + S_-(f)$, where

$$S_+(f) = \sum_{i \mid f(i) > i} (f(i) - i), \qquad S_-(f) = - \sum_{i \mid f(i) < i} (f(i) - i).$$

Since we clearly have $\sum_{i=1}^{10}(f(i) - i) = 0$, it follows that $S_+(f) = S_-(f)$ and thus $S(f) = 2S_+(f)$ is always even. This implies that the answer to the second question is 0.

As for the first question, $S(f) = 2$ implies $S_+(f) = S_-(f) = 1$, so there is exactly one index i for which $f(i) = i + 1$ and exactly one j for which $f(j) = j - 1$, whereas for all other indices k different from i, j we have $f(k) = k$. The map f must therefore exchange two consecutive numbers and keep all others fixed. There are exactly nine pairs of consecutive integers in $\{1, 2, \ldots, 10\}$, so the answer is 9.

For the third question, we must have $S_+(f) = S_-(f) = 2$. We shall distinguish three cases.

① There is exactly one i such that $f(i) = i + 2$ and exactly one j for which $f(j) = j - 2$. This case can be dealt with in the same way as question (i), by noticing that there are eight pairs of integers in $\{1, 2, \ldots, 10\}$ whose difference is 2.

② There are exactly two indices i for which $f(i) = i + 1$ and exactly two indices j for which $f(j) = j - 1$. In this case, the map f must exchange the numbers within each of two pairs of consecutive integers. In order to find the number of such maps, remark that, if the smallest pair—that is, the pair whose integers are smaller—is $\{1, 2\}$, then there are seven ways to choose the other pair; if the smallest pair is $\{2, 3\}$, then there are six ways to choose the other pair, and so on. We thus have $7 + 6 + \cdots + 1 = 28$ possibilities for f.

③ There are exactly two indices i for which $f(i) = i + 1$ and there is one index j such that $f(j) = j - 2$, or vice versa. We shall only discuss the first case, since the second is symmetric.

The permutation f can only be of the form $(i, i + 1, i + 2)$, that is, it must cycle three consecutive integers. There are 8 possible triples of consecutive integers, which, by symmetry, yields a total of 16 possibilities for f.

The answer to the third question is obtained by summing the three results above: we get $8 + 28 + 16 = 52$.

36 The sum of all integers from 1 to 100 (inclusive) is 5050.

(i) Choosing a subset A containing 96 elements is equivalent to choosing its complement B containing 4 elements. Since 5050 is even, the sum of all elements in A is even if and only if the sum of all elements in B is even. We have the following possible cases:

① The 4 numbers are all even, that is, B is a four-element subset of the set of the 50 even numbers between 1 and 100; in this case, there are $\binom{50}{4}$ choices for B.

② The four numbers are all odd; like before, there are $\binom{50}{4}$ choices for B.

③ Two of the 4 numbers are even, two are odd; in this case B consists of a two-element subset of a set of 50 even numbers and a two-element subset of a set of 50 odd numbers; this yields $\binom{50}{2} \cdot \binom{50}{2}$ possibilities for B.

There are therefore $2\binom{50}{4} + \binom{50}{2}^2$ possibilities in total.

(ii) As before, rather than choosing the subset A we shall choose its three-element complement C. Since $5050 \equiv 1 \pmod 3$, the sum of all elements in A is divisible by 3 if and only if the sum of all elements in C is congruent to 1 modulo 3.

Remark that, among the numbers from 1 to 100, there are 33 congruent to zero modulo 3, 34 congruent to 1 and 33 congruent to 2. Suppose $C = \{a, b, c\}$. Up to reordering the labels a, b, c, there are three possibilities: ① $a \equiv b \equiv 0, c \equiv 1 \pmod 3$, in which case $\{a, b\}$ can be chosen in $\binom{33}{2}$ ways and c in 34 ways; ② $a \equiv b \equiv 1, c \equiv 2 \pmod 3$, in which case $\{a, b\}$ can be chosen in $\binom{34}{2}$ ways and c in 33 ways; or, finally, ③ $a \equiv b \equiv 2, c \equiv 0 \pmod 3$, in which case $\{a, b\}$ can be chosen in $\binom{33}{2}$ ways and c in 33 ways.

We thus get a total of $\binom{33}{2} \cdot 34 + \binom{34}{2} \cdot 33 + \binom{33}{2} \cdot 33$ possibilities for A.

37 The set A is in bijection with the set of all possible five-element subsets of $\{1, \ldots, 100\}$: indeed, we can make each $f \in A$ correspond to the set of its values $\{f(1), f(2), f(3), f(4), f(5)\}$. The cardinality of A is therefore equal to the number of ways one can choose 5 elements in a set of 100 elements, that is, $\binom{100}{5}$.

The set B is obtained from the set A by excluding all maps such that $f(i + 1) = f(i) + 1$ for all $i = 1, 2, 3, 4$, that is, by subtracting the set of maps such that $f(1) = a$, $f(2) = a + 1$, $f(3) = a + 2$, $f(4) = a + 3$, $f(5) = a + 4$ with $a \in \{1, \ldots, 96\}$. Therefore, $|B| = \binom{100}{5} - 96$.

In order to find the cardinality of C, remark that the condition that defines it is equivalent to the following: the map $g(i) = f(i) - i$ is strictly increasing and takes values in $\{0, \ldots, 95\}$. Indeed, we have

$$f(i + 1) > f(i) + 1 \iff f(i + 1) - (i + 1) > f(i) + 1 - (i + 1) = f(i) - i$$

and, moreover,

$$f(1) \geq 1 \iff g(1) \geq 0 \quad \text{and} \quad f(100) \leq 100 \iff g(100) \leq 95.$$

So in order to find the cardinality of C it is enough to find the number of strictly increasing maps taking values in $\{0, \ldots, 95\}$. By the same argument used for finding the cardinality of A, we find that C has $\binom{96}{5}$ elements.

38

(i) First, consider the problem of forming an *ordered* sequence of n teams of four. In order to form the first team, we must choose four people from a set of $4n$ people, which can be done in $\binom{4n}{4}$ ways. Players for the second team will be chosen among $4n - 4$ people, so the team can be formed in $\binom{4n-4}{4}$ ways. We can iterate this argument to obtained that the ordered list of n teams can be formed in

$$\binom{4n}{4}\binom{4n-4}{4}\cdots\binom{4}{4} = \frac{(4n)!}{(4!)^n}$$

ways. Finally, since the same n teams can be ordered in $n!$ distinct ways, the answer is

$$\frac{(4n)!}{(4!)^n \cdot n!}.$$

(ii) Again, we shall first form the n teams in order. In order to form the first team, we must choose two men and two women in the respective $2n$-element sets: this can be done in $\binom{2n}{2}^2$ ways. By a procedure similar to the one from before, an ordered list of n teams can be formed in

$$\binom{2n}{2}^2\binom{2n-2}{2}^2\cdots\binom{2}{2}^2 = \frac{(2n)!^2}{2^{2n}}$$

ways, hence a non-ordered set of n teams can be formed in

$$\frac{(2n)!^2}{2^{2n}n!}$$

ways.

39 Notice that x and $f(x)$ must have the same number of divisors and the same number of multiples in the interval under consideration. It follows that 1, which is the only number with exactly one divisor, must be sent to 1, and any prime must be sent to a prime, since primes are those natural numbers that have exactly two divisors.

In general, the number of divisors of an integer m can be obtained from the prime factorisation $p_1^{a_1} \cdots p_k^{a_k}$ of m as the product $(a_1+1)\cdots(a_k+1)$; indeed, each prime p_i may appear in the prime factorisation of a divisor of m with an exponent ranging from 0 to a_i, so there are $a_i + 1$ choices for each $i = 1, \ldots, k$.

Consider in particular the case where all a_i are equal to 1, that is, $m = p_1 p_2 \cdots p_k$. In this case, m has exactly 2^k divisors. Now, since p_i divides m, its image $q_i = f(p_i)$ must be a prime dividing $f(m)$. Moreover, the q_i's must be distinct because f is injective. Therefore, $f(m)$ must be a multiple of $q_1 q_2 \cdots q_k$, and since it must have the same number of divisors as m (that is, 2^k), it must be equal to the product in question. This shows (i).

As for (ii), remark that powers of primes can be characterised as those natural numbers that do not have two distinct prime factors. Since we already showed that primes are sent to primes and that distinct primes are sent to distinct primes, it follows that any power p^n of a single prime p must be sent to some power of the corresponding prime $q = f(p)$. But we can say more: p^n must be sent to q^n, that is, maintain the same exponent n, otherwise p^n, which has $n+1$ divisors, would not have the same number of divisors as its image under f. This concludes the proof of (ii).

More generally, we shall show by induction on m that if the prime p_i is sent to $q_i = f(p_i)$ for $i = 1, \ldots, k$, then $m = p_1^{a_1} \cdots p_k^{a_k}$ is sent to $q_1^{a_1} \cdots q_k^{a_k}$.

This has just been proven for $k = 1$, so assume $k > 1$. We may also assume that exponents a_i are positive; by the induction hypothesis, $m/p_1^{a_1} = \prod_{j \neq 1} p_j^{a_j}$ is sent to $\prod_{j \neq 1} q_j^{a_j}$. Since $m/p_1^{a_1}$ divides m, $f(m/p_1^{a_1})$ must divide $f(m)$. Hence in particular $q_j^{a_j}$ divides $f(m)$ for all $j \neq 1$. A similar argument, where 1 is exchanged with another index, shows that $q_1^{a_1}$ also divides $f(m)$. Therefore, $f(m)$ must be a multiple of $q_1^{a_1} \cdots q_k^{a_k}$ and, since it must have the same number of divisors as m, must actually coincide with that number.

We have thus shown that f is actually fixed once the image of all prime numbers is fixed, that is, once we know how it permutes the primes. Let us now show that, for $n = 10$, the only possible f is the identity. It is enough to show that each prime must be sent to itself. The primes up to 10 are 2, 3, 5, 7. If 2 were sent to 3, 2^3 would have to be sent to 3^3, but this cannot be the case because $3^3 > 10$. The possibilities $f(2) = 5$ and $f(2) = 7$ are excluded in a similar way, and we have to conclude that 2 is sent to 2. Similarly, 3 is sent to 3, or we would not be able to assign an image to 3^2. The only possibilities for $f(5)$ are then 5 and 7, but, since 2 is sent to 2, $2 \cdot 5$ must be sent to $2 \cdot f(5)$, which excludes the case $f(5) = 7$. Indeed, every prime is sent to itself and f is the identity.

For $n = 13$, a similar argument shows that f can apply any permutation to the three primes 7, 11, 13 but must keep other primes fixed, so we have six possibilities.

40

(i) We may just consider the four sets of ten positions that cards of each suit will occupy, because the order of the cards in each set is fixed. There are $\binom{40}{10}$ ways to choose the ten positions occupied by coins; we are left with $\binom{30}{10}$ choices for the positions occupied by swords, then $\binom{20}{10}$ choices for the positions occupied by clubs, and finally the positions occupied by cups are all those that remain. The number we are looking for is therefore

$$\binom{40}{10} \cdot \binom{30}{10} \cdot \binom{20}{10} = \frac{40!}{(10!)^4}.$$

(ii) Let us choose the 20 positions that will be occupied by coins or swords: this can be done in $\binom{40}{20}$ ways. Now, coins must necessarily occupy the fist 10 of these positions, while swords will occupy the remaining 10. Within their assigned ten positions, both coins and swords can assume any order among the 10! available. Finally, cups and clubs can be distributed freely among the 20 remaining positions, in any of 20! possible ways. The answer is therefore

$$\binom{40}{20} \cdot (10!)^2 \cdot 20! = \frac{40! \cdot (10!)^2}{20!}.$$

41 First, suppose $n = 2m$ with $m \geq 1$; partition the n numbers into two m-element subsets E and O, containing all of the even numbers and all of the odd numbers respectively. If a subset X *does not* have at least three numbers of the same parity, then it contains at most two elements of E and at most two elements of O. Accounting for the cases of X containing 0, 1 or 2 even numbers, we find we can choose the even elements of X in

$$\binom{m}{0} + \binom{m}{1} + \binom{m}{2} = \frac{m^2 + m + 2}{2}$$

ways. The same argument can be made regarding the number of ways we can choose odd elements of X, so the number of subsets of X that do not contain three elements of the same parity is

$$\left(\frac{m^2 + m + 2}{2} \right)^2.$$

If $n = 2m - 1$ with $m \geq 1$ (that is, if n is odd), the subset of odd numbers has cardinality m while the subset of even numbers has cardinality $m - 1$. Therefore, the expression for the number of subsets of X with no three elements of the same parity becomes

$$\frac{m^2 + m + 2}{2} \cdot \frac{m^2 - m + 2}{2}.$$

We obtain the final answer by subtracting the number of subsets with no three elements of the same parity from the number of all possible subsets of $\{1, \ldots, n\}$; that is,

$$
\begin{cases}
2^{2m} - \left(\dfrac{m^2 + m + 2}{2}\right)^2 & \text{if } n = 2m, \\[4mm]
2^{2m-1} - \dfrac{m^2 + m + 2}{2} \cdot \dfrac{m^2 - m + 2}{2} & \text{if } n = 2m - 1.
\end{cases}
$$

42 Given an element $\bar{x} \in \mathbb{Z}/40\mathbb{Z}$ we shall call $c(\bar{x}) \in \{0, 1, 2, 3\}$ the *colour* of x. Partition the set $\mathbb{Z}/40\mathbb{Z}$ into ten subsets $A_0, \ldots, A_9$, each containing four elements, with A_i being the set of elements of $\mathbb{Z}/40\mathbb{Z}$ that are congruent to i modulo 10.

No conditions are imposed on the colours of elements belonging to different A_i's, so we may colour elements in each A_i independently.

Let us consider the number of possible colourings of $A_i = \{\bar{i}, \overline{i + 10}, \overline{i + 20}, \overline{i + 30}\}$ that satisfy the given requirement. There are two possibilities: either ① $\bar{i}$ and $\overline{i + 20}$ are assigned the same colour or ② $\bar{i}$ and $\overline{i + 20}$ are assigned different colours.

① In the first case, there are four ways to choose the colour of $\bar{i}$ and $\overline{i + 20}$; the other two elements must be assigned a different colour than the one chosen: we can assign any of the remaining three colours to each of the two independently. There are in this case $4 \cdot 3^2 = 36$ valid colourings.

② In the second case, there are four ways to choose the colour of $\bar{i}$ and three ways to choose the colour of $\overline{i + 20}$ (one for each colour that is, not the one chosen for $\bar{i}$). The other two elements can be coloured independently of each other, each with any one of the two remaining colours. We get $4 \cdot 3 \cdot 2^2 = 48$ valid colourings.

The total number of ways to colour a single A_i is thus $36 + 48 = 84$. Since there are ten sets A_i and each can be coloured independently, the total number of legal colourings is 84^{10}.

43

(i) There are two ways to colour a row completely white or completely black, so $2^n - 2$ ways to colour it so that it is neither. Moreover, we can choose row colourings independently of each other. Thus, the number of possible colourings is $(2^n - 2)^n$.

(ii) We need to choose n cells in such a way that there is exactly one per row and one per column, and colour them black. In other words, if in the ith row we choose the cell belonging to the $\sigma(i)$th column, we require the mapping $i \longmapsto \sigma(i)$ to be a *permutation* of the set $\{1, \ldots, n\}$. This yields $n!$ possibilities.

(iii) In each row, we shall choose $n/2$ cells which will be coloured white; all remaining cells will be coloured black. There are $\binom{n}{n/2}$ ways to colour each row. Since rows can be coloured independently of each other, the required total number of colourings is $\binom{n}{n/2}^n$.

44

Solution 1 We shall first solve the problem by means of some properties of binomial coefficients.

Both sums have a null summand given by $k = 0$: we may of course disregard it. It immediately follows from the definition of binomial coefficients that $k\binom{n}{k} = n\binom{n-1}{k-1}$. Hence, we have:

(i) $\displaystyle\sum_{k=0}^{n} k\binom{n}{k} = \sum_{k=1}^{n} k\binom{n}{k} = n \sum_{k=1}^{n} \binom{n-1}{k-1} = n \sum_{h=0}^{n-1} \binom{n-1}{h} = n2^{n-1}$.

(ii) $\displaystyle\sum_{k=0}^{n} k^2\binom{n}{k} = \sum_{k=1}^{n} kn\binom{n-1}{k-1} = n \sum_{h=0}^{n-1} (h+1)\binom{n-1}{h} = n \sum_{h=0}^{n-1} h\binom{n-1}{h} + n \sum_{h=0}^{n-1} \binom{n-1}{h}$.

By using the first identity for $n - 1 \geq 1$ and by a direct check for $n - 1 = 0$, we finally obtain

$$\sum_{k=0}^{n} k^2\binom{n}{k} = n(n-1)2^{n-2} + n2^{n-1} = (n^2 + n)2^{n-2}.$$

Solution 2 We also propose a solution by induction.

(i) Set $p(n) : \sum_{k=0}^{n} k\binom{n}{k} = n2^{n-1}$; we shall show that $p(n)$ holds for $n \geq 1$ by induction on n. For $n = 1$ we have $\sum_{k=0}^{1} k\binom{1}{k} = \binom{1}{1} = 1$ so $p(1)$ holds. Now assume that $n \geq 1$ and that $p(n)$ holds; we shall show $p(n+1)$. We have

$$\sum_{k=0}^{n+1} k\binom{n+1}{k} = \sum_{k=1}^{n} k\left(\binom{n}{k} + \binom{n}{k-1}\right) + (n+1)\binom{n+1}{n+1}$$
$$= \sum_{k=1}^{n} k\binom{n}{k} + \sum_{k=0}^{n}(k+1)\binom{n}{k} - (n+1)\binom{n}{n} + n + 1$$
$$= n2^{n-1} + \sum_{k=0}^{n} k\binom{n}{k} + \sum_{k=0}^{n} \binom{n}{k}$$
$$= n2^{n-1} + n2^{n-1} + 2^n$$
$$= (n+1)2^n$$

where we have used the fact that $p(n)$ holds, that is, $\sum_{k=0}^{n} k\binom{n}{k} = n2^{n-1}$, and that $\sum_{k=0}^{n} \binom{n}{k} = 2^n$. This concludes our proof that $p(n+1)$ holds.

(ii) Again, we shall proceed by induction on n, this time to prove the proposition $q(n) : \sum_{k=0}^{n} k^2\binom{n}{k} = n(n+1)2^{n-2}$. Again, $q(1)$ holds trivially, so we may assume that $n \geq 1$ and $q(n)$ holds. Using what was proved in (i) we have

$$\sum_{k=0}^{n+1} k^2\binom{n}{k} = \sum_{k=1}^{n} k^2\left(\binom{n}{k} + \binom{n}{k-1}\right) + (n+1)^2\binom{n+1}{n+1}$$
$$= \sum_{k=1}^{n} k^2\binom{n}{k} + \sum_{k=0}^{n}(k+1)^2\binom{n}{k} - (n+1)^2\binom{n}{n} + (n+1)^2$$
$$= n(n+1)2^{n-2} + \sum_{k=0}^{n} k^2\binom{n}{k} + 2\sum_{k=0}^{n} k\binom{n}{k} + \sum_{k=0}^{n} \binom{n}{k}$$
$$= n(n+1)2^{n-1} + n2^n + 2^n$$
$$= (n+1)(n+2)2^{n-1}$$

so $q(n+1)$ also holds.

Solution 3 Here is a more combinatorial solution.

(i) Let X be the set $\{1, 2, \ldots, n\}$ and let $\mathcal{F} \doteq \{(A, a) \mid a \in A \subseteq X\}$, that is, the set of all subsets A of X equipped with a distinguished element a of A. We shall compute the cardinality of $\mathcal{F}$ in two separate ways.

First, choose a subset A of X with cardinality k: there are $\binom{n}{k}$ ways to do this; next, there are k ways to choose a distinguished element a in A. Thus, since A can have cardinality $0, 1, \ldots, n$, the cardinality of $\mathcal{F}$ is $\sum_{k=0}^{n} k\binom{n}{k}$.

Alternatively, we can choose an element a of X in n possible ways, then choose a subset A of X containing a. Since X has cardinality n and a must be an element of A, we have 2^{n-1} possibilities for A. Thus, the cardinality of $\mathcal{F}$ is $n2^{n-1}$. This shows the required identity.

(ii) We will argue as before, this time for the set $\mathcal{G} = \{(A, a, b) \mid a, b \in A \subseteq X\}$ of all subsets A of X with two distinguished elements a and b, which may coincide.

We can choose a k-subset A of X in $\binom{n}{k}$ ways and we then have k^2 ways to choose a and b in A. So the cardinality of $\mathcal{G}$ is $\sum_{k=0}^{n} k^2 \binom{n}{k}$.

On the other hand, we can also proceed as follows. First, consider the case $a = b$: we can choose the distinguished element in n ways, after which there are 2^{n-1} ways to choose the subset A of X in such a way that it contains a. As for the case $a \neq b$, we can choose a in n ways, b in $n - 1$ ways and then A in 2^{n-2} ways (it must contain both a and b). This gives a cardinality of $n2^{n-1} + n(n - 1)2^{n-2} = n(n + 1)2^{n-2}$ for $\mathcal{G}$, hence the required identity.

Solution 4 In this final solution we will use the binomial theorem.

Let $\mathbb{R}[x]$ be the ring of polynomials in the variable x. Write $D : \mathbb{R}[x] \longrightarrow \mathbb{R}[x]$ for the operation of taking the derivative and $\mathbb{R}[x] \ni f(x) \overset{v}{\longmapsto} f(1) \in \mathbb{R}$ for evaluation in 1.

(i) Consider the polynomial $p(x) \doteq (1 + x)^n$ and evaluate its derivative in 1: we have $v(D(p(x))) = v(n(1 + x)^{n-1}) = n2^{n-1}$. But we may also write out the development given by the binomial theorem, $p(x) = \sum_{k=0}^{n} \binom{n}{k} x^k$, which yields

$$v(D(p(x))) = v(\sum_{k=1}^{n} k \binom{n}{k} x^{k-1}) = \sum_{k=1}^{n} k \binom{n}{k}$$

from which the desired identity follows.

(ii) Also consider the mapping $\mathbb{R}[x] \ni f(x) \overset{\mu}{\longmapsto} xf(x) \in \mathbb{R}[x]$ given by multiplication by x. Apply $vD\mu D$ to the polynomial $p(x) = (1 + x)^n$. We have

$$vD\mu D(1 + x)^n = vD(nx(1 + x)^{n-1})$$
$$= nv((1 + x)^{n-1} + (n - 1)x(1 + x)^{n-2})$$
$$= n(2^{n-1} + (n - 1)2^{n-2})$$
$$= n(n + 1)2^{n-2}.$$

By the binomial theorem, we also have

$$
\begin{aligned}
\nu D\mu D(1+x)^n &= \nu D\mu D(\textstyle\sum_{k=0}^{n}\binom{n}{k}x^k) \\
&= \textstyle\sum_{k=0}^{n}\binom{n}{k}\nu D\mu D(x^k) \\
&= \textstyle\sum_{k=0}^{n}\binom{n}{k}\nu D(kx^k) \\
&= \textstyle\sum_{k=0}^{n}\binom{n}{k}\nu(k^2 x^{k-1}) \\
&= \textstyle\sum_{k=0}^{n}k^2\binom{n}{k}
\end{aligned}
$$

which proves the identity.

45   Remark that the we are imposing independent and entirely symmetric requirements on digits at even places and digits at odd places. If we had N possibilities for the digits at odd places, then the cardinality we are looking would be N^2.

It is therefore enough to find the number of 15-tuples $(a_1, a_3, \ldots, a_{29})$ of 0s and 1s such that their sum is at most 2. In particular, the sum can be 0, 1 or 2. A sum of 0 is obtained only for $a_1 = a_3 = \cdots = a_{29} = 0$. In order for the sum to be 1, all digits except for exactly one must be 0, so there are 15 possibilities. Finally, in order for the sum to be 2 all digits must be 0 except for two, which must be 1 and need to be chosen among 15 places, which can be done in $\binom{15}{2}$ ways. This gives $1 + 15 + 105 = 121 = 11^2$ possibilities in total.

In conclusion, the required cardinality is 11^4.

46

(i) We wish to find the number of sets $A = \{a, b, c\} \subseteq X$ such that $a + b = 10$. The equality $a + b = 10$ with $a \neq b$ in X is satisfied if and only if $\{a, b\}$ is one of the following two-element sets: $\{1, 9\}$, $\{2, 8\}$, $\{3, 7\}$ or $\{4, 6\}$; thus there are four ways to choose the set $\{a, b\}$. The set A is obtained by choosing a set from the list above and adding an element c that does not already belong to it. This can be done in 98 ways, regardless of which of the 4 sets has been chosen. The sets constructed in this way are all distinct, so we obtain $4 \cdot 98 = 392$ in total.

(ii) Let X_5 be the subset of X consisting of all multiples of 5; clearly, $|X_5| = 20$. The sets $A = \{a, b, c\}$ we are considering either consist of 3 multiples of 5 or contain exactly 2 multiples of 5. Those of the first type are all three-element subsets of X_5: there are $\binom{20}{3}$; those of the second type can be constructed by choosing two elements in X_5 and one outside, and so there are $\binom{20}{2} \cdot 80$. The total number of sets with the required property is $\binom{20}{3} + \binom{20}{2} \cdot 80$.

47   Given an element x of $\mathbb{Z}/2^{100}\mathbb{Z}$, write it as $x = 2^\alpha x_1$ with x_1 odd and $0 \leq \alpha \leq 100$. For $\alpha = 100$ we have $x = \bar{0}$; given $\alpha < 100$, choosing x is equivalent to choosing x_1 among all odd residue classes modulo $2^{100-\alpha}$; there are $\phi(2^{100-\alpha}) = 2^{99-\alpha}$ such classes.

Now let $x = 2^{\alpha} x_1$, with x_1 odd; the equation $xy = \bar{0}$ can be rewritten as $2^{\alpha} x_1 y \equiv 0 \pmod{2^{100}}$. Solutions y are residue classes modulo 2^{100} such that $y \equiv 0 \pmod{2^{100-\alpha}}$: there are 2^{α}.

In conclusion, for $0 \le \alpha \le 99$ there are $2^{99-\alpha} \cdot 2^{\alpha} = 2^{99}$ pairs $(2^{\alpha} x_1, y)$ as required; for $\alpha = 100$, we have $x = 0$ and the equation is satisfied by all possible values of y. The total number of solutions is therefore

$$\sum_{\alpha=0}^{99} 2^{99} + 2^{100} = 100 \cdot 2^{99} + 2^{100} = 51 \cdot 2^{100}.$$

48

(i) The number of teams of 4 one can form from a set of 13 people is $\binom{13}{4}$. Among these teams, those that include any two fixed people p and q are $\binom{11}{2}$ (2 of the 4 people on the team are p and q; the 2 remaining people are selected among the other 11 available). The probability of p and q being in the chosen team is therefore $\binom{11}{2}/\binom{13}{4} = 1/13$. The probability that p and q do not both belong to the team is $12/13$.

(ii) Let n be the number we are looking for. One can form $\binom{4}{2} = 6$ possible unordered pairs of people from a team; since the total number of unordered pairs is $\binom{13}{2} = 78$, we must have $78 = 6n$, hence $n = 13$.

49 Let $X = \{0, 1, \ldots, 100\}$. We know that the positive divisors d of $2^{100} 3^{100}$ are the numbers of the form $2^x 3^y$ with $x, y \in X$; we must find out which of those satisfy the congruence

$$2^x 3^y \equiv 4 \pmod 5.$$

Remark that $3 \equiv 2^{-1} \pmod 5$, so the congruence above can be rewritten as $2^{x-y} \equiv 2^2 \pmod 5$. Its solutions are given by $x - y \equiv 2 \pmod 4$ since the multiplicative order of 2 modulo 5 is 4.

Since the set X has 26 elements congruent to 0 modulo 4 and 25 elements in each of the other classes modulo 4, the number of pairs $(x, y) \in X \times X$ such that $x \equiv y + 2 \pmod 4$ is $26 \cdot 25 + 25 \cdot 25 + 25 \cdot 26 + 25 \cdot 25 = 2550$.

50

(i) Let $A = \{x, y\}$ be a two-element subset of X. In order for $x + y$ to be divisible by 4 we must have one of the following: x and y are both divisible by 4; x and y are both congruent to 2 modulo 4; or, finally, one among x and y is congruent to 1 modulo 4 and the other to 3 modulo 4.

In the first case, A must be contained in the subset of X consisting of all multiples of 4, which has 25 elements; this leads to $\binom{25}{2}$ possibilities. Similarly, the second case also leads to $\binom{25}{2}$ possibilities.

In the last case, A can contain any of the 25 elements congruent to 1 and any of the 25 congruent to 3 modulo 4, so it can be chosen in 25^2 ways.

Summing the results obtained so far yields the answer $2 \cdot \binom{25}{2} + 25^2$.

(ii) Let $A = \{a, b, c\}$ be a three-element subset of X. We may assume $a < b < c$. There are $x = a - 1$ numbers in X that are strictly less than a; there are $y = b - a - 1$ strictly between a and b and $z = c - b - 1$ strictly between b and c; there are $t = 100 - c$ strictly larger than c.

Clearly, the subset A is entirely determined by the choice of x, y, z, t. Moreover, we must have $x + y + z + t = 97$. By assumption, y and z are both positive. Setting $x' = x + 1$ and $t' = t + 1$ we have that x' and t' are also positive; the condition on x', y, z, t' becomes $x' + y + z + t' = 99$, which has $\binom{98}{3}$ solutions in the positive integers.

51

(i) There are nine possibilities for the pair (m_f, M_f), that is, $(1, 2)$, $(2, 3)$, $\ldots$, $(9, 10)$. Each of them corresponds to maps taking only two possible values, under the condition that both values are actually attained. Once we fix the pair that a map takes values in, all we have to do is exclude the constant maps, so we have $2^{10} - 2$ possibilities. Given the nine possible choices of (m_f, M_f), the number of maps with the required property is $9 \cdot (2^{10} - 2)$.

(ii) The set of maps from $\{1, 2, \ldots, 10\}$ to itself consists of 10^{10} elements. We need to remove from this set all maps such that $M_f \neq 10$ or $m_f \neq 1$. Let $A = \{f \in X \mid M_f \neq 10\}$, $B = \{f \in X \mid m_f \neq 1\}$. Both A and B are the sets of all maps from a set of ten elements to a certain set of nine elements, while $A \cap B$ is the set of maps from a ten-element set to a set of eight elements. It follows that

$$|A \cup B| = 9^{10} + 9^{10} - 8^{10},$$

so the cardinality we want is $10^{10} - 2 \cdot 9^{10} + 8^{10}$.

52 We can rewrite the equation as $n - \phi(n) = 8$. Clearly, n cannot be equal to 1, so it must have prime divisors.

First, assume that only one prime appears in the prime factorisation of n, so $n = p^a$ for some prime p and some exponent $a \geq 1$. Then $n - \phi(n) = p^a - (p^a - p^{a-1}) = p^{a-1}$, hence $p^{a-1} = 8$, that is, $p = 2$, $a - 1 = 3$, which gives the solution $n = 2^4 = 16$.

Next, assume that $n = p^a q^b$ has exactly two distinct prime factors p and q, with $p < q$. Then $n - \phi(n) = p^a q^b - (p^a - p^{a-1})(q^b - q^{b-1}) = p^{a-1} q^{b-1}(p + q - 1)$. It follows that $p^{a-1} q^{b-1} \mid 8$.

If $p^{a-1} q^{b-1} = 1$, then $p + q - 1 = 8$ and so, since $p + q = 9$ is odd, one of the two summands must be even: we must have $p = 2$ and $q = 7$. This yields the solution $n = 14$.

If $p^{a-1}q^{b-1} = 2$, then $p + q - 1 = 4$, which has the unique solution $p = 2$, $q = 3$. This yields the solution $n = 2^2 \cdot 3 = 12$.

If $p^{a-1}q^{b-1} \geq 4$, then $p + q - 1 \leq 2$, which is clearly impossible.

Finally, assume n has at least three distinct prime factors p, q, r, with $p < q < r$. Then the set of integers $\{1 \leq x \leq n \mid p$ divides $x\}$ contains at least $qr \geq 3 \cdot 5 = 15$ numbers that are not coprime to n, so $n - \phi(n) \geq 15$.

In conclusion, the solutions of the given equation are $n = 12, 14, 16$.

53

(i) The set of fixed points is a 10-element subset of a set of 100 elements, so it can be chosen in $\binom{100}{10}$ ways. The other 90 elements must not be fixed, so each can be sent to 99 possible elements, all but the element itself. The number of maps with the required property is therefore

$$\binom{100}{10} \cdot 99^{90}.$$

(ii) There are two ways to obtain $\sum_{x \in X} |f(x) - x| = 2$: either there is only one element x which is not fixed by f and we have $|f(x) - x| = 2$, or there are exactly two elements x, y not fixed by f and we have $|f(x) - x| = |f(y) - y| = 1$.

In the first case, there are two possibilities for $f(x)$ if $x \neq 1, 2, 99, 100$ and one otherwise. In the second case, there are two possibilities for the image of each element of the pair x, y, unless at least one of the elements belongs to $A = \{1, 100\}$.

If only one of the two belongs to A, then there is one way to choose its image and there are two choices for the image of the other element. If the set $\{x, y\}$ is A then there is only one choice for the two images.

In conclusion, the number of maps we are looking for is

$$2 \cdot 100 - 4 + \binom{98}{2} \cdot 4 + 2 \cdot 98 \cdot 2 + 1 = 19601.$$

54

(i) The number of all five-element subsets of X is $\binom{20}{5}$. For each subset A containing five elements, the subset B must have exactly seven elements that do not belong to A and any number of elements in common with A. Therefore, once we fix A, the number of choices for B is $\binom{15}{7} \cdot 2^5$. The answer is thus

$$\binom{20}{5} \cdot \binom{15}{7} \cdot 2^5.$$

(ii) We can choose the set $Y = (A \cup B) \cap C$ in $\binom{20}{8}$ ways. Given an element of Y, it can belong to $A \setminus B$, $B \setminus A$ or $A \cap B$; consequently, we have 3^8 ways to distribute the elements of Y among A and B.

By a similar argument, each element of $X \setminus Y$ can be assigned to any one of five sets: $X \setminus (A \cup B \cup C)$, $A \setminus (B \cup C)$, $B \setminus (A \cup C)$, $C \setminus (A \cup B)$ or $(A \cap B) \setminus C$. This gives 5^{12} ways to place elements of $X \setminus Y$.

We have

$$\binom{20}{8} \cdot 3^8 \cdot 5^{12}.$$

choices in total.

55

(i) The strings with k even components are $\binom{10}{k} \cdot 3^k \cdot 2^{10-k}$. Indeed, they can be constructed by choosing k positions among the ten available and assigning them even values, that is, 0, 2 or 4, whereas the remaining $10 - k$ positions are assigned odd values, that is, 1 or 3. The strings we want are those with 6, 7, 8, 9 or 10 even components and are therefore

$$\sum_{k=6}^{10} \binom{10}{k} \cdot 3^k \cdot 2^{10-k} = 3^7 \cdot 2827.$$

(ii) We can interpret strings as sequences of digits in the base 5 representation of a natural number: $(a_0, \ldots, a_9) \longmapsto a_0 + a_1 5 + \cdots + a_9 5^9$. By remarking that $5 \equiv -1 \pmod 6$, we obtain that the strings we have to count are those such that

$$a_0 + a_1 5 + \cdots + a_9 5^9 \equiv \sum_{i=0}^{9} (-1)^i a_1 \equiv 0 \pmod 6,$$

that is, all those that correspond to a multiple of 6. Strings with ten components correspond to natural numbers between 0 and $5^{10} - 1$ (inclusive); among these, there are $\lceil 5^{10}/6 \rceil$ multiples of 6.

⟦Since $5^2 \equiv 1 \pmod 6$, we have $5^{10} \equiv 1 \pmod 6$. It follows that $\lceil 5^{10}/6 \rceil = 1 + (5^{10} - 1)/6 = 1 + (5^5 - 1)(5^5 + 1)/6 = 1 + (5^5 - 1)(5 + 1)(5^4 - 5^3 + 5^2 - 5 + 1)/6 = 1 + (5^5 - 1)(5^4 - 5^3 + 5^2 - 5 + 1) = 5^9 - 5^8 + 5^7 - 5^6 + 5^5 - 5^4 + 5^3 - 5^2 + 5.$⟧

56

(i) Let $A \in X$, $a = \min A$ and $b = \max A$. We have $b = 60 + a$, so, since $b \leq 100$, there are 40 possible values for a, that is, the integers from 1 to 40. Sets $A \in X$ with $\min A = a$ and $\max A = b$ are constructed by choosing $A \setminus \{a, b\}$ to be any subset of $\{a + 1, \ldots, a + 59\}$, in one of 2^{59} possible ways. We therefore have $|X| = 40 \cdot 2^{59} = 2^{62} \cdot 5$.

(ii) Let N_2 and N_5 be the subsets of N consisting of all elements that are not multiples of 2, and not multiples of 5, respectively. Set $Y_i = \{f \in Y \mid f(N) \subseteq N_i\}$ for $i = 2, 5$. We have $Y = Y_2 \cup Y_5$ and $Y_2 \cap Y_5 = \{f \in Y \mid f(N) \subseteq N_2 \cap N_5\}$, so $|Y| = |Y_2 \cup Y_5| = |Y_2| + |Y_5| - |Y_2 \cap Y_5| = |N_2|^{100} + |N_5|^{100} - |N_2 \cap N_5|^{100}$. Now $|N_2| = 50$, $|N_5| = 80$ and $|N_2 \cap N_5| = 40$, because half of the non-multiples of 5 are odd and half are even. We therefore have $|Y| = 50^{100} + 80^{100} - 40^{100}$.

57   In order to count pairs (α, β) satisfying the required conditions we shall distinguish three cases.

① The word α is spelled with only one letter: there are 26 words of this type, one for each letter in the alphabet, and in this case β is any word of length 3 formed with the 25 remaining letters, so there are 25^3 possible choices for β.

② The word α is spelled with two letters: there are $\binom{26}{2} \cdot 6$ words of this type, as many as the ways one can choose two letters from the alphabet, times the number of words of length 3 one can form using both letters. In this case, β is any word of length 3 formed with the 24 remaining letters, so we have 24^3 possible choices for β.

③ Finally, assume that the word α is spelled with three different letters: there are $\binom{26}{3} \cdot 3!$ words of this type, as many as the ways to choose three letters from the alphabet, times the number of possible permutations of three letters. In this case β is any word of length 3 formed with the 23 remaining letters, so there are 23^3 possible choices for β.

The total number of pairs satisfying the required condition is

$$26 \cdot 25^3 + \binom{26}{2} \cdot 6 \cdot 24^3 + \binom{26}{3} \cdot 3! \cdot 23^3.$$

58

(i) We can construct the required pairs (A, B) by choosing the 40 elements of $A \cup B$ in one of $\binom{100}{40}$ ways, then selecting 10 of those elements to be the elements of A, which can be done in $\binom{40}{10}$ ways, and, finally, choosing the subset of A representing the intersection with B, for which we have 2^{10} choices. It follows that the cardinality we want is $\binom{100}{40}\binom{40}{10}2^{10}$.

(ii) Call Γ the set of all subsets with the required properties and let Σ be the set of five-element subsets A of X such that $\prod_{x \in A} x \not\equiv 0 \pmod 9$. We have $|\Gamma| = \binom{100}{5} - |\Sigma|$.

In order to compute the cardinality of Σ, consider the partition given by:

$$\Sigma_3 = \{A \subseteq X \mid |A| = 5, \ \textstyle\prod_{x \in A} x \not\equiv 0 \pmod 3\}$$
$$\Sigma_9 = \{A \subseteq X \mid |A| = 5, \ \textstyle\prod_{x \in A} x \equiv 0 \pmod 3, \ \textstyle\prod_{x \in A} x \not\equiv 0 \pmod 9\}.$$

Now, $|\Sigma_3| = \binom{100-33}{5}$ since we can choose the five elements of A among the $100 - 33$ elements that are not divisible by 3. Moreover, $|\Sigma_9| = \binom{100-33}{4} \cdot 22$:

we need to choose four elements that are not divisible by 3 and a fifth that is divisible by 3 but not by 9, for which we have 22 choices (there are 33 multiples of 3, 11 of which are also multiples of 9).

We thus obtain

$$|\Gamma| = \binom{100}{5} - \binom{100-33}{5} - \binom{100-33}{4} \cdot 22.$$

59 Denote by T_k the score after the first k rolls, and by P_k the probability that T_k is divisible by 7; we wish to compute P_n. Clearly, $T_n = T_{n-1} + i$, where $i \in \{1, \ldots, 6\}$ is the score of the nth die roll, so $T_n \equiv 0 \pmod 7$ if and only if $i \equiv -T_{n-1} \pmod 7$. It follows that if $T_{n-1} \equiv 0 \pmod 7$ then it is not possible to obtain a multiple 7, whereas if $T_{n-1} \not\equiv 0 \pmod 7$ then there is a unique value of i which would yield a multiple of 7.

What we have just shown is that P_n satisfies the recurrence relation $P_n = (1 - P_{n-1})/6$.

Keeping in mind that $P_1 = 0$, one can easily show by induction that

$$P_n = \frac{1}{7}\left(1 + \frac{(-1)^n}{6^{n-1}}\right).$$

60

(i) Since $(2, 3) = 1$, the equation does have integer solutions. One particular solution of the equation $2x + 3y = 1$ is $x = -1$, $y = 1$, so a particular solution of $2x + 3y = 100$ is $x = -100$, $y = 100$, and a general solution (with x, y integers) has the form $x = -100 + 3k$, $y = 100 - 2k$, where $k \in \mathbb{Z}$.

We have $x > 0$ if and only if $-100 + 3k > 0$, that is, $3k > 100$, or equivalently $k \geq 34$. We have $y > 0$ if and only if $100 - 2k > 0$, that is, $2k < 100$, or in other words $k < 50$.

We thus have a solution in the positive integers for each k with $34 \leq k < 50$, so there are 16 such solutions.

(ii) The equation $x + y + z = 100$ where x, y, z are positive integers has $\binom{99}{2}$ solutions; these are in bijection with the subsets $\{x, x + y\}$ of $\{1, 2, \ldots, 99\}$. Naturally, all these solutions are ordered triples (x, y, z) with $x, y, z \in \{1, 2, \ldots, 100\}$, but only those such that x, y, z are all different correspond to three-element subsets of $\{1, 2, \ldots, 100\}$. To be more precise, six solutions of this type correspond to a single three-element subset, whose elements can be ordered in $3! = 6$ ways.

Solution triples with exactly two equal values are of the form (x, x, y), (x, y, x) or (y, x, x), with $2x + y = 100$. Since we must have $y = 100 - 2x$, the value of y is determined by the value of x, and the inequalities $x > 0$, $y > 0$ give $0 < x < 50$, that is, 49 possible values for x. By considering the three possible forms of the solutions, one obtains $3 \cdot 49$ solutions of this type.

There are no solutions where the three unknowns take the same value, since $3x = 100$ has no integer solutions.

Therefore, the solutions with the unknowns taking three different values are $\binom{99}{2} - 3 \cdot 49 = 96 \cdot 49$ and the number of sets we want is $96 \cdot 49/6 = 784$.

61 For $n = 1$ we have $\phi(n) = 1$, so for $a = 43$ the equation has a solution. From now on, assume $n > 1$, so that $\phi(n) < n$, that is, $a < 43$. By rearranging the formula for Euler's totient function ϕ we obtain

$$\frac{\phi(n)}{n} = \prod_{p \,|\, n} \frac{p-1}{p} = \frac{a}{43},$$

where the product ranges over all primes that divide n. The largest prime q that divides n certainly appears in the denominator of $\phi(n)/n$, so $q = 43$. If 43 is the only prime that divides n then $n = 43^k$ for some positive integer k and $\phi(n)/n = 42/43$, so $a = 42$.

Now assume there is at least another prime that divides n and let q_1 be the largest such prime. If $q_1 \nmid 42$, then q_1 appears in the denominator of $\phi(n)/n$, which gives a contradiction. We therefore have $q_1 = 2, 3$ or 7. Also remark that $5 \nmid n$, because, even if we had $q_1 = 7$, if 5 were a divisor of n then the denominator of $\phi(n)/n$ would be divisible by 5, hence a contradiction.

If follows that n can only be of the form $n = 2^x 3^y 7^z 43^t$, with $x, y, z \geq 0$ and $t > 0$. By considering the eight different cases given by each among x, y and z being either equal to or strictly greater than 0, one obtains the eight possible values of a that are less than 43, that is: $42, 36, 28, 24, 21, 18, 14, 12$.

62

(i) We can rewrite the congruence in the statement as $(a - 1)(b - 1) = ab - a - b + 1 \equiv 1 \pmod 3$. We are therefore considering all subsets $\{a, b\}$ of X with $a - 1 \equiv b - 1 \equiv 1 \pmod 3$ or $a - 1 \equiv b - 1 \equiv -1 \pmod 3$. If the first holds then $a \equiv b \equiv 2 \pmod 3$ and if the second does then $a \equiv b \equiv 0 \pmod 3$. The number of elements of X that are congruent to 2 modulo 3 is 33, the same as the number of elements of X congruent to zero modulo 3. Hence there are $\binom{33}{2}$ subsets of each of the two types and the answer is $2 \cdot \binom{33}{2} = 33 \cdot 32 = 1056$.

(ii) Let S be the family of all subsets $\{a, b\}$ of X such that $ab \equiv 0 \pmod 3$, and let T be the family of all subsets $\{a, b\}$ of X such that $a + b \equiv 0 \pmod 3$. The number we want is $|S \cup T|$; by the inclusion-exclusion principle, we have $|S \cup T| = |S| + |T| - |S \cap T|$.

The subsets in S are those that contain at least one multiple of 3, so the complement of S among all two-element subsets of X comprises all subsets that do not contain a multiple of 3. It follows that $|S| = \binom{100}{2} - \binom{67}{2} = 50 \cdot 99 - 67 \cdot 33 = 33 \cdot 83$.

The subsets in T are those for which $b \equiv -a \pmod 3$. So we must either have $a \equiv b \equiv 0 \pmod 3$, and there are $\binom{33}{2}$ subsets of this type, or one element must be congruent to 1 and the other to -1 modulo 3, and there are $34 \cdot 33$ subsets of this second type. In total, $|T| = 33 \cdot 16 + 33 \cdot 34 = 33 \cdot 50$.

The subsets in $S \cap T$ are those with both elements divisible by 3: there must be one multiple of 3 because the subset belongs to S, and therefore both elements must be multiples of 3 in order for the subset to belong to T. Hence $|S \cap T| = \binom{33}{2} = 33 \cdot 16$.

In conclusion, $|S \cup T| = 33 \cdot (83 + 50 - 16) = 33 \cdot 97 = 3201$.

63

(i) For each $\sigma \in S(X)$ we clearly have

$$\sum_{i=1}^{100} (\sigma(i) - i) = \sum_{i=1}^{100} \sigma(i) - \sum_{i=1}^{100} i = 0,$$

hence

$$\sum_{i=1}^{100} (\sigma(i) - i) \equiv 0 \cdot |X_{0,\sigma}| + 1 \cdot |X_{1,\sigma}| + 2 \cdot |X_{2,\sigma}| \equiv 0 \pmod 3,$$

that is, $|X_{1,\sigma}| \equiv -2 \cdot |X_{2,\sigma}| \equiv |X_{2,\sigma}| \pmod 3$.

(ii) Partition the set X as $X_0 \sqcup X_1$, where X_0 is the subset of even numbers in X and X_1 the subset of odd numbers. Given a permutation σ, partition X_0 and X_1 as follows: $X_0 = X_{0,0}^\sigma \sqcup X_{0,1}^\sigma$ and $X_1 = X_{1,0}^\sigma \sqcup X_{1,1}^\sigma$, where $X_{i,j}^\sigma = \{x \in X_i \mid \sigma(x) \in X_j\}$ for $i, j = 0, 1$.

It is clear that $|X_{0,1}^\sigma|$ can be any integer k with $0 \le k \le 50$ and that, if $|X_{0,1}^\sigma| = k$, then $|X_{1,0}^\sigma| = k$ and $|X_{0,0}^\sigma| = |X_{1,1}^\sigma| = 50 - k$. We therefore have $\binom{50}{k}$ independent choices for $X_{0,1}^\sigma$ and for $X_{1,0}^\sigma$, whereas the two sets $X_{0,0}^\sigma$ and $X_{1,1}^\sigma$ are fixed once the first two are chosen, as they are their complements.

Moreover, the condition in the statement is satisfied if and only if the following requirements hold: $\sigma(X_{0,1}^\sigma) = X_{1,0}^\sigma$, which corresponds to $k!$ choices; $\sigma(X_{1,0}^\sigma) = X_{0,1}^\sigma$, with $k!$ choices; $\sigma(X_{0,0}^\sigma) = X_{0,0}^\sigma$, with $(50 - k)!$ choices, and finally $\sigma(X_{1,1}^\sigma) = X_{1,1}^\sigma$, with $(50 - k)!$. It follows that the number of permutations with the required property is

$$\sum_{k=0}^{50} \binom{50}{k}^2 (k!)^2 \cdot ((50 - k)!)^2 = \sum_{k=0}^{50} (50!)^2 = 50!^2 \cdot 51!$$

64

(i) Let us write $x = 2^{\alpha_1} 5^{\beta_1}$, $y = 2^{\alpha_2} 5^{\beta_2}$, $z = 2^{\alpha_3} 5^{\beta_3}$, where $\alpha_1, \alpha_2, \alpha_3$ and $\beta_1, \beta_2, \beta_3$ are non-negative integers. The condition on x, y, z is equivalent to the system

$$\begin{cases} \alpha_1 + \alpha_2 + \alpha_3 = 100 \\ \beta_1 + \beta_2 + \beta_3 = 100. \end{cases}$$

The two equations are independent and each has as many solutions as the non-negative integer solutions of the equation $t_1 + t_2 + t_3 = 100$, that is, $\binom{102}{2}$.

Therefore the required number of triples is $\binom{102}{2}^2$.

(ii) The condition is equivalent to the system

$$\begin{cases} 2\alpha_1 + \alpha_2 + \alpha_3 = 100 \\ 2\beta_1 + \beta_2 + \beta_3 = 100 \end{cases}$$

and again the two equations are independent. By symmetry, we may count the solutions of the first equation, which is equivalent to the inequality

$$2\alpha_1 + \alpha_2 \leq 100;$$

indeed, the inequality holds for any solution of the original equation and, if α_1, α_2 satisfy the inequality, there is a unique α_3 such that $2\alpha_1 + \alpha_2 + \alpha_3 = 100$. The inequality has $101 - 2\alpha_1$ solutions for each possible value of α_1 between 0 and 50 inclusive. In total we have

$$\sum_{\alpha_1=0}^{50} (101 - 2\alpha_1) = 51^2$$

solutions, and thus the number of possible triples is $51^2 \cdot 51^2 = 51^4$.

65 We shall consider two separate types of maps: ① those for which $1 \notin \mathrm{Im}(f)$ and ② those for which $1 \in \mathrm{Im}(f)$.

① If $1 \notin \mathrm{Im}(f)$, then $f(a) \neq 1$ and $f(b) \neq 1$, so the number $f(a) \cdot f(b)$ cannot be a prime for any choice of a and b in $X \setminus \{1\}$. Since $|X \setminus \{1\}| = 9$, there are 9^{10} maps of this type.

② If $1 \in \mathrm{Im}(f)$, then there exists $c \in X$ such that $f(c) = 1$. Consequently, in order for the map f to fulfil the requirement, $f(a)$ cannot be a prime for any a in X; indeed, if this weren't the case then the number $f(a) \cdot f(c) = f(a)$ would be prime. On the other hand, if $f(a)$ is a composite number for all a in X such that $f(a) \neq 1$ then the map f does satisfy the condition.

What we have to do is find the number of all maps such that their image is contained in $X \setminus \{2, 3, 5, 7\}$ and $1 \in \mathrm{Im}(f)$. In order to do this, we can count maps from X to $X \setminus \{2, 3, 5, 7\}$ and subtract the number of maps such that $1 \notin \mathrm{Im}(f)$, that is, maps from X to $X \setminus \{1, 2, 3, 5, 7\}$. There are 6^{10} maps $X \to X \setminus \{2, 3, 5, 7\}$ and 5^{10} maps $X \longrightarrow X \setminus \{1, 2, 3, 5, 7\}$, hence $6^{10} - 5^{10}$ maps of this second type.

Summing the numbers obtained for case ① and case ② yields $9^{10}+6^{10}-5^{10}$ maps in total.

⟦In order to count maps of the second type, that is, maps $X \longrightarrow X \setminus \{2, 3, 5, 7\}$ such that $1 \in \mathrm{Im}(f)$, we may also count maps with $|f^{-1}(1)| = k$ for each $k \geq 1$ and then sum over all possible cardinalities. Indeed, the number we want is equal to

$$\sum_{k=1}^{10} |\{f : X \longrightarrow X \setminus \{2, 3, 5, 7\} \mid |f^{-1}(1)| = k\} = \sum_{k=1}^{10} \binom{10}{k} 5^{10-k}$$
$$= \sum_{k=0}^{10} \binom{10}{k} 5^{10-k} - \binom{10}{0} 5^{10}$$
$$= 6^{10} - 5^{10}.$$

⟧

66

(i) Teams that satisfy the requirement can be constructed by choosing two pairs of twins among the n available, which can be done in $\binom{n}{2}$ ways, then completing the team with two people that are not twins, selected from the remaining $n - 2$ pairs. The latter selection amounts to choosing two pairs of twins among the remaining $n - 2$, which can be done in $\binom{n-2}{2}$ ways, then choosing one twin from each, which can be done in 2^2 ways. In conclusion, there are

$$\binom{n}{2}\binom{n-2}{2}2^2 = n(n-1)(n-2)(n-3)$$

teams satisfying the requirements.

(ii) In order to find the number of valid partitions we shall first find the number of all possible partitions into four teams of six and then subtract the number of all partitions such that each pair of twins is on the same team. Forming 4 teams of 6 with 24 people is the same as partitioning a 24-element set into 4 subsets with 6 elements each. The number of such partitions is

$$\frac{1}{4!}\binom{24}{6}\binom{18}{6}\binom{12}{6}\binom{6}{6} = \frac{24!}{4!720^4}.$$

We need to subtract the number of partitions such that each pair of twins is on the same team, that is

$$\frac{1}{4!}\binom{12}{3}\binom{9}{3}\binom{6}{3}\binom{3}{3} = \frac{12!}{4!6^4}.$$

The answer is therefore $\dfrac{24!}{4!720^4} - \dfrac{12!}{4!6^4}$.

67

(i) For all $x \in \mathbb{N}$ the integers x^{100} and x have the same parity, so we must find the number of all subsets A of $X = \{1, \ldots, 100\}$ such that $\sum_{a \in A} a$ is even. A subset A satisfies this condition if and only if it contains an even number of odd elements.

Consider the partition of X given by the sets $X_0 = \{2, 4, \ldots, 100\}$ and $X_1 = \{1, 3, \ldots, 99\}$, which contain 50 elements each. As with all nonempty sets, there are as many subsets of X_1 with even cardinality as with odd cardinality; there are therefore 2^{49} subsets of X_1 with an even number of elements. It follows that the cardinality we are looking for is $2^{50} \cdot 2^{49} = 2^{99}$.

(ii) Let (A, B) be a pair of subsets of X satisfying the requirements; consider the partition of $A \cup B$ given by $A' = A \setminus (A \cap B)$, $B' = B \setminus (A \cap B)$ and $C = A \cap B$. According to the conditions on (A, B), A' and B' can be any disjoint sets, whereas the set C, which is also disjoint from A' and B', must be chosen to satisfy one of the following: ① C contains exactly one element which is a multiple of 4 but not of 8, ② C contains exactly two elements which are even but not multiples of 4.

Remark that X contains $25 = 50 - 25$ elements that are multiples of 2 but not of 4 and $13 = 25 - 12$ elements that are multiples of 4 but not of 8. Constructing a pair (A, B) is equivalent to constructing a triple (A', B', C) and, in order to find the number of such triples, we shall consider case ① and case ② separately.

In case ① we have 13 ways to choose the one multiple of 4 appearing in C. We then have three choices for each of the 49 even elements of X, each of which can be put in A', in B' or in neither of the two sets. Finally, each of the 50 odd elements of X can be chosen to be a member of A', B', C or of none of the three sets, so we have four possible choices for each. This yields $13 \cdot 3^{49} \cdot 4^{50}$ possibilities in total.

Similarly, in case ② we have $\binom{25}{2}$ choices for the pair of even elements we assign to C. Each of the remaining 48 even elements of X can be assigned to A', to B' or to neither and, finally, each of the 50 odd elements of X can be assigned to A', B', C or to none of the three sets. We have $\binom{25}{2} \cdot 3^{48} \cdot 4^{50}$ possibilities.

Summing the numbers obtained in the two cases we get the answer

$$13 \cdot 3^{49} \cdot 4^{50} + \binom{25}{2} \cdot 3^{48} \cdot 4^{50} = 113 \cdot 3^{49} \cdot 2^{100}.$$

3 Congruences

68 Remark that $13 \equiv 64 \equiv 2^6 \pmod{17}$ and thus the first equation becomes $2^{ax} \equiv 2^6 \pmod{17}$. Since the order of 2 in $(\mathbb{Z}/17\mathbb{Z})^*$ is 8, we have $ax \equiv 6 \pmod 8$. This equation implies that a and x cannot both be even; indeed, if they were we would have both $ax \equiv 0 \pmod 4$ and $ax \equiv 2 \pmod 4$, impossible. We shall distinguish the cases a even and a odd.

⓵ If a is even then (as we just showed) x is odd, so $x - 2$ is invertible modulo 4. From the second equation we get $x \equiv a \pmod{4}$ which is impossible because it implies x even.

⓶ If a is odd, then x is even, say $x = 2y$, from which the first equation becomes $2ay \equiv 6 \pmod{8}$, hence $ay \equiv 3 \pmod{4}$. Now, from a being odd we obtain $a^2 \equiv 1 \pmod{4}$, so the first equation becomes $y \equiv -a \pmod{4}$; hence y is odd. Substitute $x = 2y$ into the second equation to obtain $(2y - a)(2y - 2) \equiv 0 \pmod{4}$, that is, $y - 1 \equiv 0 \pmod{2}$, which holds because y is odd.

In conclusion, for a odd, the solution is $x \equiv -2a \pmod{8}$.

69 Since $2 \in (\mathbb{Z}/9\mathbb{Z})^*$, we must also have $a \equiv 2^x \in (\mathbb{Z}/9\mathbb{Z})^*$. So $a \not\equiv 0 \pmod{3}$, hence $x \equiv a^2 \equiv 1 \pmod{3}$.

The order of $(\mathbb{Z}/9\mathbb{Z})^*$ is 6 and 2 is a generator. Moreover, $x \equiv 1 \pmod{3}$ implies that x can be congruent to 1 or to 4 modulo 6. In the first case, $a \equiv 2^x \equiv 2 \pmod{9}$, whereas in the second case $a \equiv 2^x \equiv -2 \pmod{9}$. So the condition $a \equiv \pm 2 \pmod{9}$ is necessary.

On the other hand, if $a \equiv 2 \pmod{9}$ then $x \equiv 1 \pmod{3}$ is a solution of the system, whereas if $a \equiv -2 \pmod{9}$ then $x \equiv 4 \pmod{6}$ is a solution.

70 From the second congruence we get that $4 = (4, 24)$ divides a^2 and thus a is even, say $a = 2b$. The second congruence becomes $x^2 \equiv b^2 \pmod{6}$.

The first congruence can be rewritten as $2^x \equiv 3^{2b} \equiv (3^2)^b \equiv 2^b \pmod{7}$. It is equivalent to $x \equiv b \pmod{3}$ because 3 is the order of 2 in $(\mathbb{Z}/7\mathbb{Z})^*$. For $a = 2b$ even, the system is therefore equivalent to

$$\begin{cases} x^2 \equiv b^2 \pmod{6} \\ x \equiv b \pmod{3}. \end{cases}$$

From the first equation we get $x \equiv x^2 \equiv b^2 \equiv b \pmod{2}$; we thus must have $x \equiv b \pmod{6}$. But this condition is clearly sufficient. In conclusion, the system has solutions for all even a.

71 Remark that the order of 2 in $(\mathbb{Z}/13\mathbb{Z})^*$ is 12, that is, 2 is a generator of that group, and that $3 \equiv 2^4 \pmod{13}$. The second congruence yields that $x - 1 = 3y$ for some integer y. Then the first congruence becomes

$$(2^4)^{3y(3y+2)} = 2^{12y(3y+2)} \equiv 1 \equiv 2^a \pmod{13}.$$

Such a congruence has a solution if and only if $a \equiv 0 \pmod{12}$, in which case $x \equiv 1 \pmod{3}$ is the solution of the system.

72 First of all remark that 2 has order 3 in $(\mathbb{Z}/7\mathbb{Z})^*$, so 2^x only depends on the residue class modulo 3 of x. Moreover, the second equation is equivalent to the systems $x \equiv \pm 1 \pmod{3}$, $x \equiv \pm 1 \pmod{5}$, where signs are chosen independently. In particular, x is not congruent to 0 modulo 3. Therefore, if $x \equiv 1$

(mod 3) then $x \equiv 2^x \equiv 2$ (mod 7), whereas if $x \equiv -1$ (mod 3) then $x \equiv 2^x \equiv -3$ (mod 7).

The system is thus equivalent to the union of the four systems

$$
\begin{cases}
x \equiv 1 & \text{(mod 3)} \\
x \equiv 2 & \text{(mod 7)} \\
x \equiv \pm 1 & \text{(mod 5),}
\end{cases}
\qquad
\begin{cases}
x \equiv -1 & \text{(mod 3)} \\
x \equiv -3 & \text{(mod 7)} \\
x \equiv \pm 1 & \text{(mod 5).}
\end{cases}
$$

Straightforward computations lead to the conclusion that the original system has as solutions the classes 16, -26, 11 and -31 modulo 105.

73 From the definition of binomial coefficient we get

$$
\begin{cases}
\frac{n(n-1)(n-2)}{6} \equiv 0 & \text{(mod 2)} \\
\frac{n(n-1)(n-2)(n-3)}{24} \equiv 0 & \text{(mod 2)}
\end{cases}
$$

hence, eliminating factors that are invertible modulo 2,

$$
\begin{cases}
n(n-1)(n-2) \equiv 0 & \text{(mod 4)} \\
n(n-1)(n-2)(n-3) \equiv 0 & \text{(mod 16).}
\end{cases}
$$

The first equation is clearly solved by all n congruent to 0, 1 or 2 modulo 4, whereas n is not a solution if it is congruent to -1 modulo 4.

In order to solve the second equation, remark that exactly two among the factors $n, n - 1, n - 2, n - 3$ are even, and that they are consecutive even numbers. Given two consecutive even numbers, only one can be divisible by 4, while the other is divisible by 2 but not by 4. Therefore, if the product must be divisible by 16, then the number which is a multiple of 4 must actually be divisible by 8. It follows that the second equation is solved by n congruent to 0, 1, 2, 3 modulo 8.

Since n congruent to a modulo 8 implies n congruent to a modulo 4, the solutions of the system are given by n congruent to 0, 1, 2 modulo 8.

74   The congruence $x^2 \equiv 4$ (mod 14) is equivalent to the system of two congruences $x^2 \equiv 0$ (mod 2) and $x^2 \equiv 4$ (mod 7). Moreover, we can rewrite the first of these as $x \equiv 0$ (mod 2), because x^2 is always congruent to x modulo 2.

The second one in turn becomes $(x - 2)(x + 2) \equiv 0$ (mod 7) and, since 7 is prime, is solved by $x \equiv 2$ (mod 7) or $x \equiv -2$ (mod 7).

So the solutions of the original system are given by the union of the solutions of the two systems

$$
\begin{cases}
x \equiv 0 & \text{(mod 2)} \\
x \equiv 2 & \text{(mod 7)} \\
x \equiv 3 & \text{(mod 5),}
\end{cases}
\qquad
\begin{cases}
x \equiv 0 & \text{(mod 2)} \\
x \equiv -2 & \text{(mod 7)} \\
x \equiv 3 & \text{(mod 5).}
\end{cases}
$$

Straightforward calculations yield that the solutions of the original system are given by the residue classes -12 and -2 modulo 70.

75 Let us consider the prime factorisation $7 \cdot 11^2$ of 847 and replace the first equation with the system of congruences $x^{660} \equiv 1 \pmod{7}$ and $x^{660} \equiv 1 \pmod{11^2}$.

We shall examine the first of these. If 7 were a divisor of x, then 7 would obviously divide x^{660}, which contradicts the condition $x^{660} \equiv 1 \pmod{7}$. Conversely, if 7 does not divide x then we have $x^6 \equiv 1 \pmod{7}$ by Fermat's little theorem, hence $x^{660} = (x^6)^{110} \equiv 1^{110} \equiv 1 \pmod{7}$. We have thus shown that the first congruence is solved by all x that are not divisible by 7.

As for the second congruence, we will employ a similar argument. If 11 were a divisor of x then it would also divide x^{660}, which is not possible. So $x^{\phi(11^2)} = x^{110} \equiv 1 \pmod{11}$ and thus $x^{660} = (x^{110})^6 \equiv 1^6 \equiv 1 \pmod{11^2}$. The second congruence is therefore solved by all x that are not divisible by 11.

We can rewrite the original system as

$$\begin{cases} x \not\equiv 0 & \pmod{7} \\ x \not\equiv 0 & \pmod{11} \\ x \equiv 11 & \pmod{13}. \end{cases}$$

The solutions of the third equation are clearly given by all integers of the form $x = 11 + 13k$, where k is any integer. By imposing the condition that x is neither divisible by 7 nor by 11 we find that k can neither be congruent to 4 modulo 7 nor to 0 modulo 11.

76 By considering the factorisation $85 = 5 \cdot 17$ we obtain that the congruence is equivalent to the system

$$\begin{cases} x^3 - a^3 \equiv 0 & \pmod{5} \\ x^3 - a^3 \equiv 0 & \pmod{17}. \end{cases}$$

Now let p be either 5 or 17. If $a \equiv 0 \pmod{p}$ then we have $x^3 \equiv 0 \pmod{p}$ and so, since p is prime, $x \equiv 0 \equiv a \pmod{p}$. If $a \not\equiv 0 \pmod{p}$, then a is invertible modulo p, so we can rewrite the congruence as $(x/a)^3 \equiv 1 \pmod{p}$.

Since 0 cannot be a solution and 3 does not divide $p - 1$, the only possible solution is $x/a \equiv 1 \pmod{p}$, that is, $x \equiv a \pmod{p}$.

So the system is in any case equivalent to

$$\begin{cases} x \equiv a & \pmod{5} \\ x \equiv a & \pmod{17} \end{cases}$$

and, since $x \equiv a \pmod{85}$ is a solution, it must be the unique solution by the Chinese remainder theorem.

In conclusion, we have that for each a the congruence $x^3 - a^3 \equiv 0 \pmod{85}$ has the unique solution $x \equiv a \pmod{85}$.

77 Remark that 2 is coprime with 3^3 and so $2^{\phi(3^3)} = 2^{18} \equiv 1 \pmod{3^3}$. Since 9 and 6 are the only maximal proper divisors of 18, we shall compute 2^9 and 2^6 modulo 3^3. We find that $2^9 \equiv -1 \pmod{3^3}$ and $2^6 \equiv 10 \pmod{3^3}$. This implies that 2 has multiplicative order 18 modulo 3^3. Moreover, $2^5 = 32 \equiv 5 \pmod{3^3}$.

We can now rewrite the initial congruence as $2^x \equiv 2^5 \pmod{3^3}$ and conclude that it is equivalent to $x \equiv 5 \pmod{18}$.

We can thus rewrite the first of the two systems as

$$\begin{cases} x \equiv 5 \pmod{18} \\ x \equiv 2 \pmod{15}. \end{cases}$$

By the Chinese remainder theorem, by taking prime factorisations, one obtains

$$\begin{cases} x \equiv 5 \pmod 9 \\ x \equiv 1 \pmod 2 \\ x \equiv 2 \pmod 3 \\ x \equiv 2 \pmod 5. \end{cases}$$

Notice that the first congruence implies the third, which can thus be ignored. Straightforward computations yield the solutions $-13 + 90k$, where k is any integer.

Let us now show that the second system in the problem statement has no solutions. If $2^x \equiv 5 \pmod{3^4}$ then in particular we have $2^x \equiv 5 \pmod{3^3}$, so $x \equiv 5 \pmod{18}$ because of the argument above. Then $x \equiv -1 \pmod 3$, which is incompatible with the last congruence, as that one implies $x \equiv 0 \pmod 3$.

78

(i) If $81^x \equiv b \pmod{125}$ then we must have $81^x \equiv b \pmod 5$, that is, $b \equiv 1 \pmod 5$. Hence, in order for a solution to exist, the residue class of b modulo 125 must be one of the 25 classes congruent to 1 modulo 5. Let us show that this condition is sufficient.

Remark that 81 is coprime to 125; let us compute its multiplicative order. Since $\phi(125) = 100$, such an order must be a divisor of 100. We start by computing 81^5 modulo 125.

We have $81 = 1 + 16 \cdot 5$, so $81^5 = (1 + 16 \cdot 5)^5 \equiv 1 + 5 \cdot 16 \cdot 5 + 10 \cdot 16^2 \cdot 5^2 + \cdots \equiv 1 + 16 \cdot 5^2 \equiv 1 + 5^2 \pmod{125}$, where we omitted terms containing higher powers of 5, which are 0 modulo 125. Thus the multiplicative order of 81 modulo 125 cannot be a divisor of 5.

Similarly, we have $81^{25} = (81^5)^5 \equiv (1 + 5^2)^5 \equiv 1 + 5 \cdot 5^2 + \cdots \equiv 1 \pmod{125}$. So the order we want must be a divisor of 25 and, since it cannot be 1 or 5, it must be equal to 25.

It follows that the powers of 81 modulo 125 must take 25 different values. Since our congruence has a solution if and only if b is a power of 81 modulo 125 we just showed that the condition we found initially is sufficient.

(ii) Given a solution x_0, we have $81^{x_0} \equiv b_0 \pmod{125}$. The equation becomes $81^x \equiv 81^{x_0} \pmod{125}$ and, since 81 has multiplicative order 25 modulo 125, all solutions are given by $x_0 + 25 \cdot k$, where k is any integer.

79 First of all, let us compute the order of 2 modulo 125, which is a divisor of $\phi(125) = 100$.

Remark that if $2^n \equiv 1 \pmod{125}$ then $2^n \equiv 1 \pmod 5$. It follows that, since the multiplicative order of 2 modulo 5 is 4, the order of 2 modulo 125 is a multiple of 4, that is, one among 4, 20, 100.

We have $2^4 = 16 = 1 + 3 \cdot 5$, hence, by using the binomial development, we get $2^{20} = (2^4)^5 = (1 + 3 \cdot 5)^5 \equiv 1 + 3 \cdot 5^2 \pmod{5^3}$. Thus the multiplicative order of 2 modulo 5^3 cannot be a divisor of 20, so it must be 100.

Since $2^7 = 128 \equiv 3 \pmod{125}$, 7 is a solution of the original congruence. In particular, the congruence amounts to $2^x \equiv 2^7 \pmod{125}$ and, because of what we just showed, its solutions are given by $7 + 100k$, where k is any integer.

Let us now solve the congruence modulo $625 = 5^4$. We have already remarked that $2^{4 \cdot 5} \equiv 1 + 3 \cdot 5^2 \pmod{5^3}$, that is, $2^{4 \cdot 5} = 1 + 3 \cdot 5^2 + h \cdot 5^3$ for some integer h. Using the binomial development again, we have $2^{4 \cdot 5^2} = (1 + 3 \cdot 5^2 + h \cdot 5^3)^5 = (1 + (3 + 5h)5^2)^5 \equiv 1 + 5(3 + 5h)5^2 \equiv 1 + 3 \cdot 5^3 \pmod{5^4}$. Any solution of $2^x \equiv 3 \pmod{5^4}$ is also a solution of $2^x \equiv 3 \pmod{5^3}$, so it must be of the form $x = 7 + 100k$ because of the argument above.

By combining the two facts $2^7 = 3 + 5^3$ and $2^{100k} \equiv (1 + 3 \cdot 5^3)^k \equiv 1 + 3k5^3 \pmod{5^4}$ we obtain $2^x = 2^{7+100k} \equiv (3+5^3)(1+3k5^3) \equiv 3+(1+9k)5^3 \pmod{5^4}$. So x is a solution if and only if $k \equiv 1 \pmod 5$. In conclusion, $x \equiv 107 \pmod{500}$.

80 By examining powers of 5 modulo 11 we find that 5 has multiplicative order 5 modulo 11 and that $3 \equiv 5^2 \pmod{11}$. The first congruence can thus be rewritten as $5^x \equiv 5^2 \pmod{11}$ and is equivalent to $x \equiv 2 \pmod 5$.

By the Chinese remainder theorem, the second congruence is equivalent to the system

$$\begin{cases} x^2 \equiv 0 & \pmod 3 \\ x^2 \equiv -3 & \pmod 7. \end{cases}$$

The first of these congruences is equivalent to $x \equiv 0 \pmod 3$ because 3 is prime. The second is equivalent to $x \equiv \pm 2 \pmod 7$ because 7 is prime and $(\pm 2)^2 \equiv -3 \pmod 7$.

We can thus rewrite the original system as

$$\begin{cases} x \equiv 0 & \pmod 3 \\ x \equiv 2 & \pmod 5 \\ x \equiv \pm 2 & \pmod 7. \end{cases}$$

Straightforward computations yield that the solutions are given by the classes of 12 and 72 modulo 105.

81 The condition in the problem statement is equivalent to the congruence

$$7x^3 - 8ax^2 + 9x + 3a \equiv 0 \pmod{21}$$

which is itself equivalent to the system

$$\begin{cases} 7x^3 - 8ax^2 + 9x + 3a \equiv 0 \pmod 3 \\ 7x^3 - 8ax^2 + 9x + 3a \equiv 0 \pmod 7. \end{cases}$$

By reducing the coefficients of the above equations modulo 3 and 7 one obtains

$$\begin{cases} x^3 + ax^2 \equiv 0 \pmod 3 \\ -ax^2 + 2x + 3a \equiv 0 \pmod 7. \end{cases}$$

Since $x^3 + ax^2 = x^2(x + a)$ and 3 is prime, the solutions of the first equation are $x \equiv 0 \pmod 3$ and $x \equiv -a \pmod 3$.

As for whether or not the second equation has any solutions, we have

$x \equiv 0 \pmod 7$ is a solution $\Longrightarrow a \equiv 0 \pmod 7$;
$x \equiv 1 \pmod 7$ is a solution $\Longrightarrow a \equiv -1 \pmod 7$;
$x \equiv 2 \pmod 7$ is a solution $\Longrightarrow a \equiv -3 \pmod 7$;
$x \equiv 3 \pmod 7$ is a solution $\Longrightarrow a \equiv 1 \pmod 7$;
$x \equiv -3 \pmod 7$ is a solution $\Longrightarrow a \equiv -1 \pmod 7$;
$x \equiv -2 \pmod 7$ is a solution $\Longrightarrow a \equiv 3 \pmod 7$;
$x \equiv -1 \pmod 7$ is a solution $\Longrightarrow a \equiv 1 \pmod 7$.

The list above can be rearranged according to the possible values of a, as:

if $a \equiv 0 \pmod 7$ then $x \equiv 0 \pmod 7$;
if $a \equiv 1, -1, 3 \pmod 7$ then a solution is $x \equiv a + 2 \pmod 7$;
if $a \equiv 1, -1, -3 \pmod 7$ then a solution is $x \equiv a - 2 \pmod 7$;
if $a \equiv \pm 2 \pmod 7$ then there are no solutions.

If both equations in the system have a solution, then the system itself does, because $(3, 7) = 1$. By solving the various systems which do have solutions, we obtain

if $a \equiv 0 \pmod 7$, then $x \equiv 0, -a \pmod{21}$;
if $a \equiv 1, -1, 3 \pmod 7$, then the solutions are $x \equiv 15a + 9, 8a + 9 \pmod{21}$;
if $a \equiv 1, -1, -3 \pmod 7$, then the solutions are $x \equiv 15a + 12, 8a + 12 \pmod{21}$.

82 By the Chinese remainder theorem, the congruence is equivalent to the system

$$\begin{cases} x^{15} \equiv x^{27} \pmod 7 \\ x^{15} \equiv x^{27} \pmod{11} \end{cases}$$

which can be rewritten as

$$\begin{cases} x^{15}(x^{12} - 1) \equiv 0 \pmod{7} \\ x^{15}(x^{12} - 1) \equiv 0 \pmod{11}. \end{cases}$$

Clearly, $x \equiv 0 \pmod{7}$ and $x \equiv 0 \pmod{11}$ are solutions of the first and second equation, respectively. If $(x, 7) = 1$, then by Fermat's little theorem we have $x^6 \equiv 1$ $\pmod{7}$ and therefore $x^{12} \equiv 1 \pmod{7}$. So the first equation is satisfied for all integers x, that is, it has seven solutions modulo 7.

As for the second equation, Fermat's little theorem implies that, if $(x, 11) = 1$, then $x^{10} \equiv 1 \pmod{11}$; such values of x are solutions if and only if $x^2 \equiv 1$ $\pmod{11}$, that is, $(x+1)(x-1) \equiv 0 \pmod{11}$. Since 11 is prime, the only solutions are $x \equiv \pm 1 \pmod{11}$. Including the solution $x \equiv 0 \pmod{11}$, there are three solutions modulo 11.

The number of solutions modulo 77 is given by all possible combinations of solutions modulo 7 and solutions modulo 11 and is therefore equal to $7 \cdot 3 = 21$.

83 Because of the second condition, integers x such that $x^3 \equiv x \pmod{7}$ cannot be solutions of the system. Now, $x^3 \equiv x \pmod{7}$ is equivalent to $x(x - 1)(x + 1) \equiv 0 \pmod{7}$ and, since 7 is prime, the solutions of this equation are $x \equiv 0, 1, -1$ $\pmod{7}$.

In particular, given that the class $\overline{0}$ modulo 7 cannot be a solution of the system, we may rewrite the first equation as $x^{k-1} \equiv 1 \pmod{7}$. A class a solves the latter equation if and only if $k - 1$ is a multiple of the multiplicative order of a modulo 7. One can check that the multiplicative order of classes $\overline{2}$ and $\overline{4}$ is 3, whereas that of classes $\overline{3}$ and $\overline{5}$ is 6. Therefore, if $k \equiv 1 \pmod{6}$ then the solutions are $x \equiv 2, 3, 4, 5$ $\pmod{7}$; if $k \equiv 1 \pmod{3}$ but $k \not\equiv 1 \pmod{6}$, that is, if $k \equiv 4 \pmod{6}$, then the solutions are $x \equiv 2, 4 \pmod{7}$, and finally if $k \not\equiv 1 \pmod{3}$ then there are no solutions.

84 In order for the first equation to have solutions, it is necessary and sufficient that $(a, 25) \mid 4$, or in other words that $(a, 25) = 1$. The second equation is equivalent to the system

$$\begin{cases} x^2 + a \equiv 0 \pmod{5} \\ x^2 + a \equiv 0 \pmod{3} \end{cases}$$

and one can check that this has solutions if and only if $a \equiv 0, \pm 1 \pmod{5}$ and $a \equiv 0, -1 \pmod{3}$. Therefore, both initial equations have a solution for $a \equiv \pm 1$ $\pmod{5}$ and $a \equiv 0, -1 \pmod{3}$.

However, because the greatest common divisor of the two moduli is $(25, 15) = 5$, the original system has a solution if and only if its two equations have solutions that are congruent modulo 5.

If $a \equiv 1 \pmod{5}$ then the solutions of the first equation are congruent to $-1 \pmod{5}$, whereas those of the second equation are $\pm 2 \pmod{5}$. If $a \equiv -1$

(mod 5) then the solutions of the first equation are congruent to 1 (mod 5) and those of the second equation to ± 1 (mod 5). Consequently, the system has a solution if and only if $a \equiv -1$ (mod 5) and $a \equiv 0, -1$ (mod 3), that is, if and only if $a \equiv -1, 9$ (mod 15).

Now consider the case $a = -1$. The first equation yields $x \equiv -4$ (mod 25); this fixes the class of solutions modulo 5, that is, $x \equiv 1$ (mod 5). Moreover, we know by the argument above that solutions of the second equation satisfying this condition do exist. What is left to do is to find out which of them also satisfy $x^2 - 1 \equiv 0$ (mod 3), that is, $x \equiv \pm 1$ (mod 3). By considering both the solutions modulo 25 and those modulo 3, we find that the solutions of the system are $x \equiv -4, -29$ (mod 75).

85 Both equations have a solution if and only if $(a, 9) = 1$. Solutions of the first equation are given by $x \equiv a^{-1}$ (mod 9), so are coprime to 9. The solution of the second equation is given by $x \equiv 0$ (mod $\mathrm{ord}(a)$), where $\mathrm{ord}(a)$ is the order of a in the multiplicative group $(\mathbb{Z}/9\mathbb{Z})^*$ and so is a divisor of $\phi(9) = 6$. In order for the system to have a solution we must therefore have that $\mathrm{ord}(a)$ is not a multiple of 3, that is, $\mathrm{ord}(a) = 1, 2$.

If $\mathrm{ord}(a) = 1$ then $a \equiv 1$ (mod 9) and the solution of the first equation is $x \equiv 1$ (mod 9), while the second equation is satisfied for all x. So the solution of the system is $x \equiv 1$ (mod 9).

If $\mathrm{ord}(a) = 2$, then $a \not\equiv 1$ (mod 9) and $a^2 \equiv 1$ (mod 9), that is, $9 \mid a^2 - 1 = (a + 1)(a - 1)$. In particular, three divides one of the factors $a + 1$, $a - 1$ but cannot divide both, since their difference is 2. It follows that 9 must divide one of the factors; having excluded $a \equiv 1$ (mod 9), we must have $a \equiv -1$ (mod 9).

In conclusion, the solution of the first equation is $x \equiv -1$ (mod 9) and that of the second equation is $x \equiv 0$ (mod 2). The solution of the system is therefore $x \equiv 8$ (mod 18).

86 The first equation has a solution if and only if $(6a - 1, 21) = 1$, that is, if and only if $3 \nmid 6a - 1$ and $7 \nmid 6a - 1$. Since $3 \mid 6a$ for all a, we have $3 \nmid 6a - 1$ for all a, whereas $7 \mid 6a - 1$ if and only if $a \equiv -1$ (mod 7). So the first equation has a solution if and only if $a \not\equiv -1$ (mod 7).

The second equation always has a solution, which is expressed by the equation itself.

The greatest common divisor of the moduli of the two equations is 7, so assuming that the first equation does have a solution the system does if and only if there are solutions of the two equations that coincide modulo 7. Substituting the value for x given by the second equation into the first, one obtains the condition $(6a - 1)a \equiv 1$ (mod 7), whose solutions are $a \equiv 2, 4$ (mod 7).

Now remark that if $a \equiv 2, 4$ (mod 7) then the solution of the system is a congruence class modulo the least common multiple of the moduli, which is 105. Because of the second equation we know its congruence class modulo 35, so it is enough to glean its congruence class modulo 3 from the first equation, which clearly gives $x \equiv -1 \pmod 3$. We thus obtain the system

$$\begin{cases} x \equiv -1 \pmod 3 \\ x \equiv a \pmod{35} \end{cases}$$

whose solution is $x \equiv 36a + 35 \pmod{105}$, as one can easily find.

87 By the Chinese remainder theorem, the first equation is equivalent to a system of two congruences, one modulo 2 and one modulo 17. Since $9 \equiv 1 \pmod 2$, the congruence modulo 2 is satisfied for all possible values of a and x and can therefore be disregarded. As for the congruence modulo 17, one can check that the multiplicative order of 9 modulo 17 is 8, so the congruence is satisfied if and only if $ax \equiv 0 \pmod 8$. In particular, it has a solution for all possible values of a, and the solution is $x \equiv 0 \pmod{8/(a, 8)}$.

Similarly, we can replace the second equation by two congruences, one modulo 3 and one modulo 5. The congruence modulo 3 reduces to $x^2 \equiv 0 \pmod 3$, which does not depend on a and whose solution is $x \equiv 0 \pmod 3$. The congruence modulo 5 reduces to $x^2 + ax - 1 \equiv 0 \pmod 5$, which has a solution if and only if $a^2 + 4$ is a square modulo 5, that is, if and only if $a \equiv 0, 1, -1 \pmod 5$. The solutions are, respectively, $x \equiv \pm 1, 2, -2 \pmod 5$.

In conclusion, the system has a solution if and only if $a \equiv 0, 1, -1 \pmod 5$.

For $a = 4$ in particular, as per remarks made so far, the solution can be found via the system

$$\begin{cases} x \equiv 0 \pmod 2 \\ x \equiv 0 \pmod 3 \\ x \equiv -2 \pmod 5. \end{cases}$$

Simple calculations yield $x \equiv 18 \pmod{30}$.

88 The first equation has solutions if and only if $(3, 42) = 3 \mid a$, that is, if and only if $a = 3b$ for some $b \in \mathbb{Z}$, in which case the equation becomes $x \equiv b \pmod{14}$. Since 6 is the inverse of 6 modulo 35, the second equation is equivalent to $x \equiv 6 \pmod{35}$. By using the Chinese remainder theorem, we can turn the system into

$$\begin{cases} x \equiv b \pmod 2 \\ x \equiv b \pmod 7 \\ x \equiv 6 \pmod 7 \\ x \equiv 1 \pmod 5 \end{cases}$$

which has solutions if and only if $b \equiv 6 \pmod 7$, that is, if and only if $a \equiv 18 \pmod{21}$.

When this condition on a is satisfied, the system is equivalent to

$$\begin{cases} x \equiv b \pmod{2} \\ x \equiv 6 \pmod{35} \end{cases}$$

whose solution is $x \equiv 6 + 35b \pmod{70}$.

In conclusion, the original system has a solution if and only if $a \equiv 18 \pmod{21}$. When that is the case, its solution is $x \equiv 6 \pmod{70}$ if $a \equiv 18 \pmod{42}$ and $x \equiv 41 \pmod{70}$ if $a \equiv 39 \pmod{42}$.

89 The multiplicative order of the class of 5 modulo 2^4 is 4 and $5^2 \equiv 9 \pmod{2^4}$, so the first equation is equivalent to $x \equiv 2 \pmod{4}$.

By means of the Chinese remainder theorem, split the second equation into a congruence modulo 11 and one modulo 16. Since 11 is an odd prime, the equation $x^2 + 2x + 8 \equiv 0 \pmod{11}$ can be solved using the quadratic formula, which yields the solutions $x \equiv 1, -3 \pmod{11}$.

Now consider the subsystem

$$\begin{cases} x \equiv 2 \pmod{4} \\ x^2 + 2x + 8 \equiv 0 \pmod{16}; \end{cases}$$

if we substitute the value for x obtained from the first congruence, that is, $x = 2 + 4t$, into the second, we obtain the equation

$$(2 + 4t)^2 + 2(2 + 4t) + 8 \equiv 8t \equiv 0 \pmod{16},$$

which is satisfied if and only if $t \equiv 0 \pmod{2}$, so the solution of the subsystem is $x \equiv 2 \pmod{8}$.

So the original system is equivalent to the union of the two systems

$$\begin{cases} x \equiv 1 \pmod{11} \\ x \equiv 2 \pmod{8}, \end{cases} \qquad \begin{cases} x \equiv -3 \pmod{11} \\ x \equiv 2 \pmod{8} \end{cases}$$

whose solutions are easily computed and given by $x \equiv 34 \pmod{88}$ and $x \equiv -14 \pmod{88}$, respectively.

90   The second equation has a solution if and only if $3 \mid a$, so let $a = 3b$. The congruence $6x \equiv 3b \pmod{21}$ is equivalent to $x \equiv 4b \pmod{7}$.

The first congruence becomes $x^2 \equiv 15b \pmod{120}$ and is equivalent by the Chinese remainder theorem to a system of two congruences, one modulo 8 and one modulo 15. Keeping in mind that $x^2 \equiv 0 \pmod{15}$ if and only if $x \equiv 0 \pmod{15}$, we obtain the system

$$\begin{cases} x^2 \equiv -b \pmod{8} \\ x \equiv 0 \pmod{15} \\ x \equiv 4b \pmod{7}. \end{cases}$$

The three moduli are pairwise relatively prime, so the system has a solution if and only if each equation does. In order for that to be the case, since we already have the solutions of the second and third equation, we just need to ensure that the first equation can be solved, i.e. that $-b$ is a square modulo 8; equivalently, that $b \equiv 0, -1, 4 \pmod 8$.

In conclusion, the system has a solution if and only if $a = 3b \equiv 0, -3, 12 \pmod{24}$.

For $a = 45 \equiv -3 \pmod{24}$, that is, for $b = 15$ (where b is defined as above), the system does have a solution and becomes

$$\begin{cases} x^2 \equiv 1 \pmod 8 \\ x \equiv 0 \pmod{15} \\ x \equiv 4 \pmod 7. \end{cases}$$

It is easy to check that the subsystem given by the second and third equation has the solution $x \equiv 60 \pmod{105}$ and that the first equation has the solutions $x \equiv 1, 3, 5, 7 \pmod 8$, that is, $x \equiv 1 \pmod 2$. The system is therefore equivalent to

$$\begin{cases} x \equiv 1 \pmod 2 \\ x \equiv 60 \pmod{105} \end{cases}$$

whose solution is $x \equiv 165 \pmod{210}$.

91 By the Chinese remainder theorem, the equation in the problem statement is equivalent to the system

$$\begin{cases} x^{100} \equiv a \pmod 7 \\ x^{100} \equiv a \pmod{11}. \end{cases}$$

Let us find the number of solutions of each equation.

If $a \equiv 0 \pmod 7$, then the congruence $x^{100} \equiv a \pmod 7$ has the unique solution $x \equiv 0 \pmod 7$.

Let now $a \not\equiv 0 \pmod 7$; then $x \equiv 0 \pmod 7$ is not a solution and any potential solution will satisfy $x^6 \equiv 1 \pmod 7$. Since $100 = 16 \cdot 6 + 4$, the equation we need to solve is equivalent to $x^4 \equiv a \pmod 7$. The fourth powers in $(\mathbb{Z}/7\mathbb{Z})^*$ are $\overline{1}, \overline{2}, \overline{4}$, and so if $a \equiv 3, 5, 6 \pmod 7$ then the equation has no solutions, whereas if $a \equiv 1, 2, 4 \pmod 7$ then it has two.

For $a \equiv 0 \pmod{11}$, the congruence $x^{100} \equiv a \pmod{11}$ has the unique solution $x \equiv 0 \pmod{11}$.

Let now $a \not\equiv 0 \pmod{11}$; in this case $x \equiv 0 \pmod{11}$ is not a solution and any solution will satisfy $x^{10} \equiv 1 \pmod{11}$, hence also $x^{100} \equiv 1 \pmod{11}$. It follows that the equation has ten solutions if $a \equiv 1 \pmod{11}$ and no solutions if $a \not\equiv 0, 1 \pmod{11}$.

Let us now combine the information above.

For $a \equiv 0 \pmod 7$ and $a \equiv 0 \pmod{11}$, that is, for $a \equiv 0 \pmod{77}$, the system has the unique solution $x \equiv 0 \pmod{77}$.

For $a \equiv 0 \pmod 7$ and $a \equiv 1 \pmod{11}$, that is, for $a \equiv -21 \pmod{77}$, the first equation has one solution and the second one has 10, so there are ten solutions modulo 77.

For $a \equiv 1, 2, 4 \pmod 7$ and $a \equiv 0 \pmod{11}$, that is, for $a \equiv 22, 44, 11 \pmod{77}$, the first equation has two solutions and the second one has 1, so there are two solutions modulo 77.

Finally, for $a \equiv 1, 2, 4 \pmod 7$ and $a \equiv 1 \pmod{11}$, that is, for $a \equiv 1, 23, 67 \pmod{77}$, the first equation has two solutions and the second one has 10, so there are 20 solutions modulo 77.

If a does not belong to any of the above classes modulo 77 then the equation has no solutions.

92

(i) By the Chinese remainder theorem, we can split $x^a \equiv 1 \pmod{92}$ into a congruence modulo 4 and one modulo 23.

The congruence $x^a \equiv 1 \pmod{23}$ is solved by all elements of $(\mathbb{Z}/23\mathbb{Z})^*$ whose order divides a, and is thus equivalent to $x^d \equiv 1 \pmod{23}$, where $d = (a, \phi(23))$. Since 23 is prime, the group $(\mathbb{Z}/23\mathbb{Z})^*$ is cyclic of order $\phi(23) = 22$. We conclude that the congruence $x^a \equiv 1 \pmod{23}$ has $d = (a, 22)$ solutions modulo 23.

By the same argument, we obtain that $x^a \equiv 1 \pmod 4$ has $(a, 2)$ solutions modulo 4.

In conclusion, the equation in the problem statement has $(a, 2) \cdot (a, 22)$ solutions modulo 92.

Remark that the number of solutions can also be expressed as a function of the class of a modulo 22.

If $a \equiv 1, 3, 5, 7, 9, 13, 15, 17, 19, 21 \pmod{22}$, that is, if $(a, 2) = 1$ and $(a, 22) = 1$, then the equation has a unique solution modulo 92.

If $a \equiv 2, 4, 6, 8, 10, 12, 14, 16, 18, 20 \pmod{22}$, that is, if $(a, 2) = 2$ and $(a, 22) = 2$, then the equation has four solutions modulo 92.

If $a \equiv 11 \pmod{22}$, that is, $(a, 2) = 1$ and $(a, 22) = 11$, then the equation has 11 solutions modulo 92.

Finally, if $a \equiv 0 \pmod{22}$, that is, $(a, 2) = 2$ and $(a, 22) = 22$, then the equation has 44 solutions modulo 92.

(ii) By solving the second equation and splitting the first into an equation modulo 4 and one modulo 23 we get

$$\begin{cases} x^a \equiv 1 \pmod 4 \\ x^a \equiv 1 \pmod{23} \\ x \equiv 9 \pmod{23}. \end{cases}$$

Remark that the set of squares in $(\mathbb{Z}/23\mathbb{Z})^*$ coincides with the subgroup of order 11. Since 11 is prime, all elements of such a subgroup except for $\bar{1}$ have order

11. Now, 9 is a square and therefore its order in $(\mathbb{Z}/23\mathbb{Z})^*$ is 11. Thus, if $a \not\equiv 0$ (mod 11) then the system has no solutions, whereas if $a \equiv 0$ (mod 11) then the system is equivalent to

$$\begin{cases} x^a \equiv 1 \pmod 4 \\ x \equiv 9 \pmod{23}. \end{cases}$$

The equation $x^a \equiv 1$ (mod 4) has the unique solution $x \equiv 1$ (mod 4) if $a \equiv 1$ (mod 2) and the solutions $x \equiv \pm 1$ (mod 4) if $a \equiv 0$ (mod 2). Solving the corresponding systems, one obtains: if $a \equiv 11$ (mod 22) then the solution is $x \equiv 9$ (mod 92); if $a \equiv 0$ (mod 22) then the solutions are $x \equiv 9, 55$ (mod 92).

93 The first congruence has a solution if and only if $a \equiv 0$ (mod 2), in which case it has a unique solution modulo 11. By the Chinese remainder theorem, the second congruence is equivalent to the system

$$\begin{cases} x^2 \equiv a \pmod 3 \\ x^2 \equiv -a \pmod 4 \\ x^2 \equiv 0 \pmod 7. \end{cases}$$

The congruence $x^2 \equiv 0$ (mod 7) has the unique solution $x \equiv 0$ (mod 7) for each possible value of a.

The congruence $x^2 \equiv -a$ (mod 4) has the two solutions $x \equiv 0, 2$ (mod 4) if $a \equiv 0$ (mod 4) and has no solutions if $a \equiv 2$ (mod 4); as seen above, we need not consider the classes of 1 and 3 as we may assume a is even.

Finally, $x^2 \equiv a$ (mod 3) has the unique solution $x \equiv 0$ (mod 3) if $a \equiv 0$ (mod 3), the two solutions $x \equiv \pm 1$ (mod 3) if $a \equiv 1$ (mod 3) and no solution if $a \equiv 2$ (mod 3).

In conclusion, we have the following:

① If $a \equiv 1$ (mod 2) then the first equation, and therefore the system, has no solutions.

② If $a \equiv 0$ (mod 4) and $a \equiv 0$ (mod 3), that is, if $a \equiv 0$ (mod 12), then $x^2 \equiv 7a$ (mod 84) has two solutions modulo 84 and the equation $2x \equiv a$ (mod 22) has a unique solution modulo 11, so the system has two solutions modulo $84 \cdot 11 = 924$.

③ Finally, if $a \equiv 0$ (mod 4) and $a \equiv 1$ (mod 3), that is, if $a \equiv 4$ (mod 12), then $x^2 \equiv 7a$ (mod 84) has four solutions modulo 84 and the equation $2x \equiv a$ (mod 22) has a unique solution modulo 11, so the system has four solutions modulo $84 \cdot 11 = 924$.

In all remaining cases the system has no solutions because the second equation has none.

94 Let us consider the first congruence: remark that it has no solutions when a is even, so we must have $a \equiv 1$ (mod 2). In this case, $a^x \equiv 1$ (mod 8) if x is even and $a^x \equiv a \pmod 8$ if x is odd, so the congruence has a solution if and only if $a \equiv 3 \pmod 8$, and that solution is $x \equiv 1 \pmod 2$.

So let $a = 3 + 8b$. The second congruence becomes $x^{6+16b} \equiv 4 \pmod 9$. Note that solutions to this congruence must be sought in $(\mathbb{Z}/9\mathbb{Z})^*$: indeed, any solution will have among its powers the class of 4, which is invertible in $\mathbb{Z}/9\mathbb{Z}$. In particular, $x^6 \equiv 1 \pmod 9$, so the equation becomes $x^{4b} \equiv 4 \pmod 9$.

By again using the fact that sixth powers are 1 in $(\mathbb{Z}/9\mathbb{Z})^*$, we obtain that if $b \equiv 0 \pmod 3$ then the equation has no solutions; if $b \equiv 1 \pmod 3$ then it is equivalent to $x^4 \equiv 4 \pmod 9$ and has the solutions $x \equiv \pm 4 \pmod 9$; finally, if $b \equiv 2 \pmod 3$ then the equation is equivalent to $x^2 \equiv 4 \pmod 9$ and has the solutions $x \equiv \pm 2 \pmod 9$.

In conclusion, we have the following cases:

① If $a \equiv 11 \pmod{24}$ then the solution of the first equation is $x \equiv 1 \pmod 2$ and those of the second are $x \equiv \pm 4 \pmod 9$, so the solutions of the system are $x \equiv 5, 13 \pmod{18}$.

② If $a \equiv 19 \pmod{24}$ then the solution of the first equation is $x \equiv 1 \pmod 2$ and those of the second are $x \equiv \pm 2 \pmod 9$, so the solutions of the system are $x \equiv 7, 11 \pmod{18}$.

Finally, if $a \not\equiv 11, 19 \pmod{24}$ then one of the two equations, and therefore the system, has no solutions.

95 The problem is equivalent to finding the number of solutions modulo 100 of the equation $xy \equiv 0 \pmod{100}$, which, by the Chinese remainder theorem, is equivalent to the system

$$\begin{cases} xy \equiv 0 \pmod 4 \\ xy \equiv 0 \pmod{25}. \end{cases}$$

More generally, consider the equation $xy \equiv 0 \pmod{p^2}$, where p is a prime, that is, consider the condition $p^2 \mid xy$. The condition is satisfied when one of the two factors is divisible by p^2, that is, congruent to zero modulo p^2, and the other factor is any integer, or when both factors are divisible by p.

There are p^2 pairs of the form $(x, \overline{0})$ and p^2 of the form $(\overline{0}, y)$. But, since $(\overline{0}, \overline{0})$ is of both forms, the total number of pairs of this type is $2p^2 - 1$.

Pairs of integers that are both divisible by p and not by p^2 are represented by residue classes of the form (pa, pb) with $1 \le a, b \le p - 1$, and thus there are $(p-1)^2$ of them. We therefore have $2p^2 - 1 + (p-1)^2 = 3p^2 - 2p$ pairs in total.

By substituting the values $p = 2$ and $p = 5$, one obtains the values 8 and 65, respectively. Again by the Chinese remainder theorem, the equation $xy \equiv 0 \pmod{100}$ has $8 \cdot 65 = 520$ solutions.

96 The first congruence in the system has a solution if and only if $6 = (6, 72) \mid 4a$, that is, if and only if $3 \mid a$. If that is the case, let $a = 3b$: we may divide by 6 and obtain $x \equiv 2b \pmod{12}$.

Since $1 = (5, 39) \mid 2$, the second congruence always has the unique solution $x \equiv 16 \pmod{39}$.

However, since $(12, 39) = 3$, the system itself has a solution if and only if the solutions of the two congruences coincide modulo 3, that is, if and only if $2b \equiv 16 \pmod{3}$, or equivalently $b \equiv 2 \pmod{3}$. If this is the case, then the solution is unique modulo the least common multiple of the two moduli, that is, modulo 156.

In conclusion, there is a unique solution modulo 156 if $a \equiv 6 \pmod{9}$ and there are no solutions otherwise.

97 In order to solve the first equation, one must find the order of 8 in the group $(\mathbb{Z}/27\mathbb{Z})^*$, which will be a divisor of $\phi(27) = 18$. We have $8^2 \equiv 10$, $8^3 \equiv -1$, $8^6 \equiv 1 \pmod{27}$, so the order in question is 6. Since $8^3 \equiv -1 \pmod{27}$, the solutions of the first equation are the same as the solutions of $x^2 - 1 \equiv 3 \pmod{6}$. We can solve the equation modulo 2 and modulo 3 and obtain

$$\begin{cases} x \equiv 0 & \pmod{2} \\ x \equiv \pm 1 & \pmod{3}. \end{cases}$$

Analogously, the second equation is equivalent to the system

$$\begin{cases} x^{22} + 2x \equiv 8 & \pmod{4} \\ x^{22} + 2x \equiv 8 & \pmod{11}. \end{cases}$$

The first equation of the latter system can be rewritten as $x^{22} + 2x \equiv 0 \pmod{4}$. Since $x^{22} \equiv x \pmod{2}$, we have that x must be even. On the other hand, if x is even, then both x^{22} and $2x$ are divisible by 4, so we get a solution. The solution of the first equation is therefore $x \equiv 0 \pmod{2}$.

One can check that $x \equiv 0 \pmod{11}$ is not a solution of the second equation. But, if $x \not\equiv 0 \pmod{11}$, then by Fermat's little theorem $x^{10} \equiv x^{20} \equiv 1 \pmod{11}$, so the solutions of the second equation are the same as the solutions of $x^2 + 2x \equiv 8 \pmod{11}$.

Since this quadratic equation has solutions $2, -4$ in $\mathbb{Z}$, its solutions modulo 11 must be the classes of 2 and -4, because 11 is prime.

Finally, since the solutions of the two equations in the original system coincide modulo 2, we are left with

$$\begin{cases} x \equiv 0 & \pmod{2} \\ x \equiv \pm 1 & \pmod{3} \\ x \equiv 2, -4 & \pmod{11}, \end{cases}$$

which yields four solutions modulo $2 \cdot 3 \cdot 11 = 66$. By performing simple calculations, one finds that the solutions are $x \equiv 2, 40, 46, 62 \pmod{66}$.

98 By computing powers of 3 modulo 5, one finds that $3^3 \equiv 2$ (mod 5) and that the multiplicative order of 3 modulo 5 is 4. We can therefore rewrite the first equation as $3^x \equiv 3^{3a}$ (mod 5), whose solution is $x \equiv 3a \equiv -a$ (mod 4).

The second equation splits into two equations, one modulo 3 and one modulo 8. The equation modulo 3 always has a unique solution modulo 3, because $x^3 \equiv x$ (mod 3) for all x. As for the equation modulo 8, there are two separate possibilities.

   ① If a is even, then x must also be even; now, for all even x we have $x^3 \equiv 0$ (mod 8); so there is one solution modulo 2, that is, four solutions modulo 8, if $a \equiv 6$ (mod 8), and no solution otherwise.

   ② If a is odd, then x must also be odd; in this case, since $x^2 \equiv 1$ and $x^3 \equiv x$ (mod 8), there is always a unique solution modulo 8.

So the system has no solution for $a \equiv 0, 2, 4$ (mod 8). In all other cases, solutions to the system exist if the solution of the equation modulo 4 and that of the equation modulo 2 or 8 are compatible.

If a is odd, that is, if $a \equiv 6$ (mod 8), then the equation modulo 4 has the solution $x \equiv 2$ (mod 4), which is indeed compatible with the solution $x \equiv 0$ (mod 2). In this case, we have a unique solution modulo 12.

If a is odd then the equations are still compatible: indeed, we have remarked that the solutions modulo 8 are $x \equiv a + 2$ and $a + 2 \equiv -a$ (mod 4) for a odd. There is therefore a unique solution modulo 24.

99 The first congruence has a solution if and only if $\overline{3}$ belongs to the subrgroup of $(\mathbb{Z}/7\mathbb{Z})^*$ generated by $\overline{a}$, that is, if and only if the subgroup generated by $\overline{3}$ is contained in the subgroup generated by $\overline{a}$. Since $\overline{3}$ is a generator of the group $(\mathbb{Z}/7\mathbb{Z})^*$ (one can check that its order is 6) the argument above shows that the first congruence has a solution if and only if $\overline{a}$ is also a generator of the group.

Since $\overline{3}$ is a generator, all other generators are of the form $\overline{3}^i$ with $0 \leq i < 6$ and $(i, 6) = 1$, so they are $\overline{3}$ and $\overline{3}^5 = \overline{3}^{-1} = \overline{5}$. Any generator has order 6 and we have $3^1 \equiv 3, 5^5 \equiv 3$ (mod 7), so the first equation has the following solutions: for $a \equiv 3$ (mod 7), the solution $x \equiv 1$ (mod 6); for $a \equiv 5$ (mod 7), the solution $x \equiv 5$ (mod 6).

As for the second congruence in the system, remark that the square of an even number is divisible by 4, whereas the square of an odd number is always congruent to 1 modulo 8. Though the second congruence has solutions for $a \equiv 0, 4$ (mod 8), we may very well disregard those, because they necessarily have $x \equiv 0$ (mod 2), which is incompatible with the first congruence.

It is therefore enough to consider the case where $a \equiv 1$ (mod 8), in which the solution is $x \equiv 1$ (mod 2), which is a condition already enforced by the first congruence. In conclusion, if $a \equiv 3$ (mod 7) and $a \equiv 1$ (mod 8), that is, if $a \equiv 17$ (mod 56), then the solution is $x \equiv 1$ (mod 6); if $a \equiv 5$ (mod 7), $a \equiv 1$ (mod 8), that is, if $a \equiv 33$ (mod 56), then the solution is $x \equiv 5$ (mod 6).

For all other values of a we have no solutions.

100   The first equation has a solution if and only if $a \not\equiv 0$ (mod 5). If $a \equiv 1$ (mod 5), then all integers are solutions; if $a \equiv -1$ (mod 5), that is, if the order of a in $(\mathbb{Z}/5\mathbb{Z})^*$ is 2, then the solution is $x \equiv 0$ (mod 2); if $a \equiv 2, 3$ (mod 5), that is, if and only if the order of a in $(\mathbb{Z}/5\mathbb{Z})^*$ is 4, the solution is $x \equiv 0$ (mod 4).

The second equation has a solution if and only if $(a, 8) \mid 2$, that is, if and only if $a \not\equiv 0$ (mod 4). If a is odd, then we have $a^{-1} \equiv a$ (mod 8), so the solution is $x \equiv 2a$ (mod 8); note that in this case we have $2a \equiv \pm 2$ (mod 8). If $a \equiv 2$ (mod 4), that is, if $a = 2b$ with b odd, the solution is $x \equiv b$ (mod 4).

By comparing the solutions of the two equations, we conclude the following.

When a is odd: for $a \equiv 2, 3$ (mod 5) the two equations have no common solutions, whereas for $a \equiv \pm 1$ (mod 5), that is, for $a \equiv \pm 1$ (mod 10), solutions of the second equation also satisfy the first, so the solutions of the system are exactly those of the second equation.

When $a \equiv 2$ (mod 4): the two equations have no solutions in common except for when $a \equiv 1$ (mod 5); in this case, that is, when $a \equiv 6$ (mod 20), solutions of the second equation also satisfy the first, so they are the solutions of the original system.

101   One can immediately check that the order of 2 in the multiplicative group $(\mathbb{Z}/13\mathbb{Z})^*$ is 12 and that $5 \equiv 2^9$ (mod 13). The first equation can thus be rewritten as $2^{9(x^2-1)} \equiv 2^a$ (mod 13), which is equivalent to $9(x^2 - 1) \equiv a$ (mod 12). This implies that a must be a multiple of 3 in order for the equation to have any solutions.

So, let $a = 3b$; cancelling common factors yields $3(x^2 - 1) \equiv b$ (mod 4), that is, $x^2 \equiv 1 - b$ (mod 4). Squares of integers can only be 0 or 1 modulo 4 (0 if the integer is even and 1 if the integer is odd). Consequently, the last equation above has a solution if and only if $b \equiv 0, 1$ (mod 4), with the solution being $x \equiv 1, 0$ (mod 2), respectively.

The second equation in the system is equivalent to $64 = 2^6 \mid x^3$ and therefore to $2^2 \mid x$, that is, $x \equiv 0$ (mod 4); hence we have, in particular, $x \equiv 0$ (mod 2). In order for the system to have a solution, it is enough to have $b \equiv 1$ (mod 4), that is, $a \equiv 3$ (mod 12); when this is the case, the solution is $x \equiv 0$ (mod 4).

102   By the Chinese remainder theorem, the first equation is equivalent to the system

$$\begin{cases} a^x \equiv 1 & (\text{mod } 2) \\ a^x \equiv 4 & (\text{mod } 7). \end{cases}$$

The first of these two equations has a solution if and only if $a \equiv 1$ (mod 2), in which case all integers x are solutions.

As for the second equation in the system above, note that it has no solutions in the following cases: when $a \equiv 0$ (mod 7), because $a^x \equiv 0$ (mod 7) for all x; when $a \equiv 1$ (mod 7), because $a^x \equiv 1$ (mod 7) for all x; when $a \equiv -1$ (mod 7), because $a^x \equiv \pm 1$ (mod 7) for all x.

In all other cases, the equation does have a solution, which is listed below.

When $a \equiv 2$ (mod 7), since we have $2^2 \equiv 4$ (mod 7) and the order of 2 modulo 7 is 3, the solution is $x \equiv 2$ (mod 3).

When $a \equiv 3$ (mod 7), since we have $3^4 \equiv 4$ (mod 7) and the order of 3 modulo 7 is 6, the solution is $x \equiv 4$ (mod 6); equivalently, $x \equiv 0$ (mod 2) and $x \equiv 1$ (mod 3).

When $a \equiv 4$ (mod 7), since we have $4^1 \equiv 4$ (mod 7) and the order of 4 modulo 7 is 3, the solution is $x \equiv 1$ (mod 3).

When $a \equiv 5$ (mod 7), since we have $5^2 \equiv 4$ (mod 7) and the order of 5 modulo 7 is 6, the solution is $x \equiv 2$ (mod 6); equivalently, $x \equiv 0$ (mod 2) and $x \equiv 2$ (mod 3).

Now consider the equation $x^a \equiv 1$ (mod 9), which is satisfied for all pairs (x, a) such that $(x, 9) = 1$ (or equivalently $(x, 3) = 1$) and a is a multiple of the order of x modulo 9.

As shown before, the first equation in the original system has a solution if and only if a is odd, so the only potential solutions are integers x with an odd multiplicative order mudulo 9, that is, $x \equiv 1, 4, 7$ (mod 9). To be more specific, since $\overline{1}$ has order 1 whereas $\overline{4}$ and $\overline{7}$ have order 3, $x \equiv 1$ (mod 9) is a solution for all a, whereas $x \equiv 4, 7$ (mod 9) are solutions only if $3 \mid a$.

In any case, solutions can only be congruent to 1 modulo 3, so we must have $a \equiv 3, 4$ (mod 7) in order to ensure compatibility modulo 7.

This condition is indeed sufficient: for all a in these classes the congruence modulo 9 has among its solutions $x \equiv 1$ (mod 9), which is compatible with solutions of the congruence modulo 7.

To conclude, the system has a solution if and only if $a \equiv 1$ (mod 2) and $a \equiv 3, 4$ (mod 7), that is, if and only if $a \equiv 3, 11$ (mod 14).

103

(i) Consider the periodic sequence of all powers of 3 modulo 10:

$$3^1 = 3 \equiv 3, \quad 3^2 = 9 \equiv 9, \quad 3^3 = 27 \equiv 7, \quad 3^4 = 81 \equiv 1.$$

The sequence is periodic with period 4 and the solution of the equation in the problem statement is $x \equiv 3$ (mod 4).

(ii) Since powers of 3 are odd and 10 is even, x can be a solution only if $4 + x$ is odd, that is, if x itself is odd. Moreover, as discussed above, the residue class modulo 10 of a power of 3 only depends on the class modulo 4 of the exponent. We have the following two cases:

① If $x \equiv 1$ (mod 4) then $3^x \equiv 3$ (mod 10) and the equation reduces to $3 \equiv 4 + x$ (mod 10), that is, $x \equiv 9$ (mod 10). By combining this with the congruence modulo 4, we obtain the solution $x \equiv 9$ (mod 20).

② If $x \equiv 3$ (mod 4) then $3^x \equiv 7$ (mod 10) and the equation reduces to $7 \equiv 4 + x$ (mod 10), that is, $x \equiv 3$ (mod 10). By combining this with the congruence modulo 4, we obtain the solution $x \equiv 3$ (mod 20).

104   Clearly, the first equation cannot have a solution congruent to 0 modulo 7. But, if $x \not\equiv 0 \pmod 7$, then the exponents n for which $x^n \equiv 1 \pmod 7$ are exactly the multiples of the order of x in $(\mathbb{Z}/7\mathbb{Z})^*$. This order must be a divisor of the order of the group $(\mathbb{Z}/7\mathbb{Z})^*$, which is 6. Moreover, since the exponent $2x + 1$ is odd, the order of x must be odd, so it can only be 1 (in which case $x \equiv 1 \pmod 7$) or 3 (in which case we either have $x \equiv 2$ or $x \equiv 4 \pmod 7$).

Any $x \equiv 1 \pmod 7$ is a solution of the first equation, because $1^{2x+1} \equiv 1 \pmod 7$. In order for $x \equiv 2, 4 \pmod 7$ to be a solution, the exponent must be a multiple of 3, so we must have $2x + 1 \equiv 0 \pmod 3$, that is, $x \equiv 1 \pmod 3$.

The second equation does have a solution, because $(4, 15) \mid 7$; its solution is $x \equiv 13 \pmod{15}$, which is equivalent to the system consisting of the two congruences $x \equiv 1 \pmod 3$, $x \equiv 3 \pmod 5$.

Note that the congruence modulo 3 coming from the second equation is always compatible with the solution of the first equation, so the original system does have solutions, which are given by the following system:

$$\begin{cases} x \equiv 1, 2, 4 & \pmod 7 \\ x \equiv 1 & \pmod 3 \\ x \equiv 3 & \pmod 5. \end{cases}$$

Simple calculations yield $x \equiv 43, 58, 88 \pmod{105}$.

105

(i) The system of equations

$$\begin{cases} x \equiv 1 & \pmod 1 \\ x \equiv 1 & \pmod 2 \\ \quad \vdots \\ x \equiv 1 & \pmod{10} \end{cases}$$

does have solutions, because $x = 1$ satisfies all the required congruences. Now, if such a system of congruences has a solution, then it is given by a congruence class modulo the least common multiple of all the moduli of its equations. Let M be the least common multiple of all integers between 1 and 10, that is, $M = 2^3 \cdot 3^2 \cdot 5 \cdot 7$; the solution of the system is $x \equiv 1 \pmod M$. The number of integers between 0 and k that do satisfy this congruence is $\lceil k/M \rceil$.

(ii) If $x \equiv -1 \pmod n$ for all positive integers n then $n \mid x + 1$ for all positive integers n. However, the only number that is divisible by all positive integers is 0, so the only solution is given by $x + 1 = 0$, that is, $x = -1$.

(iii) If $x \equiv n \pmod{2n}$ then $x = n + 2hn = n(2h + 1)$ for some integer h; in particular, the requirement of the question implies that x is divisible by every integer n. As before, this implies $x = 0$; but this time 0 is not a solution, for example because we may take $n = 1$ and check that $0 \not\equiv 1 \pmod 2$. So there are no solutions in this case.

106 The pair (x, n) is a solution of the congruence if and only if there exists an integer t such that $x^n = 39 + 10xt$. In the latter equation x divides both x^n and $10xt$, so a necessary condition for it to be a solution is that x divides 39. Let us consider all possible divisors of 39.

If $x = 1$, then we get $1^n \equiv 39 \pmod{10}$ which clearly has no solutions.

If $x = 3$, then we get the equation $3^n \equiv 39 \pmod{30}$ which is equivalent to the system

$$\begin{cases} 3^n \equiv 0 \pmod 3 \\ 3^n \equiv 9 \pmod{10}. \end{cases}$$

The first equation is satisfied for all $n \geq 1$ whereas the second one is satisfied for $n \equiv 2 \pmod 4$, because 3 has order 4 in $(\mathbb{Z}/10\mathbb{Z})^*$.

If $x = 13$, then by a similar argument we obtain the system

$$\begin{cases} 13^n \equiv 0 \pmod{13} \\ 3^n \equiv 9 \pmod{10} \end{cases}$$

whose solution is again $n \equiv 2 \pmod 4$.

If $x = 39$, then the system we get is

$$\begin{cases} 39^n \equiv 0 \pmod{39} \\ 9^n \equiv 9 \pmod{10} \end{cases}$$

whose solution is $n \equiv 1 \pmod 2$.

107 By the Chinese remainder theorem, the congruence is equivalent to the system

$$\begin{cases} x^{5n} \equiv 1 \pmod 5 \\ x^{5n} \equiv 1 \pmod{11}. \end{cases}$$

Consider the first equation. By Fermat's little theorem, it is equivalent to $x^n \equiv 1 \pmod 5$. The solutions of this equation are those elements of $(\mathbb{Z}/5\mathbb{Z})^*$ whose order is a divisor of n. The possible orders of elements x of $(\mathbb{Z}/5\mathbb{Z})^*$ are

$$1, \text{ if } x = \overline{1}; \quad 2, \text{ if } x = \overline{-1}; \quad 4, \text{ if } x = \overline{\pm 2}.$$

Therefore, for $n \equiv 0 \pmod 4$ there are four solutions, because n is a multiple of the order of each element. For $n \equiv 2 \pmod 4$ there are two solutions, because n is a multiple of the order of the two elements $\pm \overline{1}$. Finally, for n odd there is a unique solution, because n is only a multiple of the order of $\overline{1}$.

Now consider the second equation. Similarly to what we just established, the solutions of this equation are those elements of $(\mathbb{Z}/11\mathbb{Z})^*$ whose order is a divisor of $5n$. Since the group $(\mathbb{Z}/11\mathbb{Z})^*$ is cyclic, its elements have order 1, 2, 5 or 10 and the number of elements of order d is $\phi(d)$. So there is a single element of order 1, a single element of order 2, and there are four elements of order 5 and four elements of order 10.

⟦It is easy to check that $\overline{1}$ is the only element of order 1, $\overline{-1}$ is the only element of order 2, and that $\overline{3}, \overline{4}, \overline{5}, \overline{9}$ are the elements of order 5 and $\overline{2}, \overline{6}, \overline{7}, \overline{8}$ are the elements of order 10.⟧

As before, an element satisfies the equation if and only if its order is a divisor of $5n$. If n is even then $5n$ is a multiple of 10, so all ten elements of $(\mathbb{Z}/11\mathbb{Z})^*$ are solutions; if n is odd, then the solutions are given by elements whose order is a divisor of 5: they are $\overline{1}$ and the elements of order 5, so we get five solutions in total.

We can summarise our findings as follows.

If $n \equiv 0 \pmod 4$ then we have four solutions modulo 5 and ten solutions modulo 11, so 40 solutions in total.

If $n \equiv 2 \pmod 4$ then we have two solutions modulo 5 and ten solutions modulo 11, so 20 solutions in total

Finally, if n is odd then there is a unique solution modulo 5 and there are five solutions modulo 11, so five solutions in total.

108

(i) Let $f(x)$ be the quadratic polynomial $x^2 - x + 43$; the congruence in the problem statement is equivalent to the system consisting of $f(x) \equiv 0 \pmod 5$ and $f(x) \equiv 0 \pmod{11}$, by the Chinese remainder theorem. Since 5 and 11 are odd primes, we can solve both congruences by means of the usual quadratic formula.

For the congruence modulo 5, we find that the discriminant of $f(x)$ is 3^2 (modulo 5), so $f(x)$ has the two roots 2 and -1 modulo 5. As for the congruence modulo 11, the discriminant is congruent to 4^2 and the two roots are -3 and 4.

Now, remark that $x_1 = 11$ and $x_2 = -10$ are solutions of the systems $x_1 \equiv 1 \pmod 5$, $x_1 \equiv 0 \pmod{11}$ and $x_2 \equiv 0 \pmod 5$, $x_2 \equiv 1 \pmod{11}$, respectively. This implies that the original equation has four solutions, which are given by

$$-3 \equiv 2x_1 - 3x_2 \pmod{55}$$
$$19 \equiv -x_1 - 3x_2 \pmod{55}$$
$$-18 \equiv 2x_1 + 4x_2 \pmod{55}$$
$$4 \equiv -x_1 + 4x_2 \pmod{55}.$$

(ii) In order to solve the system we must check that the residue classes modulo 5 of the solutions of the first equation are compatible with the second equation. The solutions of the first equation belong to the classes of 2 and -1 modulo 5. In order for a solution in the class of 2 to satisfy the second equation, we must have $2^{11^4} \equiv 2^a \pmod 5$. Since 2 has order 4 in $(\mathbb{Z}/5\mathbb{Z})^*$ and we have $11^4 \equiv (-1)^4 \equiv 1 \pmod 4$, the second equation becomes $2 \equiv 2^a \pmod 5$, which has a solution if and only if $a \equiv 1 \pmod 4$. So the classes of -3 and -18 modulo 55 are solutions of the original system if and only if $a \equiv 1 \pmod 4$.

In the case of a solution congruent to -1 modulo 5, we must have $(-1)^{11^4} = -1 \equiv (-1)^a \pmod 5$, which has a solution if and only if $a \equiv 1 \pmod 2$. So the classes of 19 and 4 modulo 55 are solutions of the original system if and only if $a \equiv 1 \pmod 2$.

We may thus conclude that if $a \equiv 0 \pmod 2$ then the system has no solutions. If $a \equiv 1 \pmod 2$ then the system does have solutions. In particular, if we split the class $a \equiv 1 \pmod 2$ into the classes $a \equiv -1 \pmod 4$ and $a \equiv +1 \pmod 4$, then we have the following: if $a \equiv -1 \pmod 4$ then the solutions are the classes of 19 and 4 modulo 55; if $a \equiv 1 \pmod 4$ then all solutions of the first equation satisfy the second, so the solutions of the system are the classes of $-3, -18, 19, 4$ modulo 55.

109

(i) The equation if equivalent to the system consisting of the two congruences $x^2 + 2x + 5 \equiv 0 \pmod 5$ and $x^2 + 2x + 5 \equiv 0 \pmod{13}$. Since both moduli are prime, each has at most two roots. The first clearly has the solutions 0 and -2 modulo 5. As for the second, we can solve it by means of the quadratic formula: the discriminant is $-4 \equiv 3^2 \pmod{13}$, so the solutions are the classes of 2 and -4 modulo 13. The solutions of the equation in the problem statement are therefore those of the four systems

$$\begin{cases} x \equiv 0 \pmod 5 \\ x \equiv 2 \pmod{13}, \end{cases} \qquad \begin{cases} x \equiv -2 \pmod 5 \\ x \equiv 2 \pmod{13}, \end{cases}$$

$$\begin{cases} x \equiv 0 \pmod 5 \\ x \equiv -4 \pmod{13}, \end{cases} \qquad \begin{cases} x \equiv -2 \pmod 5 \\ x \equiv -4 \pmod{13}. \end{cases}$$

It is easy to check that the solutions of the two systems

$$\begin{cases} x \equiv 1 \pmod 5 \\ x \equiv 0 \pmod{13}, \end{cases} \qquad \begin{cases} x \equiv 0 \pmod 5 \\ x \equiv 1 \pmod{13}. \end{cases}$$

are $x_1 \equiv 26 \pmod{65}$ and $x_2 \equiv -25 \pmod{65}$, respectively; it follows that the solutions of the four system above, and therefore the solutions of the original equation, are: $0 \cdot x_1 + 2 \cdot x_2 \equiv 15, -2 \cdot x_1 + 2 \cdot x_2 \equiv -37, 0 \cdot x_1 - 4 \cdot x_2 \equiv -30$ and $-2 \cdot x_1 - 4 \cdot x_2 \equiv -17 \pmod{65}$.

(ii) Letting $y = 3^x$, we are asking for y to be a solution of the previous equation. In particular, since $3^x \not\equiv 0$ modulo 5 and since the powers of 3 modulo 13 are 1, 3 and -4, we must have $3^x = y \equiv -2 \equiv 3 \pmod 5$ and $3^x = y \equiv -4 \pmod{13}$. The first of these two equations is equivalent to $x \equiv 1 \pmod 4$ and the second to $x \equiv 2 \pmod 3$. In conclusion, the solution of the original equation is given by the class of 5 modulo 12.

110 The first congruence in the system is equivalent to the two congruences $x^2 + 2x + 2 \equiv 0 \pmod 2$ and $x^2 + 2x + 2 \equiv 0 \pmod 5$ by the Chinese remainder theorem. The first of these two is clearly equivalent to $x \equiv 0 \pmod 2$, while the second has discriminant $\Delta = 1 - 2 = -1 \equiv 2^2$ (mod 5) and so its solutions are $-1 \pm \sqrt{\Delta} = -1 \pm 2$, that is, 1 and 2.

As for the second equation, 7 and 22 are relatively prime, so 7 is invertible modulo 22; moreover, its inverse is the class of -3. The equation can therefore be rewritten as $x \equiv -60 \equiv 6$ (mod 22), and in particular it implies $x \equiv 0$ (mod 2): since this is equivalent to the first of the two equations discussed in the previous paragraph, we may disregard that equation and only consider the one modulo 5. We thus need to solve the two systems

$$\begin{cases} x \equiv 1, 2 & \text{(mod 5)} \\ x \equiv 6 & \text{(mod 22).} \end{cases}$$

Their solutions are 6 and -38 modulo 110, respectively.

111 We can split the first congruence in the system into a congruence modulo 4 and one modulo 3, thus obtaining

$$\begin{cases} ax \equiv 2 & \text{(mod 4)} \\ ax \equiv 2 & \text{(mod 3)} \\ 9x \equiv a^2 + 2a - 3 & \text{(mod 81).} \end{cases}$$

Let us first consider the subsystem formed by the last two equations, both of which have a modulus that is a power of 3. The congruence $ax \equiv 2$ (mod 3) has a solution if and only if $a \equiv 1, 2$ (mod 3), in which case its solution is $x \equiv 2a^{-1} \equiv 2a$ (mod 3).

The congruence $9x \equiv a^2 + 2a - 3$ (mod 81) has a solution if and only if $a^2 + 2a - 3 \equiv 0$ (mod 9). Now, $a^2 + 2a - 3 = (a - 1)(a + 3) \equiv 0$ (mod 9) if and only if $a \equiv 1$ (mod 9) or $a \equiv 3$ (mod 9), or $a \equiv 1$ (mod 3) and $a \equiv 0$ (mod 3), though clearly the last condition can never be satisfied.

By imposing all conditions necessary to guarantee the existence of solution for both equations, we find that, in order for the system to have a solution, we need $a \equiv 1$ (mod 9).

So, let $a = 1 + 9k$, with $k \in \mathbb{Z}$. We have $a^2 + 2a - 3 = 9k(4 + 9k)$ and the subsystem becomes

$$\begin{cases} x \equiv 2 & \text{(mod 3)} \\ x \equiv 4k & \text{(mod 9).} \end{cases}$$

This system has a solution if and only if the two congruences are compatible, that is, if $4k \equiv 2$ (mod 3). So we have a solution if and only if $k \equiv 2$ (mod 3), that is, $a \equiv 19$ (mod 27), and that solution is unique modulo 9.

The congruence $ax \equiv 2$ (mod 4) has no solutions when $a \equiv 0$ (mod 4), has one solution modulo 4 when $a \equiv 1, 3$ (mod 4), and has one solution modulo 2 when $a \equiv 2$ (mod 4).

Since $(4, 9) = 1$, the Chinese remainder theorem allows us to reach the following conclusions.

For $a \equiv 1, 3 \pmod 4$ and $a \equiv 19 \pmod{27}$, that is, for $a \equiv 73, 19 \pmod{108}$, the system has a unique solution modulo 36, so $180/36 = 5$ solutions modulo 180.

For $a \equiv 2 \pmod 4$ and $a \equiv 19 \pmod{27}$, that is, for $a \equiv 46 \pmod{108}$, the system has a unique solution modulo 18, so $180/18 = 10$ solutions modulo 180.

Finally, for all other values of a the system has no solutions.

112 By the Chinese remainder theorem, the first equation in the system can be replaced by

$$\begin{cases} x^{131} \equiv x \pmod{11} \\ x^{131} \equiv x \pmod 5. \end{cases}$$

We can rewrite the first of the equations above as $x(x^{130} - 1) \equiv 0 \pmod{11}$, hence, since 11 is prime, we have $x \equiv 0 \pmod{11}$ or $x^{130} \equiv 1 \pmod{11}$. By Fermat's little theorem, we have $a^{10} \equiv 1 \pmod{11}$ for all $a \in \mathbb{Z}$ such that $(a, 11) = 1$. So the solutions of $x^{130} \equiv 1 \pmod{11}$ are given by all of the classes in $(\mathbb{Z}/11\mathbb{Z})^*$. It follows that the congruence $x^{131} \equiv x \pmod{11}$ is satisfied for all integers x.

Similarly, $x^{131} \equiv x \pmod 5$ if and only if $x(x^{130} - 1) \equiv 0 \pmod 5$, that is, if and only if $x \equiv 0 \pmod 5$ or $x^{130} \equiv 1 \pmod 5$. The solutions of $x^{130} \equiv 1 \pmod 5$ are the classes in $(\mathbb{Z}/5\mathbb{Z})^*$ for which $x^4 \equiv 1 \pmod 5$. We therefore have $x^{130} \equiv (x^4)^{32} x^2 \equiv x^2 \equiv 1 \pmod 5$, and the solutions are given by $x \equiv \pm 1 \pmod 5$. We conclude that the solutions of the equation $x^{131} \equiv x \pmod 5$ are $x \equiv 0, \pm 1 \pmod 5$, and the original system is equivalent to the union of the systems

$$\begin{cases} x \equiv 0 \pmod 5 \\ x(x^5 + 1) \equiv 0 \pmod{125}, \end{cases}$$

$$\begin{cases} x \equiv 1 \pmod 5 \\ x(x^5 + 1) \equiv 0 \pmod{125}, \end{cases}$$

$$\begin{cases} x \equiv -1 \pmod 5 \\ x(x^5 + 1) \equiv 0 \pmod{125}. \end{cases}$$

If $x \equiv 0 \pmod 5$ then $x^5 + 1 \in (\mathbb{Z}/125\mathbb{Z})^*$, so the solution of the first system is $x \equiv 0 \pmod{125}$.

If $x \equiv 1 \pmod 5$ then both x and $x^5 + 1$ are coprime to 5 and hence invertible in $\mathbb{Z}/125\mathbb{Z}$. So their product cannot be congruent to 0 modulo 125, and the second system has no solutions.

Suppose $x \equiv -1 \pmod 5$, that is, $x = -1 + 5y$ with $y \in \mathbb{Z}$. Such an x is invertible modulo 125, so the second equation in the third system is equivalent to $(-1 + 5y)^5 + 1 \equiv 0 \pmod{125}$. By carrying out the necessary computations we obtain

$$(-1 + 5y)^5 + 1 \equiv 5^2 y \equiv 0 \pmod{125},$$

whose solution is $y \equiv 0 \pmod 5$. So all solutions of the third system are given by integers of the form $x = -1 + 25t$ with $t \in \mathbb{Z}$, that is, by the class $x \equiv -1 \pmod{25}$

In conclusion, the solutions of the original system are $x \equiv 0 \pmod{125}$ and $x \equiv -1 \pmod{25}$.

113 By means of the Chinese remainder theorem and by inverting 13 modulo 7 and modulo 19, we may rewrite the system as

$$\begin{cases} ax \equiv 1 & (\text{mod } 11) \\ ax \equiv 5 & (\text{mod } 7) \\ x \equiv 3 & (\text{mod } 7) \\ x \equiv -1 & (\text{mod } 19). \end{cases}$$

The first equation has a solution if and only if $(a, 11) = 1$, that is, $a \not\equiv 0 \pmod{11}$, in which case its solution is $x \equiv a^{-1} \pmod{11}$. In order for the original system to have a solution, its two equations must also be compatible modulo 7, from which one obtains the condition $3a \equiv 5 \pmod 7$, that is, $a \equiv 4 \pmod 7$. The original system has a solution if and only if

$$\begin{cases} a \not\equiv 0 & (\text{mod } 11) \\ a \equiv 4 & (\text{mod } 7) \end{cases}$$

or equivalently if and only if $a \equiv 4, 18, 25, 32, 39, 46, 53, 60, 67, 74 \pmod{77}$. When a belongs to one of those classes, letting b be any representative of the class of a^{-1} modulo 11, the system becomes

$$\begin{cases} x \equiv b & (\text{mod } 11) \\ x \equiv 3 & (\text{mod } 7) \\ x \equiv -1 & (\text{mod } 19) \end{cases}$$

from which we get

$$\begin{cases} x \equiv b & (\text{mod } 11) \\ x \equiv -39 & (\text{mod } 133). \end{cases}$$

The equation that corresponds to this system is $133t - 11s = b + 39$.

We can use Euclid's algorithm to obtain that $133(1) - 11(12) = 1$; we can then multiply by $b + 39$ to find that $t = b + 39$, $s = 12b + 468$ is a particular solution of the equation. The solution of the system is therefore $x \equiv -39 + 133(b + 39) \pmod{133 \cdot 11}$, that is, $x \equiv 759 + 133b \pmod{1463}$ with $b \equiv a^{-1} \pmod{77}$ and a belonging to one of the classes listed above.

114 First of all, let us determine for which values of a each equation has a solution.

When solving the first equation, we may immediately exclude that $x \equiv 0$ (mod 7). If $x \not\equiv 0$ (mod 7), then by Fermat's little theorem we have $x^6 \equiv 1$ (mod 7). Since $80 = 13 \cdot 6 + 2$, the equation is equivalent to $x^2 \equiv 2$ (mod 7). One can check that the solutions of this equation are $x \equiv \pm 3$ (mod 7).

Thanks to the equality $80 = 11 \cdot 7 + 3$, the second equation can be rewritten as $3^x \equiv 2$ (mod 7). By examining powers of 3 modulo 7, one finds that $3^2 \equiv 2$ (mod 7) and that residue classes of powers repeat with a period of 6. So the solution of the second equation is $x \equiv 2$ (mod 6).

The third equation has a solution for all a, because $(7, 10) = 1 \mid a$. The inverse of 7 modulo 10 is 3, so its solution is $x \equiv 3a$ (mod 10).

Now let us consider whether or not the whole system has a solution. The greatest common divisor of the moduli of the last two equations is 2, so we must check that the solutions are compatible modulo 2. The solution of the second equation implies $x \equiv 0$ (mod 2), whereas the solution of the third equation implies $x \equiv a$ (mod 2). So the system has a solution if and only if $a \equiv 0$ (mod 2).

Now let $a = 2b$ and let us solve the system. The third equation can be rewritten as $x \equiv 6b$ (mod 10), or simply as $x \equiv b$ (mod 5), given that the correct congruence modulo 2 is already being enforced by the previous equation. We now have three equations whose moduli are pairwise relatively prime, so by the Chinese remainder theorem we will find solutions modulo the product of the three moduli.

Easy computations yield the solutions $10 - 42a$ and $80 - 42a$ modulo 210.

115 Since $700 = 2^2 \cdot 5^2 \cdot 7$, the Chinese remainder theorem implies that the first equation can be replaced by the following system:

$$\begin{cases} x^{41} \equiv x \pmod{4} \\ x^{41} \equiv x \pmod{25} \\ x^{41} \equiv x \pmod{7}. \end{cases}$$

Now, remark that $x^{41} - x = x \cdot (x^{40} - 1)$ and that $(x, x^{40} - 1) = 1$. Therefore, for each modulus m, the corresponding equation leads to two cases: $x \equiv 0$ (mod m) or $x^{40} - 1 \equiv 0$ (mod m).

For $m = 4$ we have the solutions $x \equiv 0$ (mod 4) and $(x, 2) = 1$. Indeed, if $(x, 2) = 1$ then $x^{\phi(4)} = x^2 \equiv 1$ (mod 4), hence $x^{40} = (x^2)^{20} \equiv 1$ (mod 4).

Similarly, for $m = 25$ we have the solutions $x \equiv 0$ (mod 25) and $(x, 5) = 1$. Indeed, if $(x, 5) = 1$ then $x^{\phi(25)} = x^{20} \equiv 1$ (mod 25) and thus $x^{40} \equiv 1$ (mod 25).

Finally, consider the case $m = 7$. As before, $x \equiv 0$ (mod 7) is a solution. If $(x, 7) = 1$ then we have $x^6 \equiv 1$ (mod 7) by Fermat's little theorem, hence $x^{36} \equiv 1$ (mod 7) and so the equation becomes $x^4 \equiv 1$ (mod 7). The solutions of this equation are the elements of $(\mathbb{Z}/7\mathbb{Z})^*$ whose order divides 4; but, because all elements of this group have an order that divides 6, the elements in question are those whose order divides $(4, 6) = 2$, that is, the solutions of $x^2 \equiv 1$ (mod 7), which are $x \equiv \pm 1$ (mod 7).

Let us now solve the second equation of the original system. Consider it modulo 4: it reduces to $x \equiv 1 \pmod 4$. When considered modulo 25, it reduces to $-5x \equiv 0 \pmod{25}$, whose solution is $x \equiv 0 \pmod 5$. Finally, it reduces modulo 7 to $3x \equiv -3 \pmod 7$, whose solution is $x \equiv -1 \pmod 7$.

By combining all conditions obtained from the first and second equation, we can turn the original system into the following:

$$\begin{cases} x \equiv 1 & \pmod 4 \\ x \equiv 0 & \pmod{25} \\ x \equiv -1 & \pmod 7. \end{cases}$$

A few computations lead to the solution $x \equiv 125 \pmod{700}$.

116 The equation can be solved only if $(x, 27) = 1$, since x must be invertible in $\mathbb{Z}/27\mathbb{Z}$; in that case, the exponent must be $x + 1 \equiv 0 \pmod k$, where k is the order of x in $(\mathbb{Z}/27\mathbb{Z})^*$.

Since $\phi(27) = 18$, the order of each element of $(\mathbb{Z}/27\mathbb{Z})^*$ is a divisor of 18.

If $x + 1 \equiv 0 \pmod{18}$, that is, if $x \equiv -1 \pmod{18}$, then x is not divisible by 3, and so by Euler's theorem $x^{x+1} \equiv 1 \pmod{27}$. So $x \equiv -1 \pmod{18}$ is a solution.

Let us now consider all other possible values for the order of an element $(\mathbb{Z}/27\mathbb{Z})^*$ in order to check whether there are any additional solutions.

If $k = 9$ then thanks to our initial remarks we must have $x + 1 \equiv 0 \pmod 9$. In particular, this equation implies that $x \equiv -1 \pmod 3$. In this case, the order of x modulo 3, hence its order modulo 27, is even. However, this contradicts our assumption that the order is 9, so in this case we have no solutions.

If $k = 6$ then we must have $x + 1 \equiv 0 \pmod 6$, that is, $x = 6a - 1$ for some $a \in \mathbb{Z}$. If $a \geq 0$, then by the binomial theorem we have $(6a - 1)^{6a} \equiv -36a^2 + 1 \pmod{27}$. So $(6a - 1)^{6a} \equiv 1 \pmod{27}$ if and only if $a \equiv 0 \pmod 3$, that is, if $x \equiv -1 \pmod{18}$. Again, we have no solutions. The case $a < 0$ can be dealt with by a similar argument, this time developing $(6a - 1)^{-6a}$.

If we have $k = 3$ then, similarly to the case $k = 9$, we must have $x \equiv -1 \pmod 3$, so the order of x must be even, that is, $x + 1 \equiv 0 \pmod 6$; by the argument above, there are no solutions.

If $k = 2$ then, since the only element of order 2 in $(\mathbb{Z}/27\mathbb{Z})^*$ is $x \equiv -1 \pmod{27}$, any solution must satisfy $x + 1 \equiv 0 \pmod 2$ and thus $x \equiv -1 \pmod{54}$. Again, this yields no additional solutions.

Finally, the only element of order 1, i.e. such that $k = 1$, is $x \equiv 1 \pmod{27}$, which is clearly a solution.

In conclusion, the solutions are $x \equiv -1 \pmod{18}$ and $x \equiv 1 \pmod{27}$.

117 The first congruence in the system has a solution if and only if $(a, 27) \mid 12$, that is, if $(a, 27) = 1, 3$. The second congruence can be split into the two congruences $a^3 x^2 \equiv 9 \pmod 3$ and $a^3 x^2 \equiv 9 \pmod{13}$. The first of those always has at least the solution $x \equiv 0 \pmod 3$. As for the second, in order for it to have a solution it is necessary that $a \not\equiv 0 \pmod{13}$.

Now, let g be a generator of the group $(\mathbb{Z}/13\mathbb{Z})^*$; let $a = g^i, x = g^j, 3 = g^k$. We must have $3i + 2j \equiv 2k \pmod{12}$, hence $i = 2i'$ must be even, that is, a must be a square modulo 13. Moreover, if $i = 2i'$ then the congruence becomes $3i' + j \equiv k \pmod 6$, whose solution is clearly $j \equiv k - 3i' \pmod 6$.

Let us now solve the system. By the Chinese remainder theorem, if each of the two congruences has a solution then the only obstruction to the existence of a solution of the system is the fact that the solutions may be incompatible modulo $(27, 39) = 3$. If $(a, 27) = 1$ then the solution of the first equation is $x \equiv 12a^{-1} \pmod{27}$, so in particular $x \equiv 0 \pmod 3$, which agrees with the second equation. If $(a, 27) = 3$ then let $a = 3b$ with $(b, 3) = 1$; the second congruence becomes $27b^3 x^2 \equiv 9 \pmod{39}$, which, when considered modulo 3, is satisfied by all integers.

In conclusion, the whole system has a solution if and only if each of the equations does, that is, if $(a, 27) = 1, 3$ (i.e. $a \not\equiv 0 \pmod 9$) and $a \equiv \pm 1, \pm 3, \pm 4 \pmod{13}$ (i.e. a is a nonzero square modulo 13).

118 First of all, let us solve each equation separately.

The first congruence has a solution if and only if $2a$ is a square modulo 5. Since the squares modulo 5 are $0, \pm 1$, the congruence has a solution if and only if $a \equiv 0 \pmod 5$ (one solution, namely, $x \equiv 0 \pmod 5$) or $a \equiv \pm 2 \pmod 5$ (two solutions in each case, namely, the two square roots of $2a$).

The second congruence has a solution if and only if it does both modulo 5 and modulo 7 and the solutions obtained for the two moduli are compatible. The two congruences modulo 5 and 7 have a solution if the class of 3 belongs to the multiplicative subgroup generated by the class of a.

Since 3 is a generator in $(\mathbb{Z}/5\mathbb{Z})^*$, the congruence $a^x \equiv 3 \pmod 5$ has a solution if and only if a itself is a generator, that is, if and only if $a \equiv \pm 2 \pmod 5$. Both cases yield a unique solution modulo $4 = \phi(5)$; moreover, the class that solves the congruence must be odd, otherwise a^x would be a square and therefore not a generator.

⟦For $a \equiv 2 \pmod 5$ we can find the solution $x \equiv 3 \pmod 4$, whereas the solution for $a \equiv -2 \pmod 5$ is $x \equiv 1 \pmod 4$.⟧

Similarly, since 3 is a generator of $(\mathbb{Z}/7\mathbb{Z})^*$, the congruence $a^x \equiv 3 \pmod 7$ has a solution if and only if a is a generator, that is, if and only if $a \equiv 3, 5 \pmod 7$. Again, we have a unique solution modulo $6 = \phi(7)$; moreover, by the same argument as before, the solution must be an odd class modulo 6.

⟦For $a \equiv 3 \pmod 7$ we can find the solution $x \equiv 1 \pmod 6$, whereas for $a \equiv 5 \pmod 7$ the solution is $x \equiv 5 \pmod 6$.⟧

It follows that the second equation has a solution if and only if $a \equiv \pm 2 \pmod 5$ and $a \equiv 3, 5 \pmod 7$, that is, $a \equiv 3, 12, 17, 33 \pmod{35}$: we have shown that this is a necessary condition, but it is also sufficient; indeed, we have remarked that solutions are odd, hence compatible modulo $2 = (4, 6)$. The solution is unique modulo 12, which is the least common multiple of 4 and 6.

To summarise, since the moduli of the two equations are relatively prime, the system has a solution if and only if both equations do, that is, if $a \equiv 3, 12, 17, 33$

(mod 35). So the system can be solved for all a in the classes listed above and we have two solutions modulo 5 for the first equation and one solution modulo 12 for the second equation, so two solutions of the system modulo 60, which is the least common multiple of 5 and 12.

119 Because 13 is an odd prime, the first equation can be solved by means of the usual quadratic formula (or even by trial and error): its solutions are $x \equiv 3, 9$ (mod 13).

The second equation has a solution if and only if $(a, 78) \mid 27$. Since $78 = 2 \cdot 3 \cdot 13$ and $27 = 3^3$, the condition is equivalent to $(a, 2) = (a, 13) = 1$. Let us split the equation into separate equations modulo 2, 13 and 3. For values of a that satisfy the conditions modulo 2 and modulo 13 we get the solutions $x \equiv 1$ (mod 2) and $x \equiv a^{-1}$ (mod 13). As for the congruence modulo 3, it imposes no conditions on a; however, if $(a, 3) = 1$ then the only solution is $x \equiv 0$ (mod 3), whereas if $3 \mid a$ then all integers x are solutions.

In order for the system to have a solution, we must have solutions of the first equation that also satisfy the second. In order for the solution $x \equiv 3$ (mod 13) to satisfy the second equation we must have $3a \equiv 27$ (mod 13), that is, $a \equiv 9$ (mod 13). In order for the solution $x \equiv 9$ (mod 13) to satisfy the second equation, we must have $9a \equiv 27$ (mod 13), that is, $a \equiv 3$ (mod 13).

In conclusion, the system has a solution if and only if $a \equiv 1$ (mod 2) and $a \equiv 9, 3$ (mod 13), that is, if and only if $a \equiv 9, 3$ (mod 26). If $(a, 3) = 1$ (that is, if $a \equiv 35, 61$ (mod 78) or $a \equiv 29, 55$ (mod 78)), then the solutions satisfy $x \equiv 1$ (mod 2), $x \equiv 3, 9$ (mod 13), $x \equiv 0$ (mod 3), so they are $x \equiv 3, 9$ (mod 78), respectively. Finally, if $3 \mid a$ (that is, if $a \equiv 9, 3$ (mod 78)) then there are no constraints on the class of x modulo 3 and the solutions are $x \equiv 3, 9$ (mod 26), respectively.

120 By the Chinese remainder theorem, we can split the first congruence into

$$\begin{cases} x^2 + x + 3 \equiv 0 & \text{(mod 5)} \\ x^2 - x \equiv 0 & \text{(mod 3)} \\ 30x \equiv -6 & \text{(mod 81)}. \end{cases}$$

Let us solve each congruence separately.

Since 5 is prime, the quadratic formula yields the solutions 1 and 3 modulo 5 for the equation $x^2 + x + 3 \equiv 0$ (mod 5).

Since 3 is prime, the solutions of $x^2 - x = x(x - 1) \equiv 0$ (mod 3) are 0 and 1, by the principle of zero products (or by direct verification).

Dividing by 6, we obtain that $30x \equiv -6$ (mod 81) is equivalent to $5x \equiv -1$ (mod 27) and, multiplying by 11 which is the inverse of 5, we get $x \equiv 16$ (mod 27).

The original system is therefore equivalent to the union of the systems

$$\begin{cases} x \equiv 1, 3 & \text{(mod 5)} \\ x \equiv 0, 1 & \text{(mod 3)} \\ x \equiv 16 & \text{(mod 27)}. \end{cases}$$

If $x \equiv 0$ (mod 3) then the subsystem consisting of the last two equations, and thus the system itself, has no solutions. The subsystem

$$\begin{cases} x \equiv 1 & \text{(mod 3)} \\ x \equiv 16 & \text{(mod 27),} \end{cases}$$

on the contrary, has the solution $x \equiv 16$ (mod 27).

The solutions of the original system are therefore those classes modulo $5 \cdot 27 = 135$ that satisfy either of the two systems

$$\begin{cases} x \equiv 1, 3 & \text{(mod 5)} \\ x \equiv 16 & \text{(mod 27).} \end{cases}$$

It is easy to check that the solutions are $x \equiv 16$ (mod 135) and $x \equiv 43$ (mod 135).

121   By the Chinese remainder theorem, the congruence in the problem statement is equivalent to the system

$$\begin{cases} x(x^{100} - 1) \equiv 0 & \text{(mod 7)} \\ x(x^{100} - 1) \equiv 0 & \text{(mod 11)} \\ x(x^{100} - 1) \equiv 0 & \text{(mod 13).} \end{cases}$$

Let p be a prime, and let us find the number of solutions of the equation $x(x^{100} - 1) \equiv 0$ (mod p). Since p is prime, the product is 0 if and only if one of the two factors is 0, so we have $x \equiv 0$ (mod p) or $x^{100} \equiv 1$ (mod p). The solutions of the equation $x^{100} \equiv 1$ (mod p) are given by the elements of $(\mathbb{Z}/p\mathbb{Z})^*$ whose order divides 100. Since the order of each element is a divisor of the order of the group, the solutions are given by those elements whose order divides $(100, p - 1)$. We know that $(\mathbb{Z}/p\mathbb{Z})^*$ is cyclic and has order a multiple of $(100, p - 1)$, so it contains exactly $(100, p - 1)$ solutions of the equation $x^{100} \equiv 1$ (mod p). Therefore, the initial equation has $(100, p - 1) + 1$ solutions.

It follows that the three equations in the system have $(100, 6) + 1 = 3$ solutions modulo 7, $(100, 10) + 1 = 11$ solutions modulo 11 and $(100, 12) + 1 = 5$ solutions modulo 13, respectively.

We obtain the solutions of the systems by combining any solution modulo 7 with any solution modulo 11 and any solution modulo 13; they are the solutions of $3 \cdot 11 \cdot 5 = 165$ systems of the form

$$\begin{cases} x \equiv a & \text{(mod 7)} \\ x \equiv b & \text{(mod 11)} \\ x \equiv c & \text{(mod 13).} \end{cases}$$

By the Chinese remainder theorem, each of these systems has a unique solution modulo $7 \cdot 11 \cdot 13 = 1001$; no two systems can have the same solution, so the original equation has 165 solutions.

122 If we consider the congruence modulo 2, we find that all solutions x must be even, so we may set $x = 2y$. The equation becomes $2^5 y^5 - 2^5 y = 2^5 y(y^4 - 1) \equiv 0$ (mod 2^{10}), hence $y(y^4 - 1) \equiv 0$ (mod 2^5). Remark that exactly one among y and $y^4 - 1$ is even, so we either have $y \equiv 0$ (mod 2^5) or $y^4 - 1 \equiv 0$ (mod 2^5).

The solutions of the congruence $y \equiv 0$ (mod 2^5) are given by $x \equiv 0$ (mod 2^6), so by 2^4 residue classes modulo 2^{10}.

Consider now the congruence $y^4 - 1 \equiv 0$ (mod 2^5). We can factor the polynomial and obtain $y^4 - 1 = (y - 1)(y + 1)(y^2 + 1) \equiv 0$ (mod 2^5).

If y is a solution then it is odd and the three factors $y - 1$, $y + 1$ and $y^2 + 1$ are all even. On the other hand, an immediate check yields that $y^2 + 1 \equiv 2$ (mod 4), that is, $y^2 + 1$ is divisible by 2 but not by 4; the congruence is therefore equivalent to $(y - 1)(y + 1) \equiv 0$ (mod 2^4). Now, since $y - 1$ and $y + 1$ are two consecutive even numbers, one is exactly divisible by 2, so the solutions are $y \equiv 1$ (mod 2^3) and $y \equiv -1$ (mod 2^3). In other words, in this case we have the solutions $x = 2y \equiv \pm 2$ (mod 2^4), that is, $2 \cdot 2^6 = 2^7$ solutions modulo 2^{10}.

The congruence in the problem statement thus has $2^4 + 2^7 = 144$ solutions modulo 2^{10}.

123 Let us start by solving the first congruence. One can check that, in $(\mathbb{Z}/17\mathbb{Z})^*$, we have ord$(2) = 8$, ord$(3) = 16$ and $2 = 3^{14}$. The first congruence in the system is thus the same as $3^{14x} \equiv 3^{x+a^2}$ (mod 17), which is equivalent to $14x \equiv x + a^2$ (mod 16). The solution is $x \equiv 5a^2$ (mod 16).

The congruence $3x \equiv a^{23}$ (mod 24) has a solution if and only if $3 = (3, 24) \mid a^{23}$, that is, if and only if $a \equiv 0$ (mod 3). When that is the case, by the Chinese remainder theorem the original system is equivalent to

$$\begin{cases} x \equiv 5a^2 \pmod{16} \\ 3x \equiv a^{23} \pmod{3} \\ 3x \equiv a^{23} \pmod{8}. \end{cases}$$

Since the second equation is always satisfied and the third has the solution $x \equiv 3a^{23}$ (mod 8), the system itself has a solution if and only if $8 = (16, 8) \mid 5a^2 - 3a^{23}$, that is, if and only if $5a^2 - 3a^{23} \equiv 0$ (mod 8).

Now, the congruence $a^2(5 - 3a^{21}) \equiv 0$ (mod 8) is satisfied if and only if $a^2 \equiv 0$ (mod 8) or $5 - 3a^{21} \equiv 0$ (mod 8), because the two factors always have different parity. Let us solve these two equations separately. For the first one, we have $a^2 \equiv 0$ (mod 8) if and only if $a \equiv 0$ (mod 4); for the second one, $5 - 3a^{21} \equiv 0$ (mod 8) if and only if $a^{21} \equiv -1$ (mod 8), which, since we have $a^2 \equiv 1$ (mod 8) because of a being odd, is equivalent to $a \equiv -1$ (mod 8).

By combining the conditions found above, we obtain that the system has a solution if and only if

$$\begin{cases} a \equiv 0 \pmod 3 \\ a \equiv 0 \pmod 4 \end{cases} \quad \text{or} \quad \begin{cases} a \equiv 0 \pmod 3 \\ a \equiv -1 \pmod 8. \end{cases}$$

In other words, the system has a solution if and only if $a \equiv 0, 12, 15 \pmod{24}$ and it has no solution if a belongs to any of the remaining residue classes modulo 24.

124 The first equation has a solution if and only if $3 = (3, 9) \,|\, a + 1$, that is, if and only if $a \equiv 2 \pmod 3$. So if $a \not\equiv 2 \pmod 3$ then the system has no solutions. Assume that $a \equiv 2 \pmod 3$ and let $a = 2 + 3b$ with $b \in \mathbb{Z}$. The first equation becomes $x \equiv b + 1 \pmod 3$.

By the Chinese remainder theorem, the second equation is equivalent to the system

$$\begin{cases} (x - 1)(x - a) \equiv 0 \pmod 3 \\ (x - 1)(x - a) \equiv 0 \pmod 5. \end{cases}$$

Since 3 and 5 are primes, by the principle of zero products the solutions of the first equation are $x \equiv 1 \pmod 3$ and $x \equiv a \equiv 2 \pmod 3$, and those of the second equation are $x \equiv 1 \pmod 5$ and $x \equiv a \equiv 2 + 3b \pmod 5$. We have found that the system is equivalent to

$$\begin{cases} x \equiv b + 1 & \pmod 3 \\ x \equiv 1, 2 & \pmod 3 \\ x \equiv 1, 3b + 2 & \pmod 5 \end{cases}$$

and so it has a solution if and only if the two equations modulo 3 are compatible. In particular, if $b \equiv 2 \pmod 3$ (that is, if $a \equiv 8 \pmod 9$) then there are no solutions; if $b \equiv 0 \pmod 3$ (that is, if $a \equiv 2 \pmod 9$) then the system is equivalent to

$$\begin{cases} x \equiv 1 & \pmod 3 \\ x \equiv 1, 3b + 2 & \pmod 5; \end{cases}$$

finally, if $b \equiv 1 \pmod 3$ (that is, if $a \equiv 5 \pmod 9$) then the system is equivalent to

$$\begin{cases} x \equiv 2 & \pmod 3 \\ x \equiv 1, 3b + 2 & \pmod 5. \end{cases}$$

Let us find the number of solutions of the first of these last two systems: if the two solutions of the equation modulo 5 coincide, that is, if $2 + 3b \equiv 1 \pmod 5$, or equivalently if $b \equiv 3 \pmod 5$, then the system has a unique solution modulo 15, so six solutions modulo 90. If $b \not\equiv 3 \pmod 5$ then the system has two solutions modulo 15, so 12 solutions modulo 90. The same argument also applies to the second system.

As for the system in the problem statement, we come to the following conclusion by combining the information above: if $a \equiv 0, 1$ (mod 3) then the first equation, and thus the system itself, has no solutions; if $a \equiv 2$ (mod 3) then there are a few cases to consider.

①　If $a \equiv 8$ (mod 9) then the equations are not compatible modulo 3, so the system has no solutions. ②　If $a \equiv 2$ (mod 9) then the system has either six solutions modulo 90 (when $a \equiv 11$ (mod 45)) or 12 solutions modulo 90 (when $a \equiv 2, 20, 29, 38$ (mod 45)). ③　Finally, if $a \equiv 5$ (mod 9) then the system has either six solutions modulo 90 (when $a \equiv -4$ (mod 45)) or 12 solutions modulo 90 (when $a \equiv 5, 14, 23, 32$ (mod 45)).

125　First of all, let us solve each equation separately. The first equation is $x^2(x^{25} - 1) \equiv 0$ (mod 144) which, by the Chinese remainder theorem, is equivalent to the system

$$\begin{cases} x^2(x^{25} - 1) \equiv 0 \pmod{16} \\ x^2(x^{25} - 1) \equiv 0 \pmod{9}. \end{cases}$$

Notice that x^2 and $x^{25} - 1$ are relatively prime for all integers x. In the first of the two equations above, this implies that we either have $x^2 \equiv 0$ (mod 16), hence $x \equiv 0$ (mod 4), or $x^{25} \equiv 1$ (mod 16), hence in particular $(x, 2) = 1$. In the latter case, since $\phi(16) = 8$, Euler's theorem implies $x^8 \equiv 1$ (mod 16), which yields $x^{25} = x^{1+3\cdot 8} \equiv x$ (mod 16) and therefore the solution $x \equiv 1$ (mod 16).

Similarly, for the second equation we either have $x^2 \equiv 0$ (mod 9), hence $x \equiv 0$ (mod 3) or, using the fact that $\phi(9) = 6$, we have $x^{25} = x^{1+6\cdot 4} \equiv x$ (mod 9), which yields the additional solution $x \equiv 1$ (mod 9).

Consider the second equation in the original system, that is, $10x \equiv a$ (mod 25) with $a \in \mathbb{Z}$. In order for this equation to have a solution we must have $5 = (25, 10) \mid a$, so $a = 5b$ for some $b \in \mathbb{Z}$. Dividing by five yields the equivalent equation $2x \equiv b$ (mod 5), hence the solution $x \equiv 2^{-1}b \equiv 3b$ (mod 5).

Finally, consider the third equation, that is, $2^{x-1} \equiv 4$ (mod 11). Since we have $2^2 \equiv 4$ (mod 11), the equation always has a solution. Let us find the multiplicative order of 2 (mod 11); we have $2^2 \equiv 4$ and $2^5 \equiv -1$ (mod 11), so $2^d \not\equiv 1$ (mod 11) for all maximal divisors of $10 = \phi(11)$, hence the order we want is 10. Therefore, the solution of the equation is $x - 1 \equiv 2$ (mod 10), that is, $x \equiv 3$ (mod 10). By the Chinese remainder theorem, this solution can be expressed as the solution of the system

$$\begin{cases} x \equiv 1 \pmod{2} \\ x \equiv 3 \pmod{5}. \end{cases}$$

Let us now turn back to the original system. Again by the Chinese remainder theorem, the solutions of its equations are certainly compatible when their resulting moduli are relatively prime. The compatibilities we need to check are the following.

The first equation implies $x \equiv 0 \pmod 4$ or $x \equiv 1 \pmod{16}$, whereas from the third we get $x \equiv 1 \pmod 2$. We must therefore exclude the case where $x \equiv 0 \pmod 4$ and, since $x \equiv 1 \pmod{16}$ implies $x \equiv 1 \pmod 2$, the condition to keep is $x \equiv 1 \pmod{16}$.

Moreover, from the second equation we obtain that $x \equiv 3b \pmod 5$, whereas $x \equiv 3 \pmod 5$ because of the third equation. The two are compatible for $b \equiv 1 \pmod 5$ and they give the condition $x \equiv 3 \pmod 5$.

To summarise: a necessary and sufficient condition for the system to have a solution is $a = 5b$ with $b \equiv 1 \pmod 5$, that is, $a \equiv 5 \pmod{25}$. When this holds, the solutions are those of the two systems

$$\begin{cases} x \equiv 1 \pmod{16} \\ x \equiv 0 \pmod 3 \\ x \equiv 3 \pmod 5 \end{cases} \qquad \begin{cases} x \equiv 1 \pmod{16} \\ x \equiv 1 \pmod 9 \\ x \equiv 3 \pmod 5. \end{cases}$$

Simple calculations yield the solutions $x \equiv 33 \pmod{240}$ and $x \equiv 433 \pmod{720}$.

126 We can simplify the first equation by dividing it by 2, which yields

$$2^{2y^2-5y+3} \equiv 1 \pmod{18}.$$

So 2^{2y^2-5y+3} must be invertible $\pmod{18}$, hence $(2^{2y^2-5y+3}, 18) = 1$. This is the case if and only if $2y^2 - 5y + 3 = (y-1)(2y-3) = 0$, which, since we are looking for integer solutions, implies $y = 1$.

Now consider the third equation. One can immediately check that $x \equiv -1 \pmod{100}$ is a solution. On the other hand, if x is a solution then we must have $(x, 100) = 1$, and so by Euler's theorem $x^{\phi(100)} = x^{40} \equiv 1 \pmod{100}$. Since the inverse of 23 modulo 40 is 7, the equation $x^{23} \equiv -1 \pmod{100}$ implies $x \equiv x^{23 \cdot 7} \equiv (-1)^7 \equiv -1 \pmod{100}$. It follows that the only solution of the third equation is $x \equiv -1 \pmod{100}$. It will be useful later to express this solution, by means of the Chinese remainder theorem, in the form of the system

$$\begin{cases} x \equiv -1 \pmod 4 \\ x \equiv -1 \pmod{25}. \end{cases}$$

Finally, consider the second equation. We can substitute the value $y = 1$, which yields $(2x^2 + 17)(2x^2 + 5x + 2)^{-1} \equiv 1 \pmod{592}$.

By factoring, we get $592 = 2^4 \cdot 37$ and $2x^2 + 5x + 2 = (x + 2)(2x + 1)$. So $2x^2 + 5x + 2$ is invertible modulo 592 if and only if $x + 2$ and $2x + 1$ are both invertible modulo 2 and modulo 37. The condition $x \equiv -1 \pmod{100}$, which implies that x is odd, ensures that both $x + 2$ and $2x + 1$ are invertible modulo 2. Invertibility modulo 37 is equivalent to $x + 2 \not\equiv 0 \pmod{37}$, that is, $x \not\equiv -2 \pmod{37}$, and $2x + 1 \not\equiv 0 \pmod{37}$, that is, $x \not\equiv 18 \pmod{37}$.

If these conditions are satisfied, then $(2x^2 + 17)(2x^2 + 5x + 2)^{-1} \equiv 1 \pmod{592}$ is equivalent to $x \equiv 3 \pmod{592}$, which is compatible with both $x \equiv -1 \pmod 4$ and $x \not\equiv -2, 18 \pmod{37}$. It follows that $(x, 1)$ is a solution of the original system if and only if x is a solution of the system of congruences

$$\begin{cases} x \equiv 3 & \pmod{16} \\ x \equiv 3 & \pmod{37} \\ x \equiv -1 & \pmod{25}. \end{cases}$$

A few calculations show that the system above is equivalent to $x \equiv 7699$ (mod 14800), so the solutions of the original system are all pairs of the form $(7699 + 14800t, 1)$ with $t \in \mathbb{Z}$.

127 The second congruence has a solution if and only if $(a, 10) = 1$, in which case its solution is $x \equiv a^{-1} \pmod{10}$; in particular, we have $(x, 10) = 1$.

The first congruence is equivalent to the system

$$\begin{cases} a^x \equiv 1 & \pmod{11} \\ a^x \equiv 1 & \pmod 7. \end{cases}$$

A necessary condition for the two equations in the system above to have a solution is that $a \not\equiv 0 \pmod{11}$ and $a \not\equiv 0 \pmod 7$. Both conditions are definitely compatible with $(a, 10) = 1$, because $(10, 77) = 1$.

Assume that $a \not\equiv 0 \pmod{11}$; by Fermat's little theorem, we have $a^{10} \equiv 1 \pmod{11}$. But, in order for the system to have a solution, we must also have $a^x \equiv 1 \pmod{11}$ for some x such that $(x, 10) = 1$. Therefore, the order of a in $(\mathbb{Z}/11\mathbb{Z})^*$ must be a divisor of $(x, 10) = 1$, so we must have $a \equiv 1 \pmod{11}$. If this holds, then all integers x satisfy the equation $a^x \equiv 1 \pmod{11}$.

Assuming that $a \not\equiv 0 \pmod 7$, the fact that solutions of the second equation must be odd implies that the order of a in $(\mathbb{Z}/7\mathbb{Z})^*$ must be a divisor of an odd integer, and therefore be odd itself. Since the order of every element of $(\mathbb{Z}/7\mathbb{Z})^*$ is a divisor of 6, we must either have $\text{ord}(a) = 1$, that is, $a \equiv 1 \pmod 7$, or $\text{ord}(a) = 3$, that is, $a \equiv 2, 4 \pmod 7$. If $a \equiv 1 \pmod 7$ then the equation $a^x \equiv 1 \pmod 7$ is satisfied for all integers x. If $a \equiv 2, 4 \pmod 7$ then the solution of the equation $a^x \equiv 1 \pmod 7$ is $x \equiv 0 \pmod 3$. This solution is guaranteed to be compatible with $x \equiv a^{-1} \pmod{10}$ because $(3, 10) = 1$.

To summarise, in order for the system to have a solution we must have: $(a, 10) = 1$, $a \equiv 1 \pmod{11}$, $a \equiv 1, 2, 4 \pmod 7$.

First, consider the case of $a \equiv 1 \pmod 7$. All integers x are solutions of the first equation, so the solution of the system is $x \equiv a^{-1} \pmod{10}$. Depending on the value of a, we have

$$a \equiv 1 \ (\mathrm{mod}\ 10) \implies a \equiv 1 \ (\mathrm{mod}\ 770), \quad x \equiv 1 \ (\mathrm{mod}\ 10),$$
$$a \equiv 3 \ (\mathrm{mod}\ 10) \implies a \equiv 463 \ (\mathrm{mod}\ 770),\ x \equiv 7 \ (\mathrm{mod}\ 10),$$
$$a \equiv 7 \ (\mathrm{mod}\ 10) \implies a \equiv 617 \ (\mathrm{mod}\ 770),\ x \equiv 3 \ (\mathrm{mod}\ 10),$$
$$a \equiv 9 \ (\mathrm{mod}\ 10) \implies a \equiv 309 \ (\mathrm{mod}\ 770),\ x \equiv 9 \ (\mathrm{mod}\ 10).$$

Now consider the case of $a \equiv 2, 4 \ (\mathrm{mod}\ 7)$. We have

$$a \equiv 1 \ (\mathrm{mod}\ 10) \implies a \equiv 331, 221 \ (\mathrm{mod}\ 770),\ x \equiv 21 \ (\mathrm{mod}\ 30),$$
$$a \equiv 3 \ (\mathrm{mod}\ 10) \implies a \equiv 23, 683 \ (\mathrm{mod}\ 770), \quad x \equiv 27 \ (\mathrm{mod}\ 30),$$
$$a \equiv 7 \ (\mathrm{mod}\ 10) \implies a \equiv 177, 67 \ (\mathrm{mod}\ 770), \quad x \equiv 3 \ \ (\mathrm{mod}\ 30),$$
$$a \equiv 9 \ (\mathrm{mod}\ 10) \implies a \equiv 639, 529 \ (\mathrm{mod}\ 770), \quad x \equiv 9 \ (\mathrm{mod}\ 30).$$

128 By the Chinese remainder theorem, we can split the equation modulo 200 into one modulo 8 and one modulo 25, thus obtaining the following system

$$\begin{cases} 7^x \equiv a \quad (\mathrm{mod}\ 8) \\ (x+a)^4 \equiv 0 \quad (\mathrm{mod}\ 8) \\ (x+a)^4 \equiv 0 \quad (\mathrm{mod}\ 25) \end{cases}$$

which is equivalent to the original one.

A necessary condition for the first equation above to have a solution is that $a \equiv \pm 1 \ (\mathrm{mod}\ 8)$; in particular, a must be odd. This implies that in the second equation x must also be odd. On the other hand, if both a and x are odd, then $2 \mid a + x$, so $16 \mid (a + x)^4$ and thus the second equation is satisfied. Going back to the first equation, x being odd implies $a \equiv 7 \ (\mathrm{mod}\ 8)$. Conversely, if $a \equiv 7 \ (\mathrm{mod}\ 8)$ then the first two equations have the unique solution $x \equiv 1 \ (\mathrm{mod}\ 2)$.

The third equation has a solution for each value of a, namely $x \equiv -a \ (\mathrm{mod}\ 5)$: as before, if $5 \mid x + a$ then $5^4 \mid (x + a)^4$.

In conclusion, the system has a solution if and only if $a \equiv 7 \ (\mathrm{mod}\ 8)$, and the solutions are the following.

If $a \equiv 0 \ (\mathrm{mod}\ 5)$, that is, if $a \equiv 15 \ (\mathrm{mod}\ 40)$, then $x \equiv 0 \ (\mathrm{mod}\ 5)$ and $x \equiv 1$ (mod 2), that is, $x \equiv 5 \ (\mathrm{mod}\ 10)$.

If $a \equiv 1 \ (\mathrm{mod}\ 5)$, that is, if $a \equiv 31 \ (\mathrm{mod}\ 40)$, then $x \equiv 4 \ (\mathrm{mod}\ 5)$ and $x \equiv 1$ (mod 2), that is, $x \equiv 9 \ (\mathrm{mod}\ 10)$.

If $a \equiv 2 \ (\mathrm{mod}\ 5)$, that is, if $a \equiv 7 \ (\mathrm{mod}\ 40)$, then $x \equiv 3 \ (\mathrm{mod}\ 5)$ and $x \equiv 1$ (mod 2), that is, $x \equiv 3 \ (\mathrm{mod}\ 10)$.

If $a \equiv 3 \ (\mathrm{mod}\ 5)$, that is, if $a \equiv 23 \ (\mathrm{mod}\ 40)$, then $x \equiv 2 \ (\mathrm{mod}\ 5)$ and $x \equiv 1$ (mod 2), that is, $x \equiv 7 \ (\mathrm{mod}\ 10)$.

If $a \equiv 4 \ (\mathrm{mod}\ 5)$, that is, if $a \equiv 39 \ (\mathrm{mod}\ 40)$, then $x \equiv 1 \ (\mathrm{mod}\ 5)$ and $x \equiv 1$ (mod 2), that is, $x \equiv 1 \ (\mathrm{mod}\ 10)$.

129 Consider the first equation. In order for it to have a solution we must have $(7a, 49) \mid a$. The greatest common divisor $(7a, 49)$ can take one of two possible values.

If $7 \nmid a$ then $(7a, 49) = 7$ and the necessary condition $7 \mid a$ is not satisfied, so we have no solutions. If $7 \mid a$ then $(7a, 49) = 49$ and the necessary condition becomes $49 \mid a$, that is, $a \equiv 0 \pmod{49}$. The equation $7ax \equiv a \pmod{49}$ becomes $0 \equiv a \pmod{49}$, which is always satisfied in this case.

To summarise, the first equation has a solution if and only if $a \equiv 0 \pmod{49}$, in which case it is satisfied for all integers.

As for the second equation, we shall distinguish three separate cases. ① If $x \equiv 0 \pmod 3$ then we have $x^a \equiv 0^a \equiv 0 \pmod 3$ for all $a > 0$; the equation cannot be satisfied for $a = 0$ or $a < 0$, because x is not invertible modulo 3. So in this case we have no solutions for any value of a. ② If $x \equiv 1 \pmod 3$ then the equation is satisfied for all $a \in \mathbb{Z}$. ③ Finally, if $x \equiv 2 \pmod 3$ then the equation becomes $2^a \equiv 1 \pmod 3$, which is satisfied if and only if $a \equiv 0 \pmod 2$.

So the solution of the second equation is $x \equiv 1 \pmod 3$ for all $a \in \mathbb{Z}$, and we have the additional solution $x \equiv 2 \pmod 3$ if $a \equiv 0 \pmod 2$.

In conclusion, the solutions of the original system can be described as follows. If $a \equiv 0 \pmod{49}$ and $a \equiv 0 \pmod 2$, that is, if $a \equiv 0 \pmod{98}$, then we have the solutions $x \equiv 1, 2 \pmod 3$. If $a \equiv 0 \pmod{49}$ and $a \not\equiv 0 \pmod 2$, that is, if $a \equiv 49 \pmod{98}$, then we have the solution $x \equiv 1 \pmod 3$. For all other values of $a \in \mathbb{Z}$ there are no solutions, because there are no solutions of the first equation.

130 Let us solve the first equation. Since we have $1000 = 8 \cdot 125$, by the Chinese remainder theorem the equation is equivalent to the system

$$\begin{cases} x^3 \equiv 0 & (\text{mod } 8) \\ x^3 \equiv 2^3 & (\text{mod } 125). \end{cases}$$

The solutions of $x^3 \equiv 0 \pmod 8$ are obviously given by all even x. As for the second equation in the system above, remark that 2 is invertible modulo 125, so x is as well. We may thus solve the equation in $(\mathbb{Z}/125\mathbb{Z})^*$ by rewriting it as $(x/2)^3 \equiv 1 \pmod{125}$; since 3 does not divide $|(\mathbb{Z}/125\mathbb{Z})^*| = \phi(125) = 100$, we get $x/2 \equiv 1 \pmod{125}$. We can therefore conclude that the equation $x^3 \equiv 8 \pmod{125}$ has the unique solution $x \equiv 2 \pmod{125}$.

We can combine the solutions via the system

$$\begin{cases} x \equiv 2 & (\text{mod } 125) \\ x \equiv 0 & (\text{mod } 2) \end{cases}$$

whose solutions are $x \equiv 2, 252, 502, 752 \pmod{1000}$. So we obtain the four systems

$$\begin{cases} x \equiv 0, 252, 502, 752 & (\text{mod } 1000) \\ x \equiv 2 & (\text{mod } 3) \end{cases}$$

with the additional constraint that $0 \le x < 3001$. Again by the Chinese remainder theorem, each system has a unique solution modulo 3000, so we have exactly four solutions.

131

(i) Let us solve the congruence $x^{36} \equiv x \pmod 9$ by distinguishing two cases.
   ① If $(x, 3) \ne 1$ then $x = 3y$ for some $y \in \mathbb{Z}$. Consequently, $x^{36} = (3y)^{36} = 3^{36}y^{36} \equiv 0 \pmod 9$, so $x \equiv x^{36} \equiv 0 \pmod 9$.
   ② If $(x, 3) = 1$ then $x \in (\mathbb{Z}/9\mathbb{Z})^*$. In this case the equation is equivalent to $x^{35} \equiv 1 \pmod 9$. By Euler's theorem we have $x^{\phi(9)} = x^6 \equiv 1 \pmod 9$ and thus $x^{35} \equiv x^{-1} \pmod 9$. The equation becomes $x^{-1} \equiv 1 \pmod 9$, that is, $x \equiv 1 \pmod 9$.
   In conclusion, the congruence has the two solutions $x \equiv 0, 1 \pmod 9$.

(ii) Let us now solve the congruence $x^2 - x = x(x - 1) \equiv 0 \pmod{64}$. Clearly, x and $x - 1$ are relatively prime, so the only solutions are $x \equiv 0 \pmod{64}$ and $x \equiv 1 \pmod{64}$.
   The solutions of the original system are therefore those of the four systems

$$\begin{cases} x \equiv 0 \text{ or } 1 \pmod 9 \\ x \equiv 0 \text{ or } 1 \pmod{64}. \end{cases}$$

Carrying out the necessary calculations yields $x \equiv 0, 1, 64, 513 \pmod{576}$.

132 If $a = 0$ then the equation is an identity and is satisfied by all classes modulo 584; we shall therefore assume that $a > 0$. We have $584 = 2^3 \cdot 73$ and $x^{a+5} - x^a - x^5 + 1 = (x^a - 1)(x^5 - 1)$. By the Chinese remainder theorem, the original equation is equivalent to the system

$$\begin{cases} (x^5 - 1)(x^a - 1) \equiv 0 \pmod 8 \\ (x^5 - 1)(x^a - 1) \equiv 0 \pmod{73}. \end{cases}$$

Consider the first equation in the system. First of all, remark that it is not satisfied for any even number. Moreover, since the square of any odd number is congruent to 1 modulo 8, assuming x is odd we have $x^5 \equiv x \pmod 8$, whereas x^a is congruent to 1 if a is even and to x if a is odd.

Therefore, if $a \equiv 0 \pmod 2$ then

$$(x^5 - 1)(x^a - 1) \equiv 0 \pmod 8 \iff x \equiv 1 \pmod 2,$$

whereas if $a \equiv 1 \pmod 2$ then

$$(x^5 - 1)(x^a - 1) \equiv 0 \pmod 8 \iff (x - 1)^2 \equiv 0 \pmod 8$$
$$\iff x \equiv 1 \pmod 4.$$

Let us now move on to the second equation. The number 73 is prime, so the principle of zero products holds modulo 73; we have

$$(x^5 - 1)(x^a - 1) \equiv 0 \pmod{73} \iff x^5 \equiv 1 \pmod{73} \text{ or } x^a \equiv 1 \pmod{73}.$$

Now, $x^5 \equiv 1 \pmod{73}$ if and only if the order x in $(\mathbb{Z}/73\mathbb{Z})^*$ divides 5. Since the order of x must also divide $\phi(73) = 72$ and $(5, 72) = 1$, we have $x^5 \equiv 1 \pmod{73}$ if and only if $x \equiv 1 \pmod{73}$. By the same argument, $x^a \equiv 1 \pmod{73}$ if and only if $x^{(a,72)} \equiv 1 \pmod{73}$. Since $(\mathbb{Z}/73\mathbb{Z})^*$ is cyclic, this equation has exactly $(a, 72)$ solutions modulo 73. In particular, the unique solution $x \equiv 1 \pmod{73}$ of $x^5 - 1 \equiv 0 \pmod{73}$ is also a solution of $x^a - 1 \equiv 0 \pmod{73}$.

Going back to the original system, its second equation has $(a, 72)$ solutions modulo 73 and its first equation has one solution modulo 2 (that is, four solutions modulo 8) if a is even, and one solution modulo 4 (that is, two solutions modulo 8) if a is odd. By the Chinese remainder theorem, we conclude that there are $4(a, 72)$ solutions modulo 584 if a is even and $2(a, 72)$ solutions if a is odd.

133   By the Chinese remainder theorem, the congruence is equivalent to the following system:

$$\begin{cases} x^5 - 4x + 400 \equiv 0 \pmod{2^{10}} \\ x^5 - 4x + 400 \equiv 0 \pmod{5^{10}}. \end{cases}$$

The first equation implies $x \equiv 0 \pmod{4}$. By letting $x = 4y$, we have $x^5 \equiv 2^{10}y^5 \equiv 0 \pmod{2^{10}}$, hence $y \equiv 25 \pmod{2^6}$, so $x \equiv 100 \pmod{2^8}$.

The second equation in the system implies $x^5 - 4x + 400 \equiv 0 \pmod{5^2}$, from which we get $x(x^4 + 1) \equiv 0 \pmod{25}$. Now, either $x^4 + 1 \equiv 1$ or $x^4 + 1 \equiv 2 \pmod 5$ by Fermat's little theorem; in particular, $x^4 + 1$ is always invertible modulo 25. It follows that we have $x(x^4 + 1) \equiv 0 \pmod{25}$ if and only if $x \equiv 0 \pmod{25}$, that is, if and only if $x = 5^2 t$ for some $t \in \mathbb{Z}$. By substituting this condition into the initial equation we obtain

$$5^{10}t^5 - 2^2 5^2 t + 2^4 5^2 \equiv -2^2 5^2 t + 2^4 5^2 \equiv 0 \pmod{5^{10}}$$

which yields $t \equiv 4 \pmod{5^8}$ and $x \equiv 100 \pmod{5^{10}}$. The initial system becomes

$$\begin{cases} x \equiv 100 \pmod{2^8} \\ x \equiv 100 \pmod{5^{10}} \end{cases}$$

whose solution is $x \equiv 100 \pmod{2^8 \cdot 5^{10}}$. In conclusion, we have $10^{10}/(2^8 \cdot 5^{10}) = 4$ solutions modulo 10^{10}.

4 Groups

134 We show that B and C are subgroups of $\mathrm{Hom}(G, G')$, whereas A and D are not.

The neutral element in $\mathrm{Hom}(G, G')$ is the zero homomorphism, that is, the homomorphism e such that $e(x) = 0$ for all $x \in G$. The additive inverse of a homomorphism f is the homomorphism $-f$ such that $(-f)x = -f(x)$ for all $x \in G$.

We have $e \in B$, because $\mathrm{Ker}(e) = G \supseteq H$. If $f, g \in B$ and $h \in H$ then $f(h) = g(h) = 0$, so $(f + g)(h) = 0$ and $f + g \in B$. Finally, if $f \in B$ and $h \in H$ then $f(h) = 0$, so $(-f)(h) = 0$ and thus $-f \in B$.

Therefore, B is a subgroup of $\mathrm{Hom}(G, G')$. Similarly, C is a subgroup of $\mathrm{Hom}(G, G')$. Indeed, we have $e \in C$, because $e(G) = 0 \in H'$. Moreover, if $f, g \in C$ and $x \in G$ then $f(x), g(x) \in H'$, hence $f(x) + g(x) \in H'$ and thus $f + g \in C$. Finally, if $f \in C$ and $x \in G$ then $f(x) \in H'$, hence $(-f)(x) = -f(x) \in H'$ and thus $-f \in C$.

Remark that, since H and H' are proper nontrivial subgroups of G and G', the zero homomorphism e of $\mathrm{Hom}(G, G')$ does not belong to A nor to D, so those cannot be subgroups of $\mathrm{Hom}(G, G')$.

135

(i) Since $\mathbb{Z}/mn\mathbb{Z} \simeq \mathbb{Z}/m\mathbb{Z} \times \mathbb{Z}/n\mathbb{Z}$ when m and n are relatively prime, we have

$$G \cong (\mathbb{Z}/2\mathbb{Z} \times \mathbb{Z}/4\mathbb{Z}) \times \mathbb{Z}/3\mathbb{Z} \times \mathbb{Z}/5\mathbb{Z}.$$

Let $G_2 = \mathbb{Z}/2\mathbb{Z} \times \mathbb{Z}/4\mathbb{Z}$, $G_3 = \mathbb{Z}/3\mathbb{Z}$, $G_5 = \mathbb{Z}/5\mathbb{Z}$. Given $(x, y, z) \in G_2 \times G_3 \times G_5$, we have $o(x, y, z) = [o(x), o(y), o(z)] = o(x)o(y)o(z)$.
Therefore, (x, y, z) has order 60 if and only if $o(x) = 4$, $o(y) = 3$, $o(z) = 5$. There are four elements of order 4 in G_2, that is, all elements of G_2 except for the four pairs (x_1, x_2) such that $2(x_1, x_2) = (0, 0)$. There are two elements of order 3 in G_3: all elements of G_3 except for the identity. Finally, there are four elements of order 5 in G_5: all of them except for the identity. So there are $4 \cdot 2 \cdot 4 = 32$ elements of order 60 in G.

(ii) As before, (x, y, z) has order 30 if and only if $o(x) = 2$, $o(y) = 3$, $o(z) = 5$. There are three pairs $(x_1, x_2) \in G_2$ of order 2, that is, the four pairs such that $2(x_1, x_2) = (0, 0)$, except for the identity. So the number of elements of G having order 30 is 24.
Now, each element of order 30 generates a cyclic subgroup of order 30. On the other hand, the same cyclic subgroup of order 30 can be generated by any of its elements of order 30, of which there are $\phi(30) = 8$. So the number of all cyclic subgroups of G whose order is 30 is $24/8 = 3$.

(iii) All homomorphisms $f : \mathbb{Z}/12\mathbb{Z} \longrightarrow G$ are of the form $f(\overline{n}) = ng$ for some $g \in G$. Indeed, if $f(\overline{1}) = g$ then by the homomorphism property we have $f(\overline{n}) = f(\overline{1 + \cdots + 1}) = g + \cdots + g = ng$. Clearly any such map, if well defined, is a homomorphism, because we have $f(\overline{m} + \overline{n}) = (m+n)g = mg + ng = f(\overline{m}) + f(\overline{n})$.

The condition for the map to be well defined is $\overline{m} = \overline{n} \Rightarrow mg = ng$. If we also ask that the resulting homomorphism is injective, the condition becomes

$$\overline{m} = \overline{n} \iff mg = ng.$$

This condition is satisfied if and only if $\text{ord}(g) = 12$. Again, (x, y, z) has order 12 if and only if $\text{ord}(x) = 4$, $\text{ord}(y) = 3$, $\text{ord}(z) = 1$. Previous calculations imply that there are eight elements of order 12 in G, so the number of all injective homomorphisms $f : \mathbb{Z}/12\mathbb{Z} \longrightarrow G$ is 8.

136

(i) The identity map $\text{Id} : \mathbb{Z}/72\mathbb{Z} \longrightarrow \mathbb{Z}/72\mathbb{Z}$, $\text{Id}(x) = x$, which is the neutral element of G, belongs to H, because $\text{Id}(\overline{12}) = \overline{12}$.
Given $f, g \in H$ of the form $f(x) = ax$ and $g(x) = bx$ with $(a, 72) = (b, 72) = 1$, their composition is $f \circ g(x) = abx$ and, since we also have $(ab, 72) = 1$ and $f \circ g(\overline{12}) = f(\overline{12}) = \overline{12}$, it does belong to H.
Given $f \in H$ of the form $f(x) = ax$ with $(a, 72) = 1$, pick a' such that $aa' \equiv 1 \pmod{72}$. We have $(a', 72) = 1$ and the map g given by $g(x) = a'x$ is the inverse of f; since $f(\overline{12}) = \overline{12}$ implies $f^{-1}(\overline{12}) = \overline{12}$, g belongs to H.
Therefore, H is a subgroup of G.
To find the order of H, remark that $f(\overline{12}) = \overline{12}$ is equivalent to $12a \equiv 12 \pmod{72}$, that is, $a \equiv 1 \pmod 6$. There are exactly 12 classes modulo 72 that are congruent to 1 modulo 6 and they are all coprime to 72, because they are coprime to 6, which has the same prime factors. So the order of H is 12.

(ii) The subgroup H is not cyclic. This is because, given any $f \in H$ of the form $f(x) = ax$ with $a \equiv 1 \pmod 6$, we have $f^6(x) = a^6 x$ and $a \equiv 1 \pmod 2$ implies $a^2 \equiv 1 \pmod 8$, hence $a^6 \equiv 1 \pmod 8$. Similarly, $a \equiv 1 \pmod 3$ implies $a^3 \equiv 1 \pmod 9$, hence $a^6 \equiv 1 \pmod 9$.
What we have just shown is that $a^6 \equiv 1 \pmod{72}$, so the order of every map in H is a divisor of 6.

137

(i) A homomorphism is injective if and only if its kernel consists of only the neutral element. The kernel of the map $f(x) = (ax, bx)$ is $\{x \in G \mid (ax, bx) = (\overline{0}, \overline{0})\}$. Let u, v be integer representatives of the residue classes a, b, respectively. We have $ax = \overline{0}$ if and only if the order of x is a divisor of $(u, 12)$, and $bx = \overline{0}$ if and only if the order of x is a divisor of $(v, 12)$. So the kernel of f is the set of elements whose order is a divisor of $(u, v, 12)$, that is, f is injective if and only if $(u, v, 12) = 1$.
Let Y be the set of pairs satisfying the desired condition and, for each divisor d of 12, let Y_d be the set of pairs (a, b) such that $d \mid u$ and $d \mid v$. We have

$$|Y| = 12^2 - |Y_2 \cup Y_3| = 12^2 - |Y_2| - |Y_3| + |Y_6| = 12^2 - 6^2 - 4^2 + 2^2 = 96.$$

(ii) Using the same notation as above, we have $g \circ f(x) = (a + b)x$ and thus $g \circ f$ is injective if and only if $(u + v, 12) = 1$. For each $a \in G$ there are exactly $\phi(12) = 4$ values of b for which $(u + v, 12) = 1$, so the answer is $12 \cdot 4 = 48$ pairs.

138 Let $G = (\mathbb{Z}/p^2\mathbb{Z})^*$; we shall show that $H = \{x \in G \mid x \equiv 1 \ (\mathrm{mod}\ p)\} = \{1 + tp \mid t = 0, 1, \ldots, p - 1\}$ is a subgroup of G. Indeed, we have $1 \in H$, and if $x, y \equiv 1 \ (\mathrm{mod}\ p)$ then we also have $xy \equiv 1 \ (\mathrm{mod}\ p)$ and $x^{-1} \equiv 1 \ (\mathrm{mod}\ p)$. The subgroup H has p elements, so all of its elements except for the identity have order p. In particular, the element $a = \overline{p + 1}$ has order p.

Let $b \in G$ be an element such that the class of b modulo p is a generator of the cyclic group $(\mathbb{Z}/p\mathbb{Z})^*$. If $b^n \equiv 1 \ (\mathrm{mod}\ p^2)$ then we also have $b^n \equiv 1 \ (\mathrm{mod}\ p)$ and so $n \equiv 0 \ (\mathrm{mod}\ p - 1)$; so the order of b in G is a multiple of $p - 1$.

But then the cyclic subgroup generated by b has order divisible by $p - 1$, so it has a cyclic subgroup of order $p - 1$. A generator of this subgroup is an element of G of order $p - 1$.

139 Let us first show that HK is a subgroup of G. First of all, we do have $e \in HK$: indeed, $e = e \cdot e$, and $e \in H, e \in K$ because H, K are subgroups of G. Now, given $hk, h'k' \in HK$, the fact that H is a normal subgroup of G implies that $kH = Hk$, so there exists $h'' \in H$ such that $kh' = h''k$; we thus have $hkh'k' = hh''kk' \in HK$ because, H and K being subgroups of G, we have $hh'' \in H$ and $kk' \in K$. Finally, if $hk \in HK$ then $(hk)^{-1} = k^{-1}h^{-1}$ and, because H is normal, we have $k^{-1}h^{-1} = h'k^{-1} \in HK$ for some $h' \in H$.

Now we show that HK is a normal subgroup of G. For all $g \in G, h \in H$ and $k \in K$ we have $ghkg^{-1} = (ghg^{-1})(gkg^{-1})$; the latter does belong to HK because, since H and K are normal, we have $ghg^{-1} \in H$ and $gkg^{-1} \in K$.

140

(i) The set $\mathrm{Ker}(f) \cap \mathrm{Im}(f)$ is a subgroup of G, so it contains the neutral element of G. Conversely, let $x \in \mathrm{Ker}(f) \cap \mathrm{Im}(f)$; the fact that $x \in \mathrm{Ker}(f)$ implies $f(x) = e$, whereas from $x \in \mathrm{Im}(f)$ we obtain that there is an element $y \in G$ such that $x = f(y)$. It follows that $x = f(y) = f \circ f(y) = f(x) = e$.

(ii) It is clear that $\mathrm{Ker}(f) \cdot \mathrm{Im}(f) \subseteq G$. Conversely, for each $x \in G$, we can write $x = xf(x^{-1}) \cdot f(x)$. We have $f(xf(x^{-1})) = f(x) \cdot f \circ f(x^{-1}) = f(x) \cdot f(x^{-1}) = e$, hence $xf(x^{-1}) \in \mathrm{Ker}(f)$. Since naturally $f(x) \in \mathrm{Im}(f)$ we have proved the desired result.

141

(i) We have $|(\mathbb{Z}/49\mathbb{Z})^*| = \phi(49) = 42$. The elements of order 2 and those of order 3 are the solutions other than $\overline{1}$ of the congruence $x^2 \equiv 1 \ (\mathrm{mod}\ 49)$ and the congruence $x^3 \equiv 1 \ (\mathrm{mod}\ 49)$, respectively.

The condition $x^2 - 1 = (x - 1)(x + 1) \equiv 0 \ (\mathrm{mod}\ 49)$ implies that one of the following must be satisfied: $x - 1 \equiv 0 \ (\mathrm{mod}\ 49)$, whose solution is class $\overline{1}$ (which we excluded), or $x + 1 \equiv 0 \ (\mathrm{mod}\ 49)$, whose solution is the class $-\overline{1}$

which does have order 2, or the system of two equations $x - 1 \equiv 0 \pmod 7$ and $x + 1 \equiv 0 \pmod 7$, which has no solutions.

In conclusion, the class $-\overline{1}$ is the only element of order 2 in $(\mathbb{Z}/49\mathbb{Z})^*$.

If $x^3 \equiv 1 \pmod{49}$ then in particular $x^3 \equiv 1 \pmod 7$, so $x \equiv 1, 2, -3 \pmod 7$ (one can check that those classes satisfy the condition and all others do not). It follows that the solutions of the original congruence modulo 49 are integers of the form $x = a + 7t$ with $a = 1$, 2 or -3 and $t \in \mathbb{Z}$. Now, remark that $(a + 7t)^3 \equiv a^3 + 21a^2 t \pmod{49}$; substituting the three possible values of a yields the solutions $x \equiv 1, -19, 18 \pmod{49}$ for the equation $x^3 \equiv 1 \pmod{49}$. In conclusion, there are exactly two elements of order 3 in $(\mathbb{Z}/49\mathbb{Z})^*$.

(ii) There are as many homomorphisms from $\mathbb{Z}/6\mathbb{Z}$ to $(\mathbb{Z}/49\mathbb{Z})^*$ as elements of $(\mathbb{Z}/49\mathbb{Z})^*$ whose order divides 6. Such elements are the solutions modulo 49 of $x^6 \equiv 1 \pmod{49}$.

If x is an integer solution of the congruence above, then $x^6 \equiv 1 \pmod 7$, that is, x is of the form $x = a + 7t$ with $a \in \{1, 2, 3, 4, 5, 6\}$ and $t \in \mathbb{Z}$. Substituting yields $(a + 7t)^6 \equiv a^6 + 42a^5 t \equiv 1 \pmod{49}$. Since $7 \mid a^6 - 1$, the equation becomes $6a^5 t \equiv b \pmod 7$, where $b = (a^6 - 1)/7$. Now, since $6a^5$ is invertible modulo 7 for all a in the set we are considering, there is a unique solution modulo 7 for t. We can thus conclude that the equation $x^6 \equiv 1 \pmod 7$ has six solutions modulo 49.

There are therefore six homomorphisms from $\mathbb{Z}/6\mathbb{Z}$ to $(\mathbb{Z}/49\mathbb{Z})^*$.

142

(i) We have $Z(H) \neq \varnothing$ because the neutral element of G does belong to $Z(H)$. If $x, y \in Z(H)$ then for all $h \in H$ we have $(xy)h = x(yh) = x(hy) = (xh)y = (hx)y = h(xy)$, so $xy \in Z(H)$; moreover, if $x \in Z(H)$ then $xh = hx$ for all $h \in H$, hence $hx^{-1} = x^{-1}xhx^{-1} = x^{-1}hxx^{-1} = x^{-1}h$ and so $x^{-1} \in Z(H)$. Therefore, $Z(H)$ is a subgroup of G.

(ii) Assume H is a normal subgroup of G and consider $x \in Z(H)$ and $g \in G$. We shall show that $gxg^{-1} \in Z(H)$. Indeed, for all h in H we have

$$(gxg^{-1})h(gx^{-1}g^{-1}) = gx(g^{-1}hg)x^{-1}g^{-1} = g(g^{-1}hg)xx^{-1}g^{-1} = h$$

because $g^{-1}hg \in H$ and $x \in Z(H)$. This shows that $Z(H)$ is a normal subgroup of G.

(iii) Given $x \in Z(H)$, we have $f(x)f(h) = f(xh) = f(hx) = f(h)f(x)$ for all $h \in H$, so the equality holds for all $f(h) \in f(H)$. In other words, $f(Z(H))$ commutes with all elements of $f(H)$, that is, $f(Z(H)) \subseteq Z(f(H))$ as required.

(iv) We can take $G = \mathbb{Z}/2\mathbb{Z}$, $G' = S_3$, $f(\overline{0}) = \text{Id}$, $f(\overline{1}) = \sigma$ where $\sigma(1) = 2$, $\sigma(2) = 1$ and $\sigma(3) = 3$, and $H = G$. One can check that $Z(f(G)) = \{\text{Id}, \sigma\} \neq G'$.

143

(i) First, let us show that H is a subgroup of G. Clearly, $e \in H$ because the neutral element has order 1. Let $a, b \in H$ and suppose $\operatorname{ord}(a) = m$, $\operatorname{ord}(b) = n$; then $(ab)^{mn} = a^{mn}b^{mn} = e$, so ab has finite order and therefore belongs to H. Finally, if $a \in H$ then $a^{-1} \in H$ because $\operatorname{ord}(a) = \operatorname{ord}(a^{-1})$, so H is indeed a subgroup.

Let $G = \mathbb{C}^*$: we have $H = \{z \in \mathbb{C}^* \mid z^n = 1 \text{ for some } n\}$. Since for all $n \in \mathbb{N}$ the polynomial $x^n - 1$ has n roots in $\mathbb{C}$ and those belong to H, we have $|H| \geq n$ for all $n \in \mathbb{N}$, so H is infinite.

(ii) Let gH be an element of finite order in G/H and let n be its order. From $g^n H = H$ we get that $g^n \in H$, so there exists an integer d such that $(g^n)^d = g^{nd} = e$, that is, $g \in H$. This implies $gH = H$, that is, only the neutral element has finite order in G/H.

(iii) Any group isomorphism sends elements of finite order to elements of finite order. We have just shown that the only element of finite order in G/H is the neutral element, so the subgroup of elements of G having finite order must be $H = \{e\}$.

(iv) Given a homomorphism $\varphi : G \longrightarrow \mathbb{Z}$, consider an element $x \in H$ such that $n = \operatorname{ord}(x)$. We must have $\operatorname{ord}(\varphi(x)) \mid n$, that is, $\varphi(x)$ is an element of $\mathbb{Z}$ of finite order. However, 0 is the only element of finite order in $\mathbb{Z}$, so necessarily $\varphi(x) = 0$, that is, $x \in \operatorname{Ker}(\varphi)$.

144

(i) Both $H_1 \times H_2$ and $G_1 \times G_2$, endowed with the appropriate componentwise operations, are groups. Since the operations of H_1 and H_2 are the restrictions of those of G_1 and G_2, the subset $H_1 \times H_2$ is a subgroup of $G_1 \times G_2$.

Given $(x, y) \in G_1 \times G_2$, we have $(x, y)(H_1 \times H_2)(x, y)^{-1} = (x, y)(H_1 \times H_2)(x^{-1}, y^{-1}) = xH_1x^{-1} \times yH_2y^{-1} = H_1 \times H_2$ because H_1 is a normal subgroup of G_1 and H_2 is a normal subgroup of G_2. Therefore, $H_1 \times H_2$ is a normal subgroup of $G_1 \times G_2$.

(ii) For all $(x, y) \in \mathcal{H}$ we have $x = \pi_1(x, y) \in \pi_1(\mathcal{H})$ and $y = \pi_2(x, y) \in \pi_2(\mathcal{H})$, so $(x, y) \in \pi_1(\mathcal{H}) \times \pi_2(\mathcal{H})$.

(iii) Given $(x, y) \in \pi_1(\mathcal{H}) \times \pi_2(\mathcal{H})$ there exist $a \in G_1$ and $b \in G_2$ such that $(x, b), (a, y) \in \mathcal{H}$. Let $h, k \in \mathbb{Z}$ be such that $hm + kn = 1$. Letting e_1 and e_2 be the neutral elements of G_1 and G_2, we have $(x, b)^{kn} = (x^{kn}, b^{kn}) = (x^{1-hm}, e_2) = (x, e_2) \in \mathcal{H}$ and $(a, y)^{hm} = (a^{hm}, y^{1-kn}) = (e_1, y) \in \mathcal{H}$. It follows that $(x, y) = (x, e_2)(e_1, y) \in \mathcal{H}$. This, together with the containment shown above, yields the desired equality.

145

(i) Because the groups are finite we have $[G_1 : H] = |G_1|/|H|$ and also $[f(G_1) : f(H)] = |f(G_1)|/|f(H)|$. The fundamental homomorphism theorem, when applied to $f : G_1 \longrightarrow G_2$ and its restriction to H, that is, $f_{|H} : H \longrightarrow G_2$, yields that $f(G_1) \simeq G_1/\operatorname{Ker}(f)$ and $f(H) \simeq H/\operatorname{Ker}(f_{|H})$.

Since Ker$(f) \subseteq H$, we have Ker$(f_{|H}) = $ Ker(f), and thus

$$[f(G_1) : f(H)] = |f(G_1)|/|f(H)|$$
$$= (|G_1|/|\text{Ker}(f)|)(\text{Ker}(f)|/|H|)$$
$$= |G_1|/|H| = [G_1 : H].$$

(ii) If Ker$(f) \nsubseteq H$ then the equality from the first question does not hold in general. For instance, consider $G_1 = G_2 = \mathbb{Z}/2\mathbb{Z}$, $H = \{0\}$ and let f be the zero homomorphism. We have $[G_1 : H] = 2$, whereas $[f(G_1) : f(H)] = [\{0\} : \{0\}] = 1$.

(iii) If $G_1 = \mathbb{Z}$ and G_2 is a finite group then we have Ker$(f) = n\mathbb{Z}$ for some $n > 0$. Letting $H = m\mathbb{Z}$, the fact that $H \supseteq$ Ker(f) implies that $m \mid n$, and in particular we obtain $|H/\text{Ker}(f)| = n/m$ as well as $[G_1 : H] = |G_1/H| = m$. Similarly to our first answer, we have

$$[f(G_1) : f(H)] = |f(G_1)|/|f(H)|$$
$$= (|G_1/\text{Ker}(f)|) : (|H/\text{Ker}(f)|)$$
$$= nm/n$$
$$= m$$
$$= [G_1 : H].$$

Therefore, the desired equality does hold in this case.

146

(i) Since N is an intersection of subgroups, it is itself a subgroup of G. Consider an element $g \in G$ and let M be a subgroup of G. The subgroup M is maximal if and only if gMg^{-1} is maximal: we have $M \subsetneq H$ if and only if $gMg^{-1} \subsetneq gHg^{-1}$, and we have $H = G$ if and only if $gHg^{-1} = G$. This implies that the set of all maximal subgroups of G is invariant by conjugation, so their intersection N must be as well, that is, N is a normal subgroup of G.

(ii) The subgroups of $\mathbb{Z}/n\mathbb{Z}$ are the groups $d\mathbb{Z}/n\mathbb{Z}$, where $d \mid n$. Since $d\mathbb{Z}/n\mathbb{Z} \subseteq m\mathbb{Z}/n\mathbb{Z}$ if and only if $m \mid d$, the maximal subgroups of $\mathbb{Z}/n\mathbb{Z}$ are the groups $p\mathbb{Z}/n\mathbb{Z}$, where p is a prime divisor of n. It follows that, assuming $n = p_1^{e_1} \ldots p_r^{e_r}$, we have

$$N = \bigcap_{i=1}^{r} p_i\mathbb{Z}/n\mathbb{Z} = p_1 \ldots p_r\mathbb{Z}/n\mathbb{Z}.$$

In particular, N is trivial if and only if $p_1 p_2 \cdots p_r = n$, that is, if and only if n is squarefree.

(iii) The answer above immediately yields that for $n = 100$ we have $N = 10\mathbb{Z}/100\mathbb{Z}$.

147 Let $\pi : G \longrightarrow G/N$ be the projection homomorphism, that is, the map given by $\pi(g) = gN$ for all $g \in G$, and set $F = \pi \circ f : G \longrightarrow G/N$.

The map F, being a composition of surjective homomorphisms, is itself a surjective homomorphism. Moreover, we have $\mathrm{Ker}(F) = \{g \in G \mid F(g) = f(g)N = N\} = N$ because $f(g) \in N$ if and only if $g \in N$.

By the fundamental homomorphism theorem applied to F, the map $\varphi : G/N \longrightarrow G/N$ given by $\varphi(gN) = f(g)N$ is well defined and is, indeed, an isomorphism.

148

(i) The group G is isomorphic to $(\mathbb{Z}/5\mathbb{Z})^* \times (\mathbb{Z}/7\mathbb{Z})^*$. Moreover, since 5 and 7 are prime, we have $G \simeq \mathbb{Z}/4\mathbb{Z} \times \mathbb{Z}/6\mathbb{Z}$ and, since $(2, 3) = 1$, we have $G \simeq H \times K$, with $H = \mathbb{Z}/4\mathbb{Z} \times \mathbb{Z}/2\mathbb{Z}$ and $K = \mathbb{Z}/3\mathbb{Z}$.

The order of any element $(h, k) \in H \times K$ is the least common multiple of the orders of h and k in H and K, respectively.

Every element $h = (h_1, h_2) \in H$ obviously satisfies $4h = (\overline{0}, \overline{0})$. The equation $2h = (\overline{0}, \overline{0})$ has four solutions: $(\overline{0}, \overline{0}), (\overline{0}, \overline{1}), (\overline{2}, \overline{0}), (\overline{2}, \overline{1})$. Finally, the neutral element $(\overline{0}, \overline{0})$ has order 1. Therefore, H has $8 - 4 = 4$ elements of order 4, $4 - 1 = 3$ elements of order 2 and one element of order 1.

The group K has two elements of order 3 and one element of order 1 (the neutral element).

Consequently, G contains $4 \cdot 2 = 8$ elements of order 12, $4 \cdot 1 = 4$ elements of order 4, $3 \cdot 2 = 6$ elements of order 6, $3 \cdot 1 = 3$ elements of order 2, $1 \cdot 2 = 2$ elements of order 3 and a single element of order 1. There are no elements of order n for $n \neq 1, 2, 3, 4, 6, 12$.

(ii) We shall first show that a subgroup C of G of order 6 must be cyclic. Indeed, let us assume by contradiction that C has no elements of order 6: the order of its elements other than the neutral element must be a divisor of 6, and therefore can only be 2 or 3.

Elements of C other than the neutral element cannot all have order 2, otherwise C would be a subgroup of $H \times \{\overline{0}\}$, which cannot be the case because $6 \nmid 8$. Similarly, they cannot all have order 3, or C would be a subgroup $\{\overline{0}\} \times K$, which has order 3.

So C must contain both an element of order 2, which is of the form $(h, \overline{0})$ with $h \in H$, and an element of order 3, of the form $(\overline{0}, k)$ with $k \in K$. But then the element $(h, k) \in C$ has order 6, which contradicts our assumption.

Since any cyclic group of order 6 contains exactly $\phi(6) = 2$ elements of order 6 and since there are six elements of order 6 in G, the group G has $6/2 = 3$ subgroups of order 6.

149

(i) Let $N = \{x \in G \mid f(x) = g(x)\}$ and let e be the neutral element of G. First of all, remark that $f(e) = g(e) = \overline{0}$, so $e \in N$. Moreover, if $x, y \in N$, that is, if $f(x) = g(x)$ and $f(y) = g(y)$, then $f(xy) = f(x) + f(y) = g(x) + g(y) = g(xy)$, hence $xy \in N$. Finally, if $x \in N$, that is, if $f(x) = g(x)$, then $f(x^{-1}) = -f(x) = -g(x) = g(x^{-1})$ and thus $x^{-1} \in N$. We have just shown that N is a subgroup of G.

Now, consider $y \in G$, $x \in N$; the fact that $\mathbb{Z}/12\mathbb{Z}$ is Abelian implies that $f(yxy^{-1}) = f(y) + f(x) - f(y) = f(x) = g(x) = g(y) + g(x) - g(y) = g(yxy^{-1})$, hence $yxy^{-1} \in N$.

⟦The proof above would still work if we replaced $\mathbb{Z}/12\mathbb{Z}$ by any Abelian group.⟧

(ii) Note that H is a normal subgroup of G, because $\langle (123) \rangle$ is normal in S_3, as it has index 2. The condition that $f(h) = \bar{0}$ for all $h \in H$ can be rewritten as $H \subseteq \mathrm{Ker}(f)$. The homomorphisms that satisfy this condition are in bijection with the homomorphisms $\varphi : G/H \longrightarrow \mathbb{Z}/12\mathbb{Z}$.

Now, we have $G/H \simeq \langle \overline{(12)} \rangle \times \langle \bar{1} \rangle \simeq \mathbb{Z}/2\mathbb{Z} \times \mathbb{Z}/2\mathbb{Z}$.

Consequently, the homomorphisms $\varphi : G/H \longrightarrow \mathbb{Z}/12\mathbb{Z}$ correspond bijectively to the homomorphisms $\psi : \mathbb{Z} \times \mathbb{Z} \longrightarrow \mathbb{Z}/12\mathbb{Z}$ whose kernel contains $2\mathbb{Z} \times 2\mathbb{Z}$.

The homomorphisms $\psi : \mathbb{Z} \times \mathbb{Z} \longrightarrow \mathbb{Z}/12\mathbb{Z}$ are exactly those of the form $\psi(m, n) = \overline{ma + nb}$ with $a, b \in \mathbb{Z}/12\mathbb{Z}$. The kernel of such a homomorphism contains $2\mathbb{Z} \times 2\mathbb{Z}$ if and only if $\psi(2, 0) = \psi(0, 2) = \bar{0}$, that is, if and only if $\overline{2a} = \overline{2b} = \bar{0}$, or equivalently $\bar{a} = \overline{6a'}, \bar{b} = \overline{6b'}$ for some $a', b' \in \{0, 1\}$. So there are four homomorphisms satisfying the requirement of the problem statement, and each can be constructed from one of the four possible pairs (a', b').

150 First of all, remark that $G \simeq H \times K = (\mathbb{Z}/p\mathbb{Z})^* \times (\mathbb{Z}/q\mathbb{Z})^*$, that is, G is the direct product of two cyclic groups whose orders are $p - 1$ and $q - 1$, respectively. Thanks to the group structure of the direct product, we have $G^{(2)} \simeq H^{(2)} \times K^{(2)}$ and $G^{(3)} \simeq H^{(3)} \times K^{(3)}$, where $H^{(2)}$, $H^{(3)}$, $K^{(2)}$, $K^{(3)}$ are defined for H and K in the same way as $G^{(2)}$, $G^{(3)}$ are defined for G.

Denote by φ and ψ the homomorphisms from H to itself given by $\varphi(x) = x^2$, $\psi(x) = x^3$: note that we shall then replicate the same argument starting with K. The subgroups $H^{(2)}$, $H^{(3)}$ are the images of these homomorphisms. As for the kernels, they consist of the solutions of $x^2 \equiv 1 \pmod{p}$ and $x^3 \equiv 1 \pmod{p}$, respectively. The solutions of the first equation are $x \equiv \pm 1 \pmod{p}$ and the second equation has either three solutions or a single solution, depending on whether H contains elements of order 3 or not, that is, on whether 3 divides $p - 1$ or not. We should also remark that for $p = 2$ the two solutions of the first equation actually coincide.

From the known relation between the kernel and the image of a homomorphism and the fact that every subgroup of a cyclic group is cyclic, we deduce that $H^{(2)}$ is a cyclic group of order $(p - 1)/2$ if $p > 2$ and of order one if $p = 2$, whereas $H^{(3)}$ is a cyclic group of order $(p - 1)/(3, p - 1)$.

We come to the following conclusions. The order of $G^{(2)}$ is $(p-1)/2 \cdot (q-1)/2$ if $p > 2$ and $(q - 1)/2$ if $p = 2$, and the order of $G^{(3)}$ is $(p - 1)/(3, p - 1) \cdot (q - 1)/(3, q - 1)$. The subgroup $G^{(2)}$, which is a direct product of cyclic groups, is itself cyclic if and only if the orders of the two groups are relatively prime, that is, if and only if $((p - 1)/2, (q - 1)/2) = 1$ if $p > 2$ and for all q if $p = 2$. Similarly, $G^{(3)}$ is cyclic if and only if the orders of its factors are relatively prime. Note, however, that those orders are both even except for the case where $p = 2$, which yields a first factor of order 1. So the group $G^{(3)}$ is cyclic for $p = 2$ and is not cyclic for $p > 2$.

151

(i) Since G is a group, it is clear that $H + K \subseteq G$. As for the opposite inclusion, consider $x \in G$ and let $\overline{x}_H$ be its projection in G/H. We have $m\overline{x}_H = \overline{e}_H$ because G/H has order m. This implies that $mx \in H$ and, similarly, $nx \in K$. Now, let a, b be integers such that $am + bn = 1$; we have $x = amx + bnx \in H + K$, which yields the desired equality.

(ii) Consider the map $f : G \longrightarrow G/H \times G/K$ given by $f(x) = (\overline{x}_H, \overline{x}_K)$, which is clearly a homomorphism. The kernel of f is the set of elements x such that $(\overline{x}_H, \overline{x}_K) = (\overline{e}_H, \overline{e}_K)$, so it is equal to $H \cap K$.

Moreover, f is surjective. This is because, given $(\overline{x}_H, \overline{y}_K) \in G/H \times G/K$ and a, b as above, we have $bn \equiv 1 \pmod{m}$, $bn \equiv 0 \pmod{n}$ and $am \equiv 0 \pmod{m}$, $am \equiv 1 \pmod{n}$, which implies $f(bnx + amy) = f(bnx) + f(amy) = (\overline{x}_H, \overline{e}_K) + (\overline{e}_H, \overline{y}_K) = (\overline{x}_H, \overline{y}_K)$.

The fundamental homomorphism theorem yields the desired isomorphism.

152 A homomorphism $f : G \longrightarrow G$ is induced by a homomorphism $g : \mathbb{Z} \times \mathbb{Z} \longrightarrow G$ such that $20\mathbb{Z} \times 8\mathbb{Z} \subseteq \mathrm{Ker}(g)$. Moreover, given x, y in G, there is a unique homomorphism $g : \mathbb{Z} \times \mathbb{Z} \longrightarrow G$ such that $g(1, 0) = x$ and $g(0, 1) = y$, which is given by $g(a, b) = ax + by$.

The condition on the kernel of g is equivalent to $g(20, 0) = 20x = (\overline{0}, \overline{0})$ and $g(0, 8) = 8y = (\overline{0}, \overline{0})$.

In order for this condition to be satisfied, x must be represented by a pair of integers (x_1, x_2) such that $20x_1 \equiv 0 \pmod{20}$ and $20x_2 \equiv 0 \pmod{8}$. The first equation is satisfied for all integers x_1, whereas the second one is satisfied if and only if $x_2 \equiv 0 \pmod{2}$, that is, for four residue classes modulo 8. The total number of possible values for x is therefore $20 \cdot 4 = 80$.

Similarly, y must be represented by an integer pair (y_1, y_2) such that $8y_1 \equiv 0 \pmod{20}$ and $8y_2 \equiv 0 \pmod{8}$. The first equation is equivalent to $y_1 \equiv 0 \pmod{5}$, so it has four solutions modulo 20, and the second equation is always satisfied. So there are $4 \cdot 8 = 32$ possible values for y.

The number of homomorphisms from G to itself is therefore $80 \cdot 32 = 2560$.

(i) The kernel of f_n consists of the pairs (x_1, x_2) such that $nx_1 \equiv 0 \pmod{20}$ and $nx_2 \equiv 0 \pmod{8}$. The solution of the first equation is $x_1 \equiv 0 \pmod{20/(n, 20)}$ and the solution of the second equation is $x_2 \equiv 0 \pmod{8/(n, 8)}$.

The kernel of f_n is thus the direct product of a cyclic group of order $(n, 20)$ and a cyclic group of order $(n, 8)$. The direct product of two finite cyclic groups is cyclic if and only if the orders of the two factors are relatively prime.

In our case, if n is odd then the order of the second factor is 1, whereas if n is even then both orders are even. Therefore, our group is cyclic if and only if n is odd.

(ii) By the fundamental homomorphism theorem, the image of f_n is isomorphic to $G/\mathrm{Ker}(f_n)$, that is, to the product of two cyclic groups of order $20/(n, 20)$ and $8/(n, 8)$, respectively. Since $(20, 8) = 4$, these orders are relatively prime if and only if 4 divides both $(n, 20)$ and $(n, 8)$, that is, if and only if $4 \mid n$.

153

(i) Let $x \in G$ be such that its projection $\bar{x}$ in G/H is a generator of G/H. In particular, we have $\mathrm{ord}(\bar{x}) = n$, so the fact that $\mathrm{ord}(\bar{x}) \mid \mathrm{ord}(x)$ implies $\mathrm{ord}(x) = nk$ for some positive integer k.

Now, let y be a generator of H and let $z = x^k$. We have $\mathrm{ord}(y) = m$, $\mathrm{ord}(z) = n$ and we wish to show that $\mathrm{ord}(yz) = mn$, that is, that yz is a generator of G. Since $yz \in G$, it is clear that $\mathrm{ord}(yz) \mid mn = |G|$.

Let d be a positive integer such that $(yz)^d = e$. Since G is Abelian, we can write $y^d z^d = e$ as well as $y^d = z^{-d}$.

Notice that y^d belongs to H and z^{-d} belongs to the subgroup generated by z, and that the intersection of these two groups, their orders being relatively prime, consists of the identity only. So the equality above implies $y^d = e$ and $z^d = e^{-1} = e$. Since the orders of y and z are m and n, respectively, we must have $m \mid d$ and $n \mid d$; since $(m, n) = 1$, this implies that $mn \mid d$, hence the fact that G is cyclic.

(ii) A family of examples is given by groups of the form $\mathbb{Z}/m\mathbb{Z} \times \mathbb{Z}/n\mathbb{Z}$ with $(m, n) > 1$. By setting $H = \mathbb{Z}/m\mathbb{Z} \times \{e\}$ we have that $G/H \cong \mathbb{Z}/n\mathbb{Z}$. Indeed, $\mathbb{Z}/n\mathbb{Z}$ is the image of the canonical projection of G onto the second factor, whose kernel is precisely H. So H and G/H are cyclic but G, thanks to the assumption that $(m, n) > 1$, is not.

In particular, we can give the single example $G = \mathbb{Z}/2\mathbb{Z} \times \mathbb{Z}/2\mathbb{Z}$.

154

(i) Since in a group of order n every element's nth power is the neutral element, we have $f_{n-1}(x) = x^{n-1} = x^{-1}$. If f_{n-1} is a homomorphism, then for all $x, y \in G$ we have $(xy)^{-1} = x^{-1}y^{-1}$. Taking the inverse of both sides of the equality, we get $xy = yx$.

(ii) If f_8 is a homomorphism, then $f_8 \circ f_8 = f_{64}$ is as well. Now, the fact that for all $x \in G$ we have $x^{62} = e$ implies that $f_{64}(x) = f_2(x) = x^2$. If f_2 is a homomorphism, then for all $x, y \in G$ we have $(xy)^2 = x^2y^2$, that is, $xyxy = xxyy$. Multiplying by x^{-1} on the left and by y^{-1} on the right yields $yx = xy$.

(iii) Let $G = S_3$ and $k = 2$. In this case we have $f_2((12)(13)) = f_2((132)) = (132)^2 = (123)$, whereas $f((12))f((13)) = (12)^2(13)^2 = e \cdot e = e$.

155 We shall use the additive notation and denote by 0 the neutral element of G.

(i) First of all, we show that G_p is a subgroup of G. The neutral element belongs to G_p because $\mathrm{ord}(0) = 1$.

Given $x, y \in G_p$, suppose $\text{ord}(x) = p^k$, $\text{ord}(y) = p^h$. Setting $m = \max\{k, h\}$, we have $p^m(x + y) = p^m x + p^m y = 0$, so the order of $x + y$ is a divisor of p^m hence a power of p.

Finally, for each integer r we have $rx = 0$ if and only if $r(-x) = 0$. We have thus proven that G_p is a subgroup.

Now, let q be a prime that divides the order of G_p. By Cauchy's theorem, there exists an element $x \in G_p$ of order q, but then the definition of G_p implies that $q = p$, so the order of G_p must be a power of p.

(ii) Let $x + G_p$ be an element of the quotient group G/G_p and assume $\text{ord}(x) = r$. Write r in the form $r = p^k m$, with $(m, p) = 1$. We have $p^k(mx) = 0$, hence $mx \in G_p$. It follows that $m(x + G_p) = mx + G_p = G_p$ is the class of the identity in G/G_p, that is, $\text{ord}(x + G_p)$ is a divisor of m, which implies it must be coprime to p.

(iii) We have $|G_p| = p^{a'}$ and $|G_q| = q^{b'}$ for some $a', b' \in \mathbb{N}$. Since G_p and G_q are subgroups of G, we have $a' \leq a$, $b' \leq b$. Consider the homomorphism $f : G \to G$ given by $f(x) = p^a x$. We show that $\text{Ker}(f) = G_p$: if x belongs to $\text{Ker}(f)$ then its order is a divisor of p^a and, conversely, if $x \in G_p$ then $p^{a'} x = 0$ and thus $p^a x = 0$. By the fundamental homomorphism theorem, the image of f is isomorphic to G/G_p, which, by (ii), only contains elements whose orders are powers of q. Therefore, $\text{Im}(f) \leq G_q$. In particular, $|\text{Im}(f)|$ is a power of q, say $|\text{Im}(f)| = q^{b''}$ for some $b'' \in \mathbb{N}$. We have

$$p^a q^b = |G| = |\text{Ker}(f)| \cdot |\text{Im}(f)| = p^{a'} q^{b''},$$

hence $a = a'$ and $b = b''$. Since $b'' \leq b' \leq b$, we also have $b'' = b' = b$ and thus $G/G_p \simeq \text{Im}(f) = G_q$.

156 Let $q = a/b$ be a rational number with $(a, b) = 1$ and $b = 2^r 5^s h$ for some $r, s \geq 0$ and some $h > 0$ such that $(h, 10) = 1$. From the definition of L we have that $mq \in L$ if and only if $h \mid m$, since the factor m must simplify with every prime factor other than 2 and 5 appearing in the denominator of q. In particular, $x = q + L$ has order k in $\mathbb{Q}/L$ if and only if $h = k$; moreover, x is a solution of $kx = 0$ if and only if $h \mid k$.

We shall first endeavour to answer (i) and (ii) in the case where $(k, 10) = 1$.

Note that $1/k + L$ has order k in $\mathbb{Q}/L$ by our previous remarks. Moreover, all elements of the cyclic group G generated by $1/k + L$ in $\mathbb{Q}/L$ satisfy the equation $kx = 0$.

Conversely, let x be a solution; we wish to show that $x \in G$. By our previous remarks, we have $x = q + L$, where $q = a/(2^r 5^s h)$ and $h \mid k$.

Consider the congruence $ht + a \equiv 0 \pmod{2^r 5^s}$ in the indeterminate t. Since $(h, 10) = 1$, there exists a solution $t_0 \in \mathbb{Z}$. Suppose $ht_0 + a = c 2^r 5^s$ for some $c \in \mathbb{Z}$. We have

$$q - \frac{c}{h} = \frac{a}{2^r 5^s h} - \frac{ht_0 + a}{2^r 5^s h} = \frac{t_0}{2^r 5^s} \in L$$

and so $x = q + L = c/h + L$ belongs to the subgroup of G generated by $1/h + L$. We have thus proven that G is the set of solutions of $kx = 0$ in $\mathbb{Q}/L$, so there are k such solutions. Moreover, the number of elements of order k is $\phi(k)$, because G is a cyclic group.

Finally, let us consider the case where $(k, 10) > 1$. We can write $k = 2^r 5^s k_1$ with $(k_1, 10) = 1$. Our initial discussion about the order of elements in $\mathbb{Q}/L$ implies that every element has order coprime to 10. In particular, there are no elements of order k and the equation $kx = 0$ has the same solutions as $k_1 x = 0$; it therefore has k_1 solutions.

157

(i) Yes, there are subgroups of order 3: one is given by the rotations by a multiple of $2\pi/3$ around some fixed axis passing through opposite vertices of the cube.

(ii) There are two types of transformations in G: those that preserve the figure and those that rotate all the line segments within the faces by $\pi/2$. It is easy to see that, if one of the line segments is rotated, those of adjacent faces must be rotated as well, because the endpoints of line segments within adjacent faces never coincide.

The subgroup H is that of transformations of the first type. Note that transformations of the second type do exists: for instance, the rotation by $\pi/2$ around an axis joining the centres of two opposite faces. Moreover, transformations of the second type form a coset of H. Indeed, given two transformations of the second type α and β, the transformation $\alpha \circ \beta^{-1}$ does send the figure to itself, so $\alpha \circ \beta^{-1} \in H$.

Therefore, the index of H in G is 2.

(iii) We know that H is a subgroup of index 2 in G, so H is a normal subgroup of G.

158

(i) Every element of G generates a cyclic subgroup. Moreover, a cyclic subgroup of order d can be generated by $\phi(d)$ distinct elements. It follows that the number of cyclic subgroups of a finite group G is given by the expression

$$\sum_{d \mid |G|} \frac{1}{\phi(d)} \cdot (\text{number of elements of order} d).$$

In our case (that is, $G = \mathbb{Z}/6\mathbb{Z} \times \mathbb{Z}/6\mathbb{Z}$) every $x \in G$ satisfies $6x = 0$, so the order of an element can only be 1, 2, 3, or 6. The only element of order 1 is the identity. Any element $x = (a, b)$ of order 2 must satisfy $(2a, 2b) = (0, 0)$, whose solutions are $a = \bar{0}, \bar{3}, b = \bar{0}, \bar{3}$. If we exclude the neutral element, which is the only solution that does not have order 2, we find that there are $4 - 1 = 3$ elements of order 2.

Elements of order 3 are counted analogously: there are $9 - 1 = 8$ in total. All other $36 - 1 - 3 - 8 = 24$ elements have order 6. The expression above becomes

$$\frac{1}{\phi(1)} + \frac{3}{\phi(2)} + \frac{8}{\phi(8)} + \frac{24}{\phi(6)} = 1 + 3 + 2 + 12 = 18.$$

(ii) Clearly, if $\mathrm{ord}(x) = 1, 2, 3, 6$ then $|G/\langle x \rangle|$ is equal to 36, 18, 12, 6, respectively. Suppose that $G/\langle x \rangle$ is cyclic.

Remember that, if we let $\pi : G \longrightarrow G/\langle x \rangle$ be the canonical projection, we have $\mathrm{ord}(\pi(y)) \mid \mathrm{ord}(y)$ for all $y \in G$ (simply because π is a homomorphism). Since G has no elements of order 36, 18, 12, the only possibility is $\mathrm{ord}(x) = |G/\langle x \rangle| = 6$. If this is the case, then $G/\langle x \rangle$ is necessarily cyclic, since it contains an element a of order 2 and an element b of order 3, and their sum $a + b$ has order 6. The answer is therefore given by the number of elements of G whose order is 6, which, as seen above, is 24.

159

(i) Partition $\mathbb{Z}/60\mathbb{Z}$ into the twenty disjoint sets $A_0, A_1, \ldots, A_{19}$, where $A_h = \{\overline{h}, \overline{h+20}, \overline{h+40}\}$ for $h = 1, 2, \ldots, 19$.

The set G consists of all permutations of $\mathbb{Z}/60\mathbb{Z}$ that send A_h to itself for $h = 1, 2, \ldots, 19$. From this description of the set G, it is clear that G is a group isomorphic to the direct product $S(A_0) \times S(A_1) \times \cdots \times S(A_{19}) \simeq S_3^{\times 20}$, whose factors are the groups of permutations of $A_0, A_1, \ldots, A_{19}$. In particular, G has 6^{20} elements.

(ii) It is clear that G has no subgroups of order 10, because 10 is not a divisor of 6^{20}.

Remark that the order of an element in a direct product is the least common multiple of the orders of its components. Since the elements of S_3 have order 1, 2, or 3, G does have cyclic subgroups of order 6 but contains no cyclic subgroups of order 8 or 12.

There are, however, (non-cyclic) subgroups of order 8: let H be a subgroup of order 2 in S_3; the subgroup $H \times H \times H \times \mathrm{Id} \times \mathrm{Id} \times \cdots \times \mathrm{Id}$ has order 8 in $S_3^{\times 20}$. Similarly, given a subgroup K of order 3 in S_3, the subgroup $K \times H \times H \times \mathrm{Id} \times \cdots \times \mathrm{Id}$ of $S_3^{\times 20}$ has order 12.

160 Let $\varphi : G \longrightarrow 3G$ be the map given by $\varphi(x) = 3x$ for all $x \in G$. Remark that φ is a homomorphism: since G is Abelian, we have $\varphi(x + y) = 3(x + y) = 3x + 3y = \varphi(x) + \varphi(y)$.

Let H be a subgroup of G such that $(|H|, 3) = 1$; consider the map $\psi = \varphi_{|H}$, that is, the restriction to H of the homomorphism φ. The kernel of ψ is the set of elements $h \in H$ such that $3h = 0$, that is, the set of elements of H whose order is a divisor of 3. Since 3 does not divide the order of H, the group H does not contain any elements of order 3, so the kernel of ψ contains the identity only. It follows that ψ is injective, hence H is isomorphic to its image $\psi(H)$. But $\psi(H)$ is a subgroup of a cyclic group, so it is cyclic, and thus H itself must be cyclic.

161 A homomorphism $f : \mathbb{Z}/n\mathbb{Z} \longrightarrow \mathbb{Z}/10\mathbb{Z} \times \mathbb{Z}/20\mathbb{Z}$ is determined by the choice of $f(1) = (a, b) \in \mathbb{Z}/10\mathbb{Z} \times \mathbb{Z}/20\mathbb{Z}$, provided the condition $\mathrm{ord}(a, b) \mid n = \mathrm{ord}(1)$ is satisfied. What we need to find is therefore the number of elements $(a, b) \in \mathbb{Z}/10\mathbb{Z} \times \mathbb{Z}/20\mathbb{Z}$ such that $n(a, b) = 0$.

For $n = 0$, that is, when counting homomorphisms from $\mathbb{Z}$ to $\mathbb{Z}/10\mathbb{Z} \times \mathbb{Z}/20\mathbb{Z}$, all elements (a, b) satisfy the requirement, so there are 200 homomorphisms. None of them is injective because $\mathbb{Z}$ is infinite whereas the codomain is finite.

Assume $n \geq 1$. We need to count the elements $(a, b) \in \mathbb{Z}/10\mathbb{Z} \times \mathbb{Z}/20\mathbb{Z}$ such that $na = \overline{0}$ and $nb = \overline{0}$, that is, the solutions of the congruences $na \equiv 0 \pmod{10}$ and $nb \equiv 0 \pmod{20}$. We have

$$na \equiv 0 \pmod{10} \iff a \equiv 0 \left(\mathrm{mod}\ \frac{10}{(n, 10)}\right)$$

and the latter equation has exactly $(n, 10)$ solutions in $\mathbb{Z}/10\mathbb{Z}$. Similarly, the equation $nb \equiv 0 \pmod{20}$ has exactly $(n, 20)$ solutions in $\mathbb{Z}/20\mathbb{Z}$. This yields a total of $(n, 10)(n, 20)$ pairs (a, b) satisfying the requirement, so we have $(n, 10)(n, 20)$ homomorphisms in total.

In order for the homomorphism given by $f(1) = (a, b) \in \mathbb{Z}/10\mathbb{Z} \times \mathbb{Z}/20\mathbb{Z}$ to be injective, we must have $\mathrm{ord}(a, b) = \mathrm{ord}(1) = n$: there are as many injective homomorphisms as there are elements of order n in $\mathbb{Z}/10\mathbb{Z} \times \mathbb{Z}/20\mathbb{Z}$. A necessary condition for the existence of injective homomorphism is therefore that $n \mid 20$: we must have $n = 1, 2, 4, 5, 10, 20$.

Let us assume n takes one of those values and find the number d_n of elements $(a, b) \in \mathbb{Z}/10\mathbb{Z} \times \mathbb{Z}/20\mathbb{Z}$ of order n. Remember that $\mathrm{ord}(a, b)$ is the least common multiple of $\mathrm{ord}(a)$ and $\mathrm{ord}(b)$ and that, if $d \mid m$, then there are $\phi(d)$ elements of order d in $\mathbb{Z}/m\mathbb{Z}$.

The only element of order 1 in any group is the neutral element, so $d_1 = 1$.

All elements (a, b) of order 2 satisfy $2(a, b) = 0$. We have two possible choices for a and two for b, so four solutions in total; the neutral element is the only solution we need to exclude, so $d_2 = 3$.

Similarly, we exclude the neutral element from the 25 solutions of the equation $5(a, b) = 0$ and obtain $d_5 = 24$ elements of order 5.

In order to compute d_{10}, we find the number of solutions of $10(a, b) = 0$ and subtract the number of elements of order 1, 2 and 5: we have $d_{10} = 100 - d_1 - d_2 - d_5 = 100 - 1 - 3 - 24 = 72$.

An element (a, b) has order 4 if and only if b has order 4 and $\mathrm{ord}(a) \mid (4, 10) = 2$. There are two possible choices for the element b, because there are $\phi(4) = 2$ elements of order 4 in $\mathbb{Z}/20\mathbb{Z}$, and there are two possible choices for a, since there are two elements of order 1 or 2 in $\mathbb{Z}/10\mathbb{Z}$. We thus have $d_4 = 4$.

Finally, we can obtain d_{20} as the difference $d_{20} = 10 \cdot 20 - d_1 - d_2 - d_4 - d_5 - d_{10} = 200 - 1 - 3 - 4 - 24 - 72 = 96$.

⟦We can also compute d_{20} by remarking that (a, b) has order 20 if and only if b has order 20, or b has order 4 and a has order 5 or 10. There are $\phi(20) = 8$ possibilities for b and 10 for a in the first case, whereas in the second case we have two possible values for b and $8 = \phi(5) + \phi(10)$ for a, so 16 possibilities. We find $d_{20} = 80 + 16 = 96$.⟧

162

(i) Let $g = (g_1, g_2, g_3)$ be an element of G with $g_1 \in \mathbb{Z}/5\mathbb{Z}$, $g_2 \in \mathbb{Z}/10\mathbb{Z}$ and $g_3 \in \mathbb{Z}/36\mathbb{Z}$. The element g is in the kernel of f if and only if $f(g) = 78(g_1, g_2, g_3) = (3g_1, -2g_2, 6g_3) = (0, 0, 0)$; in other words, $f(g) = (0, 0, 0)$ if and only if $g_1 \equiv 0 \pmod 5$, $g_2 \equiv 0 \pmod 5$ and $g_3 \equiv 0 \pmod 6$. We therefore conclude that $\mathrm{Ker}(f)$ has $1 \cdot 2 \cdot 6 = 12$ elements. This implies that $\mathrm{Im}(f) \simeq G/\mathrm{Ker}(f)$ has $|G|/|\mathrm{Ker}(f)| = 5 \cdot 10 \cdot 36/12 = 150$ elements.

(ii) Given $g \in G$, we have $\mathrm{ord}(f(g)) \mid (|\mathrm{Im}(f)|, \mathrm{ord}(g)) = (150, \mathrm{ord}(g))$. Moreover, if $g = (g_1, g_2, g_3) \in G$ then we have

$$\mathrm{ord}(g) = [\mathrm{ord}(g_1), \mathrm{ord}(g_2), \mathrm{ord}(g_3)] \mid [5, 10, 36] = 180.$$

This implies that $\mathrm{ord}(f(g)) \mid (180, 150) = 30$ for all $g \in G$.
On the other hand, for $g = (1, 1, 1)$ we have $f(g) = (3, -2, 6)$ and therefore $\mathrm{ord}(f(g)) = [5, 5, 6] = 30$, so 30 is the largest possible order.

163

(i) Remark that $0_G = 2 \cdot 0_G \in Q$. Moreover, if $x_1, x_2 \in Q$ then $x_1 = 2g_1$ and $x_2 = 2g_2$ for some $g_1, g_2 \in G$, so $x_1 + x_2 = 2g_1 + 2g_2 = 2(g_1 + g_2) \in Q$; finally, $-x_1 = -2g_1 = 2(-g_1) \in Q$. This shows that Q is a subgroup of G.
⟦Another way to show that Q is a subgroup is to remark that, since G is Abelian, the map $\varphi : G \longrightarrow G$ given by $\varphi(g) = 2g$ is a homomorphism and that $Q = \mathrm{Im}(\varphi)$.⟧

(ii) The map $G \ni g \overset{\varphi}{\longmapsto} 2g \in G$ is a homomorphism because G is Abelian; moreover, we have $Q = \mathrm{Im}(\varphi)$ and $|G/Q| = |\mathrm{Ker}(\varphi)|$. Now, $\mathrm{Ker}(\varphi) = \{g \in G \mid 2g = 0_G\}$, so all elements in $\mathrm{Ker}(\varphi)$ other than 0_G have order 2. By Cauchy's theorem, if $|G/Q| = |\mathrm{Ker}(\varphi)|$ is finite then it is a power of 2. This shows that we cannot have $m = 3$.
Let us now give examples of groups G for which $|G/Q|$ is 1, 2 and 4. Set $G = \mathbb{Z}/3\mathbb{Z}$; since $\mathbb{Z}/3\mathbb{Z}$ has no elements of order 2, we have $\mathrm{Ker}(\varphi) = \{\bar{0}\}$, so $|G/Q| = 1$. For $G = \mathbb{Z}/2\mathbb{Z}$ we have $\mathrm{Im}(\varphi) = \bar{0}$, so $m = 2$. Finally, consider $G = \mathbb{Z}/2\mathbb{Z} \times \mathbb{Z}/2\mathbb{Z}$; all elements of G have order 1 or 2, so φ is the zero homomorphism and $|G/Q| = |G| = 4$.

164

(i) Given $(a, b) \in G$, we have that $\mathrm{ord}(a, b) = [\mathrm{ord}(a), \mathrm{ord}(b)]$ is 11 if and only if $\mathrm{ord}(a) \mid 11$, $\mathrm{ord}(b) \mid 11$ and $(a, b) \neq (0, 0)$. Since for each divisor d of m there are d elements in $\mathbb{Z}/m\mathbb{Z}$ whose order divides d, the number of pairs (a, b) satisfying the requirements is $11 \cdot 11 - 1 = 120$.
Since 11 is prime, all subgroups of order 11 are cyclic; their number is given by the number of elements of order 11, divided by the number of generators in a cyclic group with 11 elements, which is $\phi(11)$. So we have 12 subgroups of order 11.

(ii) For every subgroup H of order 11 we have $|G/H| = |G|/|H| = 3^3 \cdot 11 = 297$; if G/H were cyclic, it would contain an element of order 297. But this cannot be the case, because every element of G has an order that divides 99 and for all $x \in G$ we have $\text{ord}(x + H) \mid \text{ord}(x)$.

(iii) There are no surjective homomorphisms from G to $\mathbb{Z}/121\mathbb{Z}$. If f were one, there would be an element $x \in G$ such that $\text{ord}(f(x)) = 121$; but this is not possible because $\text{ord}(f(x)) \mid \text{ord}(x)$ and for every element x of G we have $\text{ord}(x) \mid 99$.

165 Let $\pi : G \ni x \longmapsto xH \in G/H$ be the projection onto the quotient. Since π is a homomorphism we have $\text{ord}(xH) \mid \text{ord}(x)$. Assume that G/H contains an element xH of order m; we have $\text{ord}(x) = mk$ and thus $\text{ord}(x^k) = m$.

Conversely, let $x \in G$ be an element of order m; we have $\text{ord}(xH) = d \mid m$, hence $(xH)^d = H$ and thus $x^d \in H$. By Lagrange's theorem, we have $x^{dn} = e$, so $m = \text{ord}(x) \mid nd$. The fact that $(n, m) = 1$ implies that $m \mid d$ hence, finally, $d = m$.

166

(i) All elements $x = (a, b)$ of a subgroup of order 4 must have an order that divides 4, so they must satisfy $4x = (4a, 4b) = (\bar{0}, \bar{0})$. The solutions of these congruences are given by $a \equiv 0 \pmod 2$, that is, $a \in \{\bar{0}, \bar{2}, \bar{4}, \bar{6}\}$, and $b \equiv 0 \pmod 3$, that is, $b \in \{\bar{0}, \bar{3}, \bar{6}, \bar{9}\}$.

Among the $4 \cdot 4 = 16$ elements above, 4 also satisfy the equation $(2a, 2b) = (\bar{0}, \bar{0})$—that is, the elements $(\bar{0}, \bar{0}), (\bar{0}, \bar{6}), (\bar{4}, \bar{0}), (\bar{4}, \bar{6})$—whereas the other 12 elements have order exactly 4. Each of these 12 elements generates a cyclic subgroup of order 4, and each cyclic subgroup of order 4 contains two elements of order 4, so the total number of cyclic subgroups of order 4 is $12/2 = 6$. We can list them explicitly by choosing one generator from each, for example: $(\bar{0}, \bar{3}), (\bar{2}, \bar{0}), (\bar{2}, \bar{3}), (\bar{2}, \bar{6}), (\bar{2}, \bar{9}), (\bar{4}, \bar{3})$.

The set of solutions of $(2a, 2b) = (\bar{0}, \bar{0})$ is also a subgroup of order 4, though it is not cyclic. Indeed, $(\bar{0}, \bar{0})$ belongs to the set; moreover, if (a, b) and (a', b') are solutions, then $(a + b, a' + b')$ is a solution, because $2(a + b, a' + b') = 2(a, b) + 2(a', b') = (\bar{0}, \bar{0})$; finally, if (a, b) is a solution then $-(a, b)$ is a solution, because $2(-(a, b)) = -2(a, b) = (\bar{0}, \bar{0})$.

(ii) Let H be a subgroup of G of order 48. Since G is Abelian, H is necessarily a normal subgroup of G, so the quotient G/H has a group structure, and it is clearly isomorphic to $\mathbb{Z}/2\mathbb{Z}$. Since we have $2(gH) = (2g)H = H$ for all $g \in G$, we have that $2G \subseteq H$.

The subgroups that contain $2G$ are in bijection with the subgroups of $G/2G = (\mathbb{Z}/8\mathbb{Z} \times \mathbb{Z}/12\mathbb{Z})/(2\mathbb{Z}/8\mathbb{Z} \times 2\mathbb{Z}/12\mathbb{Z}) \simeq \mathbb{Z}/2\mathbb{Z} \times \mathbb{Z}/2\mathbb{Z}$. Now, $2G$ has order $8/2 \times 12/2 = 24$, so any subgroup H of G whose order is 48 corresponds to a subgroup of $G/2G$ whose order is 2, which must consist of the identity and one of the three elements of order 2 in $G/2G$. It follows that G has exactly three subgroups of order 48, each given by the union of $2G$ with a coset $x + 2G$ of order 2 in $G/2G$. We may choose as representatives of these three cosets in $G/2G$ the elements $x = (\bar{1}, \bar{0}), (\bar{0}, \bar{1}), (\bar{1}, \bar{1})$.

167 The answer to both questions is yes: let us see why.

(i) Let $\pi : (\mathbb{Z}/p^2\mathbb{Z})^* \to (\mathbb{Z}/p\mathbb{Z})^*$ be the map given by $[x]_{p^2} \longmapsto [x]_p$, which is well defined because if $x \equiv y \pmod{p^2}$ then *a fortiori* $x \equiv y \pmod{p}$. The map π is a homomorphism because $[xy]_{p^2} = [x]_{p^2}[y]_{p^2}$ and thus $[xy]_p = [x]_p[y]_p$. Finally, π is surjective, because if $(x, p) = 1$ then $(x, p^2) = 1$.

(ii) The group G_1 is cyclic of order $p - 1$, because p is prime. Let x be a generator of G_1, let π be the map defined above and let $y \in G_2$ be such that $\pi(y) = x$. We have that $p - 1 = \operatorname{ord}(x) \mid \operatorname{ord}(y)$. Assume that $\operatorname{ord}(y) = k(p - 1)$; this implies $\operatorname{ord}(y^k) = p - 1$ and so the map $G_1 \ni x^h \mapsto y^{kh} \in G_2$ is an isomorphism between the two cyclic groups $G_1 = \langle x \rangle$ and $\langle y^k \rangle \subseteq G_2$. In particular, it is an injective homomorphism from G_1 to G_2.

168

(i) Let us check that H is a subgroup of G: $0 \in H$ because $p^a 0 = 0$; if $x, y \in H$, then $p^a(x + y) = p^a x + p^a y$ because G is Abelian, so $p^a(x + y) = 0 + 0 = 0$; finally, if $x \in H$ then $p^a(-x) = -p^a x = 0$, so $-x \in H$.

(ii) Suppose by contradiction that $x + H \in G/H$ is an element of order p. In particular, we have $p(x + H) = px + H = H$, that is, $px \in H$. By definition of H, this implies $p^a(px) = 0$, so the order of x is a divisor of p^{a+1}. On the other hand, the order of x must be a divisor of the order G, so it must also divide $(p^{a+1}, |G|) = p^a$. It follows that $p^a x = 0$, that is, $x \in H$ and the class $x + H$ is the class H which has order 1, but this contradicts our assumption.

(iii) The order of every element of H is a divisor of p^a. Therefore, if the order of $x \in H$ is prime, then $\operatorname{ord}(x) = p$. By Cauchy's theorem, the order of H must be a power of p: if this were not the case, there would be a different prime q that divides $|H|$ and thus there would be an element of order q in H. So, let $|H| = p^b$ for some natural number b; we then have that $p^b \mid |G|$ because, by Lagrange's theorem, the order of a subgroup divides the order of the group. Therefore, $b \le a$ and, if we had $b < a$, the order of the group G/H would be a multiple of p; again by Cauchy's theorem, this would imply the existence of an element of order p in that group, which contradicts what we showed above.

169

(i) Given two homomorphisms f, g from G to $\mathbb{C}^*$, the map fg is a homomorphism because $(fg)(x + y) = f(x + y)g(x + y) = f(x)f(y)g(x)g(y) = f(x)g(x)f(y)g(y) = (fg)(x) \cdot (fg)(y)$. We can therefore define the operation $(f, g) \longmapsto fg$ in $\operatorname{Hom}(G, \mathbb{C}^*)$.

This operation is associative because the multiplication of $\mathbb{C}^*$ is. The set $\operatorname{Hom}(G, \mathbb{C}^*)$ has a neutral element, that is, the map $e : G \ni x \longmapsto 1 \in \mathbb{C}^*$. Finally, every homomorphism $f \in \operatorname{Hom}(G, \mathbb{C}^*)$ has an inverse: letting $f^{-1} : G \to \mathbb{C}^*$ be the map given by $f^{-1}(x) = f(x)^{-1}$ for all $x \in G$, we have $f^{-1}(xy) = (f(xy))^{-1} = f(x)^{-1}f(y)^{-1} = f^{-1}(x)f^{-1}(y)$, so $f^{-1} \in \operatorname{Hom}(G, \mathbb{C}^*)$. Moreover, $(ff^{-1})(x) = (f^{-1}f)(x) = f(x)^{-1}f(x) = 1 = e(x)$ for all $x \in G$, so $ff^{-1} = f^{-1}f = e$.

(ii) If φ is an injective homomorphism in $\mathrm{Hom}(G, \mathbb{C}^*)$ then $\varphi(G) \simeq G$. But $\varphi(G)$ is a multiplicative subgroup of a field, so it is cyclic. It follows that G must be cyclic, that is, we must have $(m, n) = 1$.

On the other hand, if $(m, n) = 1$ then $G = \mathbb{Z}/m\mathbb{Z} \times \mathbb{Z}/n\mathbb{Z}$ is cyclic of order mn. Let g be a generator of G and let $\zeta = \zeta_{mn}$ be an mnth primitive root of unity in $\mathbb{C}^*$. Consider the map $f : G \to \mathbb{C}^*$ given by $f(g^k) = \zeta^k$.

This map is well defined because, if $k \equiv h \pmod{mn}$, then $\zeta^k = \zeta^h$. It is a homomorphism because $f(g^k g^h) = f(g^{k+h}) = \zeta^{k+h} = \zeta^k \zeta^h = f(g^k)f(g^h)$. Moreover, it is injective because $\zeta^k = 1$ implies $k \equiv 0 \pmod{mn}$ and thus $g^k = id$.

170

(i) It is clear that pG and qG are both contained in G, so $pG + qG \subseteq G$. Conversely, let $x \in G$ and let a, b be integers such that $pa + qb = 1$. We have $x = pax + qbx \in pG + qG$, so $G \subseteq pG + qG$.

(ii) If $pq \nmid n$ then at least one among p, q does not divide n; by symmetry, let us assume that $p \nmid n$. The map $f : G \to G$ given by $f(x) = px$ is an injective homomorphism, because there are no elements of order p in G. Since G is a finite group, f must also be surjective: its image pG is equal to G. This of course implies that $G = pG + qG$.

Now assume that $pq \mid n$, that is, $p \mid n$ and $q \mid n$. By Cauchy's theorem, G has an element of order p and an element of order q, so the kernels of the homomorphisms $x \longmapsto px$ and $x \longmapsto qx$ are both nontrivial. It follows that the images pG and qG of these homomorphisms are both different from G. But the union of two subgroups of a given group is itself a subgroup if and only if one of the two subgroups contains the other. In this case, since both subgroups are proper, even if their union were indeed a subgroup, it would not be all of G.

(iii) As above, if $pqr \nmid n$ then at least one of the primes, say p, does not divide n, so $G = pG$. On the other hand, if $pqr \mid n$ then G has an element x of order p, so the homomorphism $x \longmapsto px$ has a kernel of order at least p and its image pG has at most n/p elements. Similarly, $|qG| \le n/q$ and $|rG| \le n/r$, so

$$|pG \cup qG \cup rG| \le \left(\frac{1}{p} + \frac{1}{q} + \frac{1}{r}\right) n \le \left(\frac{1}{3} + \frac{1}{5} + \frac{1}{7}\right) n < n = |G|.$$

It follows that we cannot have $pG \cup qG \cup rG = G$.

171 We shall use the additive notation. If the order of $x \in G$ is a prime p, then this prime must be a divisor of 200, that is, $p = 2, 5$. By Cauchy's theorem, G contains both elements of order 2 and elements of order 5. An element x of order 2 generates a subgroup of order 2 where x itself is the only element of order 2; an element y of order 5 generates a subgroup of order 5 that contains four elements of order 5, namely, $y, 2y, 3y, 4y$. Consequently, every group G of order 200 contains at least $1 + 4 = 5$ elements whose order is prime. This minimum is actually achieved by the cyclic group $\mathbb{Z}/200\mathbb{Z}$, which contains $\phi(d)$ elements of order d for each divisor d of 200. The desired minimum is therefore 5.

Any element of order 2 will belong to the subgroup $G_2 = \{x \in G \mid 2x = 0\}$ whose order is, by Cauchy's theorem, a power of 2. Moreover, since the order of G_2 must divide the order of G, we have that $|G_2|$ must divide 8. Any group of order m has at most $m - 1$ elements of order 2—all elements except for the identity—so G can have at most seven elements of order 2.

Similarly, any element of order 5 must belong to the subgroup $G_5 = \{x \in G \mid 5x = 0\}$, which has at most 25 elements and therefore at most 24 elements of order 5.

It follows that the maximum possible number of elements whose order is prime in a group G of order 200 cannot be more than $7 + 24 = 31$. If we have $G \cong \mathbb{Z}/2\mathbb{Z} \times \mathbb{Z}/2\mathbb{Z} \times \mathbb{Z}/2\mathbb{Z} \times \mathbb{Z}/5\mathbb{Z} \times \mathbb{Z}/5\mathbb{Z}$ then we do have equality, so 31 is the desired maximum.

172 By Cauchy's theorem, G contains both elements of order p and elements of order q. Let x be an element of order p and let y be an element of order q. The element xy has order pq: indeed, we have $(xy)^{pq} = x^{pq} y^{pq} = e$ and for proper divisors of pq we clearly have $(xy)^1 = xy \neq 1$, $(xy)^p = y^p \neq 1$, $(xy)^q = x^q \neq 1$. On the other hand, any subgroup of order pq must contain both an element of order p and an element of order q; it follows that it must be cyclic and is generated by two such elements.

Set $H = \langle x \rangle$ and $K = \langle y \rangle$, so that $HK = \langle xy \rangle$. The map $(H, K) \longmapsto HK$ is a bijection between the set of pairs (H, K) of subgroups of orders p and q, respectively, and the set of subgroups of order pq. The fact that this map is surjective follows from the fact that every subgroup of order pq contains a cyclic subgroup of order p and a cyclic subgroup of order q. Moreover, the map is injective, because each subgroup of order pq contains *exactly one* subgroup of order p and *exactly one* subgroup of order q. Therefore, $h_{pq} = h_p h_q$.

As for the second equality, it is enough to remark that $m_{pq} = \phi(pq)h_{pq}$, $m_p = \phi(p)h_p$, $m_q = \phi(q)h_q$. Substituting and using the identity $\phi(pq) = \phi(p)\phi(q)$ yields the desired formula.

173

(i) By assumption, the groups G/H and G/K have order p, so they are isomorphic to $\mathbb{Z}/p\mathbb{Z}$. In order to show the desired isomorphism, it is enough to show that $G \simeq G/H \times G/K$.

Let $\varphi : G \longrightarrow G/H \times G/K$ be the map given by $g \longmapsto (gH, gK)$, which is a homomorphism because

$$\begin{aligned}
\varphi(xy) &= (xyH, xyK) \\
&= (xHyH, xKyK) \\
&= (xH, xK)(yH, yK) \\
&= \varphi(x)\varphi(y).
\end{aligned}$$

Since $\mathrm{Ker}(\varphi) = \{g \in G \mid (gH, gK) = (H, K)\} = H \cap K = \{e\}$, the homomorphism φ is injective.

In order to show that φ is surjective, it is enough to show that $|G| = p^2$. Since G embeds into $G/H \times G/K$ which has cardinality p^2, all we have to do is show that we cannot have $|G| = p$, which is true because G has two distinct subgroups of index p.

(ii) Since p is prime, all subgroups of order p are cyclic, so their number is equal to the number of elements of order p in G, divided by $\phi(p)$.

Moreover, since isomorphisms preserve the order of elements, we may as well count elements of order p in $\mathbb{Z}/p\mathbb{Z} \times \mathbb{Z}/p\mathbb{Z}$. Clearly, all elements of $\mathbb{Z}/p\mathbb{Z} \times \mathbb{Z}/p\mathbb{Z}$ other than the identity have order p, so their number is $p^2 - 1$.

We can therefore conclude that G contains $(p^2 - 1)/\phi(p) = p + 1$ subgroups of order p.

⟦Since $\mathbb{Z}/p\mathbb{Z} \times \mathbb{Z}/p\mathbb{Z} \simeq \mathbb{F}_p^2$ as Abelian groups, the number of subgroups of order p in $\mathbb{F}_p^2$, and therefore in G, is the number of vector subspaces of dimension 1, that is, the number of lines through the origin in $\mathbb{F}_p^2$. These are clearly $p + 1$.⟧

174

(i) We first show that G^k is a subgroup of G. We have $e = e^k \in G^k$; for all $a^k, b^k \in G^k$ we have $a^k b^k = (ab)^k$ because G is Abelian, so $a^k b^k = (ab)^k \in G^k$; finally, $a^k \in G^k$ implies $(a^k)^{-1} = (a^{-1})^k \in G^k$. Moreover the fact that G is Abelian implies that G^k is normal in G.

Consider the quotient G/G^k and let xG^k be one of its elements. We have $(xG^k)^k = x^k G^k = G^k$, so the order of any element in G/G^k is a divisor of k and therefore finite.

(ii) Assuming $G \cong \mathbb{Z}/n\mathbb{Z}$, we have that $G^k \simeq \langle [k]_n \rangle$ is a cyclic group of order $\mathrm{ord}([k]_n) = n/(n, k)$. It follows that $|G/G^k| = |G|/|G^k| = (n, k)$.

⟦We also know that G/G^k must be cyclic because it is a quotient of a cyclic group, so $G/G^k \simeq \mathbb{Z}/(n, k)\mathbb{Z}$.⟧

(iii) Consider the group $G = \mathbb{Z}/2\mathbb{Z} \times \mathbb{Z}/10\mathbb{Z}$; we have $G^{10} = \{(0, 0)\}$ because for all $(a, b) \in G$ we have $10(a, b) = (10a, 10b) = (0, 0)$. In this case we clearly have $G/G^{10} = G = \mathbb{Z}/2\mathbb{Z} \times \mathbb{Z}/10\mathbb{Z}$.

⟦More generally, if $G = \mathbb{Z}/m\mathbb{Z} \times \mathbb{Z}/n\mathbb{Z}$ then we have $G^k = \{(ka, kb) \mid (a, b) \in G\} = \mathbb{Z}/(m/(m, k))\mathbb{Z} \times \mathbb{Z}/(n/(n, k))\mathbb{Z}$ and $G/G^k \simeq \mathbb{Z}/(m, k)\mathbb{Z} \times \mathbb{Z}/(n, k)\mathbb{Z}$. It follows that, for $G = \mathbb{Z}/m\mathbb{Z} \times \mathbb{Z}/n\mathbb{Z}$, we have $G/G^{10} \cong \mathbb{Z}/2\mathbb{Z} \times \mathbb{Z}/10\mathbb{Z}$ if and only if $(m, 10) = 2$ and $(n, 10) = 10$.⟧

175

(i) Since $\mathbb{Z}/m\mathbb{Z}$ is cyclic and generated by $\bar{1}$, a homomorphism $\varphi : \mathbb{Z}/m\mathbb{Z} \longrightarrow \mathbb{Z}/n\mathbb{Z}$ is completely determined once one sets $\varphi(\bar{1}) = \bar{a}$ for some $\bar{a}$ such that $\mathrm{ord}(\bar{a}) \mid m$; the homomorphism φ is then given by $\bar{k} \longmapsto \varphi(\bar{k}) = k\bar{a}$. Since $\bar{a} \in \mathbb{Z}/n\mathbb{Z}$, the requirement that $\mathrm{ord}(\bar{a}) \mid m$ is equivalent to the condition $\mathrm{ord}(\bar{a}) \mid (m, n) = d$. The number of homomorphisms is therefore the number of elements of $\mathbb{Z}/n\mathbb{Z}$ whose order divides d. There are thus d homomorphisms, all of them of the form $\varphi(\bar{k}) = k\bar{a}$ for some $\bar{a} \in \mathbb{Z}/n\mathbb{Z}$ such that $\mathrm{ord}(\bar{a}) \mid d$.

It follows that the image of the map $\Phi : \text{Hom}(\mathbb{Z}/m\mathbb{Z}, \mathbb{Z}/n\mathbb{Z}) \longrightarrow \mathbb{Z}/n\mathbb{Z}$ given by $\Phi(\varphi) = \varphi(\overline{1})$ is the subgroup of order d in $\mathbb{Z}/n\mathbb{Z}$. In order to show the desired statement, it is enough to show that Φ is an injective homomorphism. The fact that Φ is a homomorphism is clear: for all pairs φ_1, φ_2 in $\text{Hom}(\mathbb{Z}/m\mathbb{Z}, \mathbb{Z}/n\mathbb{Z})$ we have $\Phi(\varphi_1 + \varphi_2) = (\varphi_1 + \varphi_2)(\overline{1}) = \varphi_1(\overline{1}) + \varphi_2(\overline{1}) = \Phi(\varphi_1) + \Phi(\varphi_2)$. To show that Φ is injective, remark that $\Phi(\varphi) = \overline{0}$ if and only if $\varphi(\overline{1}) = \overline{0}$, that is, if and only if $\varphi(\overline{k}) = k\overline{0} = \overline{0}$ for all $\overline{k} \in \mathbb{Z}/m\mathbb{Z}$, which is equivalent to φ being the zero homomorphism.

(ii) Since $12 \mid (360, 420) = 60 = |\text{Hom}(\mathbb{Z}/360\mathbb{Z}, \mathbb{Z}/420\mathbb{Z})|$, the proof above shows that there is a unique subgroup of order 12, which is the one that corresponds via the isomorphism Φ to the subgroup of order 12 in $\mathbb{Z}/420\mathbb{Z}$, that is, to $\langle \overline{35} \rangle$. The required subgroup thus consists of the homomorphisms $\varphi : \mathbb{Z}/360\mathbb{Z} \longrightarrow \mathbb{Z}/420\mathbb{Z}$ given by $\varphi(\overline{1}) = \overline{a}$ with $a \equiv 0 \pmod{35}$.

176

(i) Let $e \in G$ be the neutral element; (e, e) is the neutral element of $G \times G$ and belongs to Δ. Given (x, x) and $(y, y) \in \Delta$, clearly $(x, x)(y, y) = (xy, xy) \in \Delta$. Finally, given $(x, x) \in \Delta$, its inverse in $G \times G$ is the element (x^{-1}, x^{-1}), which is again an element of Δ. This shows that Δ is a subgroup of $G \times G$.

(ii) If G is Abelian then $G \times G$ is also Abelian, so all of its subgroups are normal. Conversely, assume Δ is normal in $G \times G$; then for all $g \in G$ and for all $x \in G$ we have $(g, e)(x, x)(g^{-1}, e) = (gxg^{-1}, x) \in \Delta$, which implies that for all $g, x \in G$ we have $gxg^{-1} = x$, that is, G is Abelian.

(iii) Consider the map $\varphi : G \times G \longrightarrow G$ given by $\varphi(x, y) = xy^{-1}$. If G is Abelian then φ is a homomorphism, since $\varphi((x, y)(u, v)) = \varphi(xu, yv) = xu(yv)^{-1} = xuv^{-1}y^{-1} = xy^{-1}uv^{-1} = \varphi(x, y)\varphi(u, v)$. Moreover, φ is surjective because we have $\varphi(x, e) = x$ for all $x \in G$. Remark that $\text{Ker}(\varphi) = \{(x, y) \mid xy^{-1} = e\} = \Delta$; by the fundamental homomorphism theorem, we obtain that $(G \times G)/\Delta \simeq G$.

177

(i) First of all, we have $0 \in G_p$ because $p^0 0 = 1 \cdot 0 = 0$. Consider $x, y \in G_p$ and $k, h \in \mathbb{N}$ such that $p^k x = 0$, $p^h y = 0$; if we set $m = \max\{k, h\}$ we have $p^m(x + y) = p^m x + p^m y = 0 + 0 = 0$, so $x + y \in G_p$. Finally, if $x \in G_p$ and $p^k x = 0$ then we also have $p^k(-x) = -p^k x = 0$, so $-x \in G_p$. We have shown that G_p is a subgroup of G.

(ii) If $\text{ord}(x) = a$ then $kax = 0$ for all $k \in \mathbb{N}$. Similarly, if $\text{ord}(y) = b$ then $hby = 0$ for all $h \in \mathbb{N}$. So $ab(x + y) = abx + aby = 0 + 0 = 0$, and therefore $\text{ord}(x + y) \mid ab$.

Now assume that $s(x + y) = 0$, that is, $sx = -sy$. The left hand side is an element of the subgroup generated by x, while the right hand side is an element of the subgroup generated by y. But the intersection of these subgroups is 0, as it must be a subgroup whose order divides both a and b, and thus also divides $1 = (a, b)$. Consequently, we must have $sx = 0$, so $a \mid s$, and $sy = 0$, so $b \mid s$. Since $(a, b) = 1$, this implies $ab \mid s$, so $ab \mid \text{ord}(x + y)$.

(iii) One implication is clear: if G is cyclic then all its subgroups are cyclic, and in particular all the groups G_p are cyclic. As for the opposite implication, we shall first show that, if p^a is the largest power of p that divides n, then the order of G_p is p^a.

By Cauchy's theorem, the order of G_p must be a power of p. Consider the quotient G/G_p; we show that it cannot contain any elements of order p. Indeed, assume by contradiction that $x + G_p \in G/G_p$ is an element of order p; we have $p(x + G_p) = G_p$, that is, $px = y \in G_p$, so there is an integer k such that $p^k y = p^{k+1} x = 0$, hence $x \in G_p$. It follows that the coset $x + G_p$ is actually the class of the neutral element in the quotient and therefore has order 1, which is a contradiction.

Now, since G/G_p does not have any elements of order p, again by Cauchy's theorem we have that p cannot divide the order of G/G_p. But, since $|G| = |G_p| \cdot |G/G_p|$, this yields the desired equality $|G_p| = p^a$.

Now, let $n = p_1^{e_1} \cdots p_k^{e_k}$ be the prime factorisation of n and let $x_i \in G$, for $i = 1, \ldots, k$, be an element of G_{p_i} of order $p_i^{e_i}$. We show by induction on k that $x_1 + \cdots + x_k$ has order $p_1^{e_1} \cdots p_k^{e_k}$. If $k = 0$ there is nothing to prove. Assuming the desired result for $k - 1$, set $x = x_1 + \cdots + x_{k-1}$, $y = x_k$, $a = p_1^{e_1} \cdots p_{k-1}^{e_{k-1}}$, $b = p_k^{e_k}$. Because of our argument above, the order of $x + y$ is $ab = n$, that is, it is the same as the order of G, which implies that G is cyclic.

178

(i) We start by showing that $f + g : G \longrightarrow G'$ is a homomorphism: we have $(f+g)(u+v) = f(u+v)+g(u+v) = f(u)+f(v)+g(u)+g(v) = f(u)+g(u)+f(v)+g(v) = (f+g)(u)+(f+g)(v)$ because G' is Abelian. Moreover, the operation $(f, g) \longmapsto f + g$ on $\mathrm{Hom}(G, G')$ is associative because the addition of G' is. The neutral element is the homomorphism $G \ni u \longmapsto 0 \in G'$. The additive inverse of a homomorphism $G \ni u \longmapsto f(u) \in G'$ is the map $G \ni u \longmapsto -f(u) \in G'$, which one can immediately show to be a homomorphism.

A homomorphism $f : G \longrightarrow G'$ induces two restricted homomorphisms $f_1 : \mathbb{Z}/18\mathbb{Z} \longrightarrow G'$ and $f_2 : \mathbb{Z}/12\mathbb{Z} \longrightarrow G'$ given by $f_1(x) = f(x, \bar{0})$ and $f_2(y) = f(\bar{0}, y)$. Conversely, given two homomorphisms $f_1 : \mathbb{Z}/18\mathbb{Z} \longrightarrow G'$ and $f_2 : \mathbb{Z}/12\mathbb{Z} \longrightarrow G'$ we can construct a homomorphism $G \ni (x, y) \longmapsto f(x, y) = f_1(x) + f_2(y) \in G'$. It follows that the set of homomorphisms from G to G' is in bijection with $\mathrm{Hom}(\mathbb{Z}/18\mathbb{Z}, \mathbb{Z}/36\mathbb{Z}) \times \mathrm{Hom}(\mathbb{Z}/12\mathbb{Z}, \mathbb{Z}/36\mathbb{Z})$ and thus has cardinality $18 \cdot 12 = 216$, because $18 = (18, 36)$ and $12 = (12, 36)$.

(ii) Suppose $f(\bar{1}, \bar{0}) = \bar{r}$ and $f(\bar{0}, \bar{1}) = \bar{s}$. We must have $\bar{r} = 2\overline{r_1}$ and $\bar{s} = 3\overline{s_1}$, because the order of $\bar{r}$ and the order of $\bar{s}$ must divide 18 and 12, respectively. The homomorphism f is given by $f(x, y) = \bar{r} \cdot x + \bar{s} \cdot y$ and is surjective if and only if there are x, y such that $\bar{r} \cdot x + \bar{s} \cdot y$ is a generator of $\mathbb{Z}/36\mathbb{Z}$, that is, if and only if there are x, y such that $(2r_1 x + 3s_1 y, 36) = 1$.

In order for this to hold, a necessary condition is that $3 \nmid r_1$ and $2 \nmid s_1$, as otherwise all numbers of the form $2r_1 x + 3s_1 y$ would be multiples of 3 or of 2.

But this condition is also sufficient because, if $3 \nmid r_1$ and $2 \nmid s_1$, then $2r_1 + 3s_1$ neither divisible by 3 nor by 2, so it is coprime to 36 and thus its residue class modulo 36 is a generator of $\mathbb{Z}/36\mathbb{Z}$.

We have shown that the number of surjective homomorphisms is the number of pairs $(2\overline{r_1}, 3\overline{s_1}) \in G' \times G'$ such that $3 \nmid r_1$ and $2 \nmid s_1$ and is therefore equal to $12 \cdot 6 = 72$.

(iii) We have $\varphi_{(a,b)}(f + g) = (f + g)(a, b) = f(a, b) + g(a, b) = \varphi_{(a,b)}(f) + \varphi_{(a,b)}(g)$, so $\varphi_{(a,b)}$ is a homomorphism. Using the same notation as above, set $\overline{r} = -\overline{2}$ and $\overline{s} = \overline{3}$; we have that $-\overline{2} + \overline{3} = \overline{1}$ is in the image of the homomorphism. Since $\overline{1}$ is a generator of $\mathbb{Z}/36\mathbb{Z}$, the homomorphism is surjective. Finally, by the fundamental homomorphism theorem its image is isomorphic to $\mathrm{Hom}(G, G')/\mathrm{Ker}(\varphi_{(\overline{1},\overline{1})})$, so we have $|\mathrm{Ker}(\varphi_{(\overline{1},\overline{1})})| = |\mathrm{Hom}(G, G')|/|G'| = 216/36 = 6$.

179

(i) It is clear that the second bullet point implies the first, but we shall prove it independently for the sake of completeness.

Let $x_1 + H, \ldots, x_m + H$ be three cosets of H and let $y_1 + K, \ldots, y_n + K$ be the cosets of K. Clearly,

$$G = \bigcup_{i=1}^{m}(x_i + H) = \bigcup_{j=1}^{n}(y_j + K) = \bigcup_{\substack{i=1,\ldots,m \\ j=1,\ldots,n}} (x_i + H) \cap (y_j + K).$$

The last expression above is a union of mn subsets; if we showed that each of them is contained in a single coset of $H \cap K$ then the desired inequality would follow. We shall now prove that, for all i and j and all elements $a, b \in (x_i + H) \cap (y_j + K)$, we have $a + H \cap K = b + H \cap K$. Indeed, we have $a - b \in H$, $a - b \in K$, so $a - b \in H \cap K$, which is equivalent to the claim.

(ii) Consider the map $f : G \longrightarrow G/H \times G/K$ given by $f(x) = (x + H, x + K)$, which is a homomorphism because its components are the projections of G onto G/H and G/K. The kernel of f is $\{x \in G \mid x \in H, \ x \in K\} = H \cap K$. By the fundamental homomorphism theorem, f induces an *injective* homomorphism from $G/(H \cap K)$ to $G/H \times G/K$, so $d = |G/(H \cap K)|$ divides $|G/H| \cdot |G/K| = mn$.

(iii) The statement is equivalent to showing that the homomorphism f defined above is surjective if and only if $H + K = G$. Suppose that $H + K = G$ and consider an element $(a + H, b + K) \in G/H \times G/K$. By assumption we can write $a - b = h + k$ for some $h \in H, k \in K$. Now, set $g = a - h = b + k$. We have $f(g) = (g + H, g + K) = (a + H, b + K)$, so f is indeed surjective. Conversely, assume that f is surjective; then for all $x \in G$ there is $g \in G$ such that $f(g) = (x + H, K)$. This implies $g + H = x + H$, that is, $x - g = h \in H$, and $g + K = K$, that is, $g \in K$. It follows that $x = h + g \in H + K$, so $G \subseteq H + K$. The reverse inclusion is clear.

180

(i) Given any two cyclic subgroups of G that are isomorphic to $\mathbb{Z}$, say $H_1 = \langle a_1/b_1 \rangle$ and $H_2 = \langle a_2/b_2 \rangle$, they have the element $a_1 a_2 = a_2 b_1 \cdot a_1/b_1 = a_1 b_2 \cdot a_2/b_2$ in common; this element is nonzero because a_1/b_1 and a_2/b_2 are nonzero. Therefore, there cannot be a subgroup of G that is isomorphic to $\mathbb{Z} \times \mathbb{Z}$.

(ii) Take for example the subgroups

$$H_m = 2^m G = \{2^m \frac{a}{b} \mid \frac{a}{b} \in G\} \qquad \text{with } m \geq 0$$

and consider the quotients $G_m = G/H_m$.

The coset $1 + H_m$ has order 2^m, because $k \cdot 1 \in H_m$ if and only if k is of the form $2^m a/b$ with $(b, 10) = 1$, that is, if and only if kb is a multiple of 2^m. Since $(b, 2^m) = 1$, this is equivalent to the condition that $2^m \mid k$.

Moreover, every coset $a/b + H_m$ is a multiple of $1 + H_m$, since if c is an integer such that $cb \equiv a \pmod{2^m}$ then $c - a/b \in H_m$, so $c(1 + H_m) = c + H_m = a/b + H_m$.

It follows that G_m is cyclic of order 2^m.

(iii) Suppose by contradiction that G has a cyclic quotient of order 3, that is, that there is a subgroup H of G such that $|G/H| = 3$. Every element of G/H must have an order that divides 3: for all $x \in G$ we must have $3(x + H) = H$, that is, $3x \in H$, so ultimately we get $3G \subseteq H$. However, we have $3G = G$ because for all $y \in G$ we have $y = 3 \cdot y/3 \in 3G$, hence a contradiction.

181

(i) Assume that G is finite; we shall show that the conditions in the problem statement are necessary.

If f is an injective homomorphism then $f(G) \simeq G$ and, since $f(G)$ is a subgroup of H, we must have that H is finite and that $a \mid b$.

Now, let $b = ac$. Suppose by contradiction that $(a, c) > 1$; then there is a prime p and there are positive integers α, β with $\alpha < \beta$ such that p^α and p^β are the largest powers of p that divide a and b, respectively. Let $G_p \simeq \mathbb{Z}/p^\alpha \mathbb{Z}$ and $H_p \simeq \mathbb{Z}/p^\beta \mathbb{Z}$ be the unique subgroups of G and H of order p^α and p^β, respectively, and set $\gamma = \beta - \alpha$. We necessarily have that $f(G_p)$ is the unique subgroup of H_p of order p^α, that is, $f(G_p) = p^\gamma H_p$. In particular, given a generator x of G_p, its image $f(x)$ is not a generator of H_p, so $f(x) = py$ for some $y \in H_p$. But then $g \circ f(x) = pg(y)$ cannot be a generator of G_p, which contradicts the assumption that $g \circ f$ is an isomorphism.

We now show that the conditions are sufficient. If $b = ac$ and $(a, c) = 1$ then $H \simeq \mathbb{Z}/a\mathbb{Z} \times \mathbb{Z}/c\mathbb{Z}$ and $f(G)$ is the unique subgroup of H of order a, that is, $\mathbb{Z}/a\mathbb{Z} \times \{\bar{0}\}$. Setting g to be the canonical projection of $\mathbb{Z}/a\mathbb{Z} \times \mathbb{Z}/c\mathbb{Z}$ onto $\mathbb{Z}/a\mathbb{Z}$, we have $g \circ f(G) = G$, so $g \circ f$ is surjective.

Since $g \circ f$ is a surjective map between two sets of the same cardinality it is also injective, so it is an isomorphism.

(ii) Assume now that G is infinite; first of all, we prove that the condition in the problem statement is necessary. Remark that there are no injective maps from an infinite set into a finite one, so H must be a cyclic infinite group, that is, it must be isomorphic to $\mathbb{Z}$. Assume by contradiction that f is not surjective; we would then have $f(1) = k$ for some $k \neq \pm 1$. But then $g(k) = kg(1)$ would be divisible by k and so $g(k) = g \circ f(1) \neq \pm 1$. It follows that the map $g \circ f$ would send the generator 1 of $\mathbb{Z}$ to an element that is not a generator, so the map would not be an isomorphism.

Let us now show that the condition is sufficient. If $H \simeq \mathbb{Z}$ then without loss of generality we can assume $G = H = \mathbb{Z}$, $f(x) = \pm x$ and, setting $g(x) = \pm x$, we have that $g \circ f(x) = x$ is an isomorphism.

182 Given a subgroup H of a group G, we shall call *intermediate* between H and G any subgroup L such that $H \subsetneqq L \subsetneqq G$.

(i) The projection $G \ni g \overset{\pi}{\longmapsto} gK \in G/K$ induces an inclusion-preserving bijection between the set of subgroups of G that contain K and the set of subgroups of G/K. Since the subgroups of G that contain M also contain K, there is an intermediate subgroup between M and G if and only if there is an intermediate subgroup between M/K and G/K, hence the conclusion.

(ii) Suppose by contradiction that the statement does not hold, that is, that G has subgroups that are not contained in any maximal subgroup. Let H be such a subgroup of G. Since the order of G is finite, we can assume that H is of maximal order among all such subgroups.

Since H cannot be maximal, there is an intermediate subgroup L between H and G. But $|L| > |H|$ and so L must be contained in a maximal subgroup M, hence $H \subseteq M$, which contradicts our assumption.

(iii) We shall use the statement we proved above and the fact that all subgroups of an Abelian group are normal.

If $[G : K] = p$ is a prime then K/K is a maximal subgroup of $G/K \cong \mathbb{Z}/p\mathbb{Z}$ because the only subgroups of a group of order p are the trivial subgroup and the group itself.

If $[G : K] = |G/K| = m$ is not prime then, given a prime p that divides m, by Cauchy's theorem there is a cyclic subgroup of order p in G/K, which is clearly intermediate between K/K and G/K. But then there is an intermediate subgroup between K and G, so K is not maximal.

183

(i) The group $\mathbb{Z}/12\mathbb{Z}$ is a cyclic group of order 12 generated by $\overline{1}$. We know that a homomorphism $\varphi : \mathbb{Z}/12\mathbb{Z} \longrightarrow \mathbb{Z}/4\mathbb{Z} \times S_3$, since $\mathbb{Z}/12\mathbb{Z}$ is a cyclic group, is completely determined by the image of $\overline{1}$, which only needs to satisfy the condition $\mathrm{ord}(\varphi(\overline{1})) \mid \mathrm{ord}(\overline{1}) = 12$. Therefore, the image of $\overline{1}$ can be any element whose order is a divisor of 12 in $\mathbb{Z}/4\mathbb{Z} \times S_3$.

The elements of the group $\mathbb{Z}/4\mathbb{Z} \times S_3$ are the pairs $(\overline{a}, \sigma)$, where $\overline{a} \in \mathbb{Z}/4\mathbb{Z}$ and $\sigma \in S_3$. We know that $\mathrm{ord}(\overline{a}, \sigma) = [\mathrm{ord}(\overline{a}), \mathrm{ord}(\sigma)]$. But $\mathrm{ord}(\overline{a}) \mid 4$ and ord$(b) \mid 6$, so ord$(\bar{a}, \sigma) \mid [4, 6] = 12$. We can therefore choose any element of $\mathbb{Z}/4\mathbb{Z} \times S_3$ to be the image of $\bar{1}$, hence there are 24 homomorphisms in total. The homomorphism φ is injective if and only if ord$(\varphi(\bar{1})) = $ ord$(\bar{1}) = 12$; counting injective homomorphisms is the same as counting elements of order 12 in $\mathbb{Z}/4\mathbb{Z} \times S_3$. The order of an element in $\mathbb{Z}/4\mathbb{Z}$ can be 1, 2 or 4, while the order of an element in S_3 can be 1, 2 or 3. Consequently, a pair $(\bar{a}, \sigma)$ has order 12 if and only if ord$(\bar{a}) = 4$ and ord$(\sigma) = 3$. Since $\mathbb{Z}/4\mathbb{Z}$ contains $\phi(4) = 2$ elements of order 4, namely, $\bar{1}$ and $\bar{3}$, and S_3 contains two elements of order 3, namely, the 2 three-cycles (123) and (132), there are $2 \cdot 2 = 4$ injective homomorphisms in total.

(ii) Because of our previous remarks, a homomorphism φ is completely determined by the image of $\bar{1}$. Let $\varphi(\bar{1}) = (\bar{a}, \sigma)$; we have $\varphi(\overline{10}) = 10\varphi(\bar{1}) = (\overline{10a}, \sigma^{10})$. It follows that the order of $\varphi(\overline{10})$ is equal to 3 if and only if we have $[\text{ord}(\overline{10a}), \text{ord}(\eta^{10})] = 3$. Since the order of an element in $\mathbb{Z}/4\mathbb{Z}$ is 1, 2 or 4 and the order of an element in S_3 is 1, 2 or 3, in order for the order of $\varphi(\overline{10})$ to be 3 we must have ord$(\overline{10a}) = 1$ and ord$(\sigma^{10}) = 3$.

Now, we have $10a \equiv 0 \pmod 4$ if and only if $2a \equiv 0 \pmod 4$, that is, $a \equiv 0 \pmod 2$. The two solutions of this equation are $\bar{a} = \bar{0}, \bar{2}$. As for $\sigma \in S_3$, remark that ord$(\sigma^{10}) = 3$ implies $3 \mid$ ord(σ) and, since $\sigma \neq e$, the element σ must be one of the 2 three-cycles of S_3, which clearly satisfy ord$(\sigma^{10}) = 3$.

It follows that the homomorphisms φ such that ord$(\varphi(\overline{10})) = 3$ are exactly those that send $\bar{1}$ to an element $(\bar{a}, \sigma)$ with $\bar{a} = \bar{0}$ or $a = \bar{2}$ and $\sigma = (123)$ or $\sigma = (132)$. That is, there are four such homomorphisms.

184

(i) By the Chinese remainder theorem, since $1000 = 8 \cdot 125$ and $(8, 125) = 1$, we know that

$$(\mathbb{Z}/1000\mathbb{Z})^* \cong (\mathbb{Z}/8\mathbb{Z})^* \times (\mathbb{Z}/125\mathbb{Z})^*.$$

Therefore, the group G has a subgroup isomorphic to $(\mathbb{Z}/8\mathbb{Z})^*$. Since $(\mathbb{Z}/8\mathbb{Z})^* = \{\pm\bar{1}, \pm\bar{3}\} \simeq \mathbb{Z}/2\mathbb{Z} \times \mathbb{Z}/2\mathbb{Z}$ is not cyclic, G itself cannot be cyclic, because every subgroup of a cyclic group is cyclic.

(ii) Clearly, the identity belongs to H, because it has order $1 = 2^0$. Given $g, h \in H$, letting $2^a = \max\{\text{ord}(g), \text{ord}(h)\}$, we have $(gh)^{2^a} = g^{2^a} h^{2^a} = e$, so the order of gh is a power of 2 as it divides 2^a. It follows that $gh \in H$. Since H is finite, this is enough to show that H is a subgroup of G.

Let us now compute the order of H. By Lagrange's theorem, $|H|$ divides $|G| = \phi(1000) = 2^4 \cdot 5^2$. Moreover, since the order of every element in H is a power of 2, by Cauchy's theorem 2 is the only prime divisor of the order of H, so $|H|$ divides 2^4. We now show that $|H| = 2^4$ by arguing that $|G/H| = 2^4 \cdot 5^2 / |H|$ is odd.

Indeed, if $|G/H|$ were even, Cauchy's theorem would yield a coset gH of order 2 in G/H, that is, such that $gH \neq H$ and $(gH)^2 = H$, or equivalently that $g \notin H$ and $g^2 \in H$. But this is not possible, because it would imply that the order of g^2 is a power of 2, and so the order of g would be a power of 2 and g would belong to H. This concludes the proof that $|H| = 2^4$.

(iii) It is enough to show that $(\mathbb{Z}/125\mathbb{Z})^*$ contains an element of order 25, that is, that there are solutions of the congruence $x^{25} \equiv 1 \pmod{125}$ that are not solutions of $x^5 \equiv 1 \pmod{125}$. For example, set $x = \overline{6} \in \mathbb{Z}/125\mathbb{Z}$; we have $x^{25} = (1 + 5)^{25} \equiv 1 + 25 \cdot 5 \equiv 1 \pmod{125}$, whereas $x^5 = (1 + 5)^5 \equiv 1 + 5 \cdot 5 = 26 \not\equiv 1 \pmod{125}$ and so x does not satisfy $x^5 \equiv 1 \pmod{125}$. We have thus shown that the order of x is 25.

We know that $|G/H| = 25$; in order to show that G/H is cyclic, it is enough to remark that, letting $x = \overline{6}$, the coset xH has order 25 in G/H. Let $d = \operatorname{ord}(xH)$ be the smallest positive integer such that $x^d \in H$. By the definition of H, the order of every element in H is a power of 2, so $\operatorname{ord}(x^d) = 25/(d, 25) = 2^k$ for some k. But this is possible only if $k = 0$ and $d = 25$.

185

(i) Since G is Abelian, for all $g, h \in G$ we have $g^k h^k = (gh)^k$, so the map $\varphi_k : G \longrightarrow G$ given by $\varphi_k(g) = g^k$ is a group homomorphism and we have $G^k = \varphi_k(G)$. Since the image of a homomorphism is always a subgroup, G^k is a subgroup of G.

(ii) Again, consider the homomorphism $G \ni g \xmapsto{\varphi_k} g^k \in G$. We have $G^k = G$ if and only if φ_k is surjective; since G is finite, this is equivalent to φ_k being injective. We therefore need to identify the values of k for which $\operatorname{Ker}(\varphi_k) = \{e\}$.

First, let us suppose that $(n, k) > 1$. Let p be a prime that divides (n, k); by Cauchy's theorem, there is an element g of order p in G, and since $k = pd$ for some integer d we have $g^k = (g^p)^d = e$, so the kernel of φ_k is nontrivial and the homomorphism is not surjective.

Now suppose that $(n, k) = 1$. We have $g^k = e$ if and only if the order of g divides k, but since k must also divide the order of the group n, the only possibility is $g = e$. So φ_k is injective and surjective. In conclusion, $G^k = G$ if and only if $(k, n) = 1$.

(iii) Consider for example $G = \mathbb{Z}$. Clearly, G^k is the subgroup $k\mathbb{Z}$ of all multiples of k. But then 1 is not an element of G^k for any $k > 1$, so $G^k \neq G$ for all $k > 1$.

⟦Another easy example is given by the multiplicative group $\mathbb{Q}^*$, where $2 \notin \mathbb{Q}^{*k}$ if $k > 1$, as one can show by generalising the proof of the fact that $\sqrt{2} \notin \mathbb{Q}$.⟧

(iv) Consider the additive group $\mathbb{Q}$ of rational numbers: for all $k \geq 1$ and for all rationals a/b we have

$$\frac{a}{b} = k\frac{a}{kb} \in k\mathbb{Q},$$

so $k\mathbb{Q} = \mathbb{Q}$.

⟦One can give other examples by letting G be the set of all roots of unity or the set of points of the unit circle in $\mathbb{C}$ (that is, complex numbers of modulus 1), endowed with the multiplication of $\mathbb{C}^*$. In both cases it is easy to show that G is a group and that every $g \in G$ is a square of some element of G, the cube of some element of G, the fourth power of some element of G, and so on. ∎

186

(i) The order of G is $3 \cdot 6 = 18$, so by Lagrange's theorem the order of any subgroup of G must divide 18.

On the other hand, if H is a subgroup of $\mathbb{Z}/3\mathbb{Z}$ and K is a subgroup of S_3 then $H \times K$ is a subgroup of G of order $|H| \cdot |K|$. We can take H to be the trivial subgroup, which has one element, or the group $\mathbb{Z}/3\mathbb{Z}$, which has three. The group S_3 has a one-element subgroup (the trivial subgroup), a two-element subgroup (any subgroup generated by a transposition), a three-element subgroup (generated by a three-cycle) and a six-element subgroup (the whole group S_3). We can therefore choose K of order 1, 2, 3 and 6. This allows us to conclude that we can construct a subgroup of G of order d for all divisors d of 18.

(ii) In order to find the number of cyclic subgroups of some fixed order n, we count the elements of order n and divide the result by $\phi(n)$. This is because every cyclic subgroup of order n contains $\phi(n)$ elements of order n, and if two subgroups have an element of order n in common then, since that element is a generator for both, they must coincide.

Each element of G has an order that divides 6, since the order of a pair in a direct product is the least common multiple of the orders of its components. It follows that there cannot be any cyclic subgroups of order 9 or 18.

The elements of order 2 are those of the form $(\overline{0}, \sigma)$ where σ is one of the transpositions (12), (13), (23) in S_3. There are therefore $3 = 3/\phi(2)$ cyclic subgroups of order 2.

The elements of G whose order is 3 are those of the form $(\overline{a}, \sigma) \neq (0, e)$, where $a = 0, 1, 2$ and σ is one of the 2 three-cycles (123) and (132), or the neutral element. We therefore have $3 \cdot 3 - 1 = 8$ elements of order 3 in total, hence $4 = 8/\phi(3)$ cyclic subgroups of order 3.

Finally, there are 3 cyclic subgroups of order 6: there 6 elements of order 6, namely, the elements of the form $(\overline{a}, \sigma)$, where $a = 1, 2$ and σ is one of the three transpositions of S_3, and we have $\phi(6) = 2$.

In conclusion, G has $1 + 3 + 4 + 3 = 11$ cyclic subgroups.

5 Rings and Fields

187 First of all, let us factor $f(x)$ as a product of irreducible polynomials. One can check, using Ruffini's theorem, that $f(x)$ is divisible by both $x - 2$ and $x - 3$, hence by their product. By performing the division we obtain

$$f(x) = (x - 2)(x - 3)(x^2 - x + 3)$$

where the last factor, which is of degree 2 and has no roots in $\mathbb{F}_7$, is irreducible. Elements of $\mathbb{F}_7[x]/(f(x))$ can be written in the form $\overline{g(x)}$, where $g(x)$ is a polynomial of degree at most 3. The element $\overline{g(x)}$ is a zero divisor if and only if $(g(x), f(x)) \neq 1$ and is invertible if and only if $(g(x), f(x)) = 1$.

The set of zero divisors is therefore the union of the sets of multiples of $x - 2$, multiples of $x - 3$, and multiples of $x^2 - x + 3$. The multiples of degree at most 3 of a polynomial of degree d are as many as the polynomials of degree at most $3 - d$, that is, 7^{4-d}. By the inclusion-exclusion principle, this implies that the number of zero divisors is

$$7^3 + 7^3 + 7^2 - 7^2 - 7 - 7 + 1 = 673.$$

By the remark above, $\overline{x + 1}$ is invertible. In order to find its inverse, note that

$$\overline{0} = \overline{x^4 + x^3 - 3} = \overline{x + 1} \cdot \overline{x^3 - 3}.$$

We therefore have $\overline{x + 1} \cdot \overline{x^3} = \overline{3}$, hence $\overline{x + 1} \cdot \overline{5x^3} = \overline{1}$, and so $\overline{5x^3}$ is the inverse of $\overline{x + 1}$.

188 First of all, remark that $x^4 - 25 = (x^2 - 5)(x^2 + 5)$ and that the two polynomials $x^2 - 5$ and $x^2 + 5$ are irreducible in $\mathbb{Q}[x]$, because their degree is 2 and they have no rational roots.

The splitting field of $x^4 - 25$ over $\mathbb{Q}$ is $\mathbb{F} = \mathbb{Q}(\sqrt{5}, \sqrt{-5})$. Since $[\mathbb{Q}(\sqrt{5}) : \mathbb{Q}] = [\mathbb{Q}(\sqrt{-5}) : \mathbb{Q}] = 2$ and $\sqrt{-5} \notin \mathbb{Q}(\sqrt{5})$ (because $\sqrt{-5} \notin \mathbb{R}$ whereas $\mathbb{Q}(\sqrt{5}) \subseteq \mathbb{R}$), we have $[\mathbb{F} : \mathbb{Q}] = 4$. Moreover, a basis of $\mathbb{F}$ as a vector space over $\mathbb{Q}$ is given by $1, \sqrt{5}, \sqrt{-5}, \sqrt{-25} = 5i$, or alternatively simply by $1, \sqrt{5}, i, i\sqrt{5}$.

Let $\mathbb{K}$ be the splitting field of $f_m(x) = (x^2 - m)(x^4 - 25)$. If m is of the form $\pm n^2$ or of the form $\pm 5n^2$ for some $n \in \mathbb{N}$, then $\sqrt{m} = \pm n, \pm in, \pm\sqrt{5}n, \pm i\sqrt{5}n$ belongs to $\mathbb{F}$, so $\mathbb{K} = \mathbb{F}$ has degree 4 over $\mathbb{Q}$.

Assume now that m is not of one of the two forms in the previous paragraph. The field $\mathbb{K}$ contains both $\sqrt{m}$ and $i\sqrt{m}$: therefore, it contains $\sqrt{|m|}$, where $|m| \neq n^2, 5n^2$. Let us show that $\sqrt{|m|} \notin \mathbb{F}$. If we had $\sqrt{|m|} \in \mathbb{F}$ we would have $\sqrt{|m|} = a + b\sqrt{5} + ci + di\sqrt{5}$ for some $a, b, c, d \in \mathbb{Q}$.

Since $\sqrt{|m|} \in \mathbb{R}$, we must have $c = d = 0$, and therefore $\sqrt{|m|} = a + b\sqrt{5}$. Taking the square of both sides, $|m| = a^2 + 5b^2 + 2ab\sqrt{5}$, hence

$$\begin{cases} |m| = a^2 + 5b^2 \\ 2ab = 0 \end{cases}$$

because 1 and $\sqrt{5}$ are linearly independent over $\mathbb{Q}$.

The second equation implies that $a = 0$ or $b = 0$. If $a = 0$ then $|m| = 5b^2$ and if $b = 0$ then $|m| = a^2$, contradicting our assumptions on n.

We can thus conclude that, if m is neither of the form $\pm n^2$ nor of the form $\pm 5n^2$ for any $n \in \mathbb{Z}$, then $\mathbb{K} = \mathbb{F}(\sqrt{|m|})$ and $\sqrt{|m|} \notin \mathbb{F}$, so $[\mathbb{K} : \mathbb{Q}] = 8$.

189

(i) The polynomial $f(x)$ is irreducible in $\mathbb{Q}[x]$: its degree is 3, so it is irreducible if and only if it has no rational roots; since any rational root must have a numerator that divides the constant coefficient and a denominator that divides the leading coefficient, it is enough to check that $f(\pm 1) \neq 0$. We thus have $[\mathbb{Q}(\alpha) : \mathbb{Q}] = 3$. Clearly, $1/(\alpha + 2) \in \mathbb{Q}(\alpha)$ and $\alpha \in \mathbb{Q}(1/(\alpha + 2))$, so $\mathbb{Q}(1/(\alpha + 2)) = \mathbb{Q}(\alpha)$. It follows that the minimal polynomial of $1/(\alpha + 2)$ over $\mathbb{Q}$ has degree 3.
Set $\beta = \alpha + 2$ and $\gamma = 1/\beta = 1/(\alpha + 2)$. Since α is a root of $f(x)$ we have that β is a root of $f(x - 2) = (x - 2)^3 - 3(x - 2) + 1 = x^3 - 6x^2 + 15x - 15$, that is, $\beta^3 - 6\beta^2 + 15\beta - 15 = 0$. Multiplying by on both sides $1/(15\beta^3) \neq 0$ yields

$$\gamma^3 - \gamma^2 + \frac{2}{5}\gamma - \frac{1}{15} = 0,$$

so γ is a root of the monic polynomial with rational coefficients

$$x^3 - x^2 + \frac{2}{5}x - \frac{1}{15}.$$

Since this polynomial has degree 3, it must be the minimal polynomial of γ over $\mathbb{Q}$.

(ii) Suppose β is a common root of $f(x)$ and $g(x)$ in an algebraic closure of $\mathbb{F}_p$. The fact that $\beta^2 = 2$ implies that $\beta^3 + 3\beta - 1 = 5\beta - 1 = 0$, so $5\beta = 1$ and $1 = (5\beta)^2 = 25\beta^2 = 25 \cdot 2 = 50$. But $1 = 50$ implies $p \mid 50 - 1 = 49$, that is, $p = 7$. Conversely, if $p = 7$ then we have the common root $\beta = 3$.

190

(i) First of all, remark that $f(x)$ is irreducible in $\mathbb{Z}[x]$ by Eisenstein's criterion for the prime 2. By Gauss's lemma, $f(x)$ is then irreducible in $\mathbb{Q}[x]$. We have $\alpha^6 + 2 = -4\alpha^3$ and, by taking the square of both sides, $\alpha^{12} + 4\alpha^6 + 4 = 16\alpha^6$, so α^2 is a root of the polynomial $g(x) = x^6 - 12x^3 + 4$.
Let us show that $g(x)$ is irreducible. Since $\mathbb{Q}(\alpha^2) \subseteq \mathbb{Q}(\alpha)$ and $[\mathbb{Q}(\alpha) : \mathbb{Q}(\alpha^2)] \leq 2$, we have that $[\mathbb{Q}(\alpha^2) : \mathbb{Q}] = 3, 6$, so it is enough to check that $g(x)$ does not factor as the product of two polynomials of degree 3.
By Gauss's lemma, we may assume by contradiction that $g(x) = p(x)q(x)$ for some monic polynomials p and q of degree 3 with integer coefficients.
Considering the reduction modulo 2 of the equality above, we necessarily have that $\overline{p} = \overline{q} = x^3$, so all coefficients of p and q except for the leading coefficient must be even. In particular, the constant coefficients must both be equal to 2 or both be equal to -2. If we write $p(x) = x^3 + ux^2 + vx \pm 2$, $q(x) = x^3 + u'x^2 + v'x \pm 2$ and equate the corresponding coefficients of $g(x)$ and $p(x)q(x)$ we immediately obtain that $u' = -u$, $v' = -v$ (from the terms of

degree 5 and 1) and $u^2 = v^2 = 0$ (from the terms of degree 4 and 2), hence a contradiction.

We thus have $\mathbb{Q}(\alpha^2) = \mathbb{Q}(\alpha)$ and so $\mathbb{Q}(1/\alpha^2) = \mathbb{Q}(\alpha^2)$ has degree 6 over $\mathbb{Q}$. Since $4(1/\alpha^2)^6 - 12(1/\alpha^2)^3 + 1 = 0$, the minimal polynomial of $1/\alpha^2$ over $\mathbb{Q}$ is

$$h(x) = x^6 - 3x^3 + \frac{1}{4}.$$

(ii) If we consider the coefficients as elements of $\mathbb{F}_7$, we have $f(x) = (x^3 - 1)(x^3 - 2)$. We can check for roots of the degree 3 factors and obtain that $x^3 - 1 = (x - 1)(x - 2)(x - 4)$, while $x^3 - 2$ is irreducible. Since the least common multiple of the degrees of the irreducible factors of $f(x)$ is 3, the splitting field of $f(x)$ is $\mathbb{F}_{7^3}$.

191 We have $f(x) = x^6 - 4 = (x^3 + 2)(x^3 - 2)$. Note that α is a root of the first factor if and only if $-\alpha$ is a root of the second factor, so the splitting field of $f(x)$ is the same as the splitting field of $g(x) = x^3 - 2$.

The polynomial $g(x)$ is irreducible in $\mathbb{Q}[x]$ by Eisenstein's criterion; it has one real root, namely, $\sqrt[3]{2}$, and two complex conjugate roots. The degree of the splitting field $\mathbb{K}$ of $g(x)$ over $\mathbb{Q}$ is at most $3! = 6$ and is a multiple of $[\mathbb{Q}(\sqrt[3]{2}) : \mathbb{Q}] = 3$. Moreover, it cannot be 3 because $\mathbb{Q}(\sqrt[3]{2}) \subseteq \mathbb{R}$ and $\mathbb{K} \not\subseteq \mathbb{R}$, so $[\mathbb{K} : \mathbb{Q}] = 6$.

The polynomial $g(x)$ does not have multiple roots in $\mathbb{F}_{11}$, because its roots are nonzero and $g'(x) = 3x^2$ is zero only for $x = 0$. The multiplicative group $\mathbb{F}_{11}^*$ is cyclic and has ten elements. Since $(3, 10) = 1$, the map $x \longmapsto x^3$ is an isomorphism. In particular, it is bijective. It follows that there is a unique element a such that $a^3 = 2$, that is, a unique root of $g(x)$ in $\mathbb{F}_{11}$, which is simple. Therefore, the irreducible factors of $g(x)$ are a polynomial of degree 1 and a polynomial of degree 2, so the degree of the splitting field of $g(x)$ over $\mathbb{F}_{11}$ is 2.

192 Given a root α of $f(x) = x^6 + 1$, we have $\alpha^6 = -1$ and $\alpha^{12} = 1$, so the order $\mathrm{ord}(\alpha)$ of α in $\mathbb{F}_p^*$ is a divisor of 12.

If $\mathrm{ord}(\alpha) = 1$, then $1^6 + 1 = 0$ and so $p = 2$.

Assume now that $p \neq 2$: we have $-1 \neq 1$ in $\mathbb{F}_p$, so $\mathrm{ord}(\alpha) \nmid 6$. We thus have the two possibilities $\mathrm{ord}(\alpha) = 4$ and $\mathrm{ord}(\alpha) = 12$. But, if there exists $\alpha \in \mathbb{F}_p$ of order 4 then α is a root of the polynomial $x^2 + 1 = (x^4 - 1)/(x^2 - 1)$, which divides $f(x)$, so α is also a root of $f(x)$. Similarly, if there exists $\alpha \in \mathbb{F}_p$ of order 12 then α is a root of the polynomial $f(x) = (x^{12} - 1)/(x^6 - 1)$.

So $f(x)$ has a root in $\mathbb{F}_p$ if and only if $\mathbb{F}_p^*$, which is a cyclic group of order $p - 1$, contains an element of order 4 or an element of order 12, that is, if and only if $4 \mid p - 1$ or $12 \mid p - 1$. Since the second condition implies the first, the primes satisfying these requirements are 2 and all primes p with $p \equiv 1 \pmod 4$.

193

(i) By using the fact that α is a root of $f(x) = x^4 - 2x^3 + x - 1$, we obtain that

$$\alpha^4 - 2\alpha^3 + \alpha = 1$$
$$(\alpha^4 - 2\alpha^3 + \alpha)^2 = \alpha^8 - 4\alpha^7 + 4\alpha^6 + 2\alpha^5 - 4\alpha^4 + \alpha^2 = 1$$
$$\alpha^2(\alpha^6 - 4\alpha^5 + 4\alpha^4 + 2\alpha^3 - 4\alpha^2 + 1) = 1.$$

It follows that $g(x) = x^6 - 4x^5 + 4x^4 + 2x^3 - 4x^2 + 1$ has the required property. ⟦The polynomial $g(x)$ given above is not an example with minimal degree: we have $1/\alpha^2 \in \mathbb{Q}(\alpha)$ and every element of $\mathbb{Q}(\alpha)$ can be expressed as a polynomial in α with integer coefficients of degree ≤ 3.⟧

(ii) The polynomial $f(x)$ is irreducible in $\mathbb{Q}[x]$. In order to show this, by Gauss's lemma it is enough to show that it is irreducible in $\mathbb{Z}[x]$. By reducing modulo 2, we find that the polynomial $x^4 + x + 1$ is irreducible in $\mathbb{F}_2[x]$, because it has no roots in $\mathbb{F}_2$ and is not the square of the unique irreducible polynomial of degree 2 with coefficients in $\mathbb{F}_2$, namely, $x^2 + x + 1$. So $f(x)$ is irreducible in $\mathbb{Z}[x]$.

Therefore, $f(x)$ is the minimal polynomial of α, we have $[\mathbb{Q}(\alpha) : \mathbb{Q}] = 4$ and, since $\beta = \alpha^2 + k\alpha \in \mathbb{Q}(\alpha)$, $d = [\mathbb{Q}(\beta) : \mathbb{Q}]$ is a divisor of 4.

The degree d cannot be 1, otherwise we would have $\beta \in \mathbb{Q}$ and α would be the root of a polynomial with rational coefficients of degree $2 < 4$.

We have $d = 2$ if and only if there are rationals a, b such that $\beta^2 + a\beta + b = 0$, that is,

$$\alpha^4 + 2k\alpha^3 + (k^2 + a)\alpha^2 + ka\alpha + b = 0.$$

This is the case if and only if $x^4 + 2kx^3 + (k^2 + a)x^2 + kax + b$ is a multiple of $f(x)$. Since both polynomials are monic and have the same degree, this is equivalent to $2k = -2$, $k^2 + a = 0$, $ka = 1$, $b = 1$, that is, $k = -1$, $a = -1$, $b = 1$.

In conclusion, the degree $[\mathbb{Q}(\alpha^2 + k\alpha) : \mathbb{Q}]$ is 2 for $k = -1$ and 4 for all other values of k.

194 First of all, remark that $f(x) = x^4 + 2x^2 + 2$ is irreducible over $\mathbb{Q}$ by Eisenstein's criterion for the prime 2, so $[\mathbb{Q}(\alpha) : \mathbb{Q}] = 4$.

The minimal polynomial of α^2 is $x^2 + 2x + 2$, because it is monic, has α^2 as a root and is irreducible. So the minimal polynomial of $\alpha^2 + 1$ is $(x - 1)^2 + 2(x - 1) + 2 = x^2 + 1$.

The polynomial $(x - 2)^4 + 2(x - 2)^2 + 2 = x^4 - 8x^3 + 26x^2 - 40x + 26$ has $\alpha + 2$ as a root, so its reciprocal polynomial $26x^4 - 40x^3 + 26x^2 - 8x + 1$ has $1/(\alpha + 2)$ as a root.

Remark that $\mathbb{Q}(1/(\alpha + 2)) = \mathbb{Q}(\alpha + 2) = \mathbb{Q}(\alpha)$; therefore, the minimal polynomial of $1/(\alpha + 2)$ has degree 4. We can thus conclude that this minimal polynomial is

$$x^4 - \frac{20}{13}x^3 + x^2 - \frac{4}{13}x + \frac{1}{26}.$$

195 The polynomial $x^3 - 2$ is irreducible in $\mathbb{Q}[x]$ because its degree is 3 and it has no rational roots. Its roots in $\mathbb{C}$ are $\sqrt[3]{2}, \sqrt[3]{2}\zeta, \sqrt[3]{2}\zeta^2$, where $\zeta = \frac{-1}{2} + \frac{\sqrt{-3}}{2}$ is a primitive third root of unity.

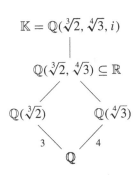

The polynomial $x^4 - 3$ is irreducible over $\mathbb{Q}$ by Eisenstein's criterion for the prime 3 and by Gauss's lemma, and its roots in $\mathbb{C}$ are $\sqrt[4]{3}, i\sqrt[4]{3}, -\sqrt[4]{3}, -i\sqrt[4]{3}$. It is easy to check that the splitting field of $f(x) = (x^3 - 2)(x^4 - 3)$ over $\mathbb{Q}$ is thus $\mathbb{K} = \mathbb{Q}(\sqrt[3]{2}, \sqrt[4]{3}, i)$.

We know that $[\mathbb{Q}(\sqrt[3]{2}) : \mathbb{Q}] = 3$ and that $[\mathbb{Q}(\sqrt[4]{3}) : \mathbb{Q}] = 4$, so $12 = [3, 4]$ divides $[\mathbb{Q}(\sqrt[3]{2}, \sqrt[4]{3}) : \mathbb{Q}] \leq 12$. It follows that $[\mathbb{Q}(\sqrt[3]{2}, \sqrt[4]{3}) : \mathbb{Q}] = 12$. Moreover, $[\mathbb{K} : \mathbb{Q}(\sqrt[3]{2}, \sqrt[4]{3})] = 2$ because $\mathbb{K}$ is obtained by adding i to the real field $\mathbb{Q}(\sqrt[3]{2}, \sqrt[4]{3})$. In conclusion, $[\mathbb{K} : \mathbb{Q}] = 24$.

In $\mathbb{F}_3$ we have $f(x) = (x - 2)^3 x^4$, so the splitting field is $\mathbb{F}_3$ itself. In $\mathbb{F}_{11}$ the polynomial $x^3 - 2$ factors as the product of a polynomial of degree 1 and an irreducible polynomial of degree 2: to see this, it is enough to remark that the endomorphism of $\mathbb{F}_{11}^*$ given by $a \mapsto a^3$ is an automorphism. On the other hand, $x^4 - 3$ factors as $(x^2 - 5)(x^2 + 5)$ so, independently of whether these factors are irreducible or not, the least common multiple of the degrees of the irreducible factors of $f(x)$ is 2. The splitting field of $f(x)$ is therefore $\mathbb{F}_{11^2}$.

196

(i) A simple calculation yields $\alpha^2 - 4 = 2i\sqrt{5}$, so $(\alpha^2 - 4)^2 = -20$, that is, $\alpha^4 - 8\alpha^2 + 36 = 0$. So α is a root of the polynomial $f(x) = x^4 - 8x^2 + 36$. The polynomial $f(x)$ is irreducible in $\mathbb{Q}[x]$: it has α and $\overline{\alpha} = \sqrt{5} - i$ as roots and, since its monomials all have even degree, $-\alpha = -\sqrt{5} - i$ and $-\overline{\alpha} = -\sqrt{5} + i$ are also roots $f(x)$.

It follows that $f(x)$ has no rational roots and that it has no factors of degree 2 with rational coefficients, because its degree 2 factors with real coefficients are $(x - \alpha)(x - \overline{\alpha}) = x^2 - 2\sqrt{5}x + 6$ and $(x + \alpha)(x + \overline{\alpha}) = x^2 + 2\sqrt{5}x + 6$, which are not in $\mathbb{Q}[x]$. So $f(x)$ is the minimal polynomial of α.

(ii) The splitting field of $f(x)$ over $\mathbb{Q}$ is $\mathbb{K} = \mathbb{Q}(\alpha, \bar{\alpha}, -\alpha, -\bar{\alpha})$.

One can immediately check that $\mathbb{K} = \mathbb{Q}(\sqrt{5}, i)$: clearly, $\mathbb{K} \subseteq \mathbb{Q}(\sqrt{5}, i)$, and moreover $\sqrt{5} = (\alpha + \bar{\alpha})/2$, $i = (\alpha - \bar{\alpha})/2 \in \mathbb{K}$. Remark that $[\mathbb{Q}(\sqrt{5}) : \mathbb{Q}] = 2$ because $x^2 - 5$ is irreducible by Eisenstein's criterion and has $\sqrt{5}$ as a root. Moreover, $[\mathbb{Q}(\sqrt{5})(i) : \mathbb{Q}(\sqrt{5})] = 2$ because i is a root of $x^2 + 1$, but $\mathbb{Q}(\sqrt{5}) \subseteq \mathbb{R}$ and $i \notin \mathbb{R}$. Therefore, $[\mathbb{K} : \mathbb{Q}] = [\mathbb{Q}(\sqrt{5})(i) : \mathbb{Q}(\sqrt{5})][\mathbb{Q}(\sqrt{5}) : \mathbb{Q}] = 2 \cdot 2 = 4$.

$$\mathbb{K} = \mathbb{Q}(\sqrt{5}, i)$$
$$2 \,|$$
$$\mathbb{Q}(\sqrt{5}) \subseteq \mathbb{R}$$
$$2 \,|$$
$$\mathbb{Q}$$

In $\mathbb{F}_7[x]$ we have $f(x) = x^4 - x^2 + 1 = (x^2 - 3)(x^2 - 5)$. Since the squares in $\mathbb{F}_7$ are $0, 1, 2, 4$, the polynomials $x^2 - 3$ and $x^2 - 5$ are irreducible, so the splitting field of $f(x)$ is $\mathbb{F}_{7^2}$.

197 The polynomial $f(x) = x^4 - x - 1$ is irreducible in $\mathbb{F}_2[x]$: it has no roots because it is not zero when evaluated at the classes of 0 and 1, and it is not the product of irreducible factors of degree 2 because it is not the square of $x^2 + x + 1$, which is the unique irreducible polynomial of degree 2 in $\mathbb{F}_2[x]$. It follows that $f(x)$ is irreducible in $\mathbb{Z}[x]$ and hence, by Gauss's lemma, in $\mathbb{Q}[x]$. We can therefore conclude that $f(x)$ is the minimal polynomial of α over $\mathbb{Q}$ and that $[\mathbb{Q}(\alpha) : \mathbb{Q}] = 4$.

Since $\mathbb{Q}(\alpha) = \mathbb{Q}(2\alpha - 1)$, the minimal polynomial of $2\alpha - 1$ over $\mathbb{Q}$ has degree 4. So the polynomial $g(x) = 2^4 f((x + 1)/2) = x^4 + 4x^3 + 6x^2 - 4x - 23$, which has $2\alpha - 1$ as a root, is monic and has degree 4, is the minimal polynomial of $2\alpha - 1$.

Let $\beta = \alpha^2$; we have $\mathbb{Q} \subseteq \mathbb{Q}(\beta) \subseteq \mathbb{Q}(\alpha)$ and we can easily compute $\beta^2 = \alpha + 1$, $\beta^3 = \alpha^3 + \alpha^2$ and $\beta^4 = \alpha^2 + 2\alpha + 1$. Since $1, \alpha, \alpha^2, \alpha^3$ are linearly independent over $\mathbb{Q}$, it is immediate to check that $1, \beta, \beta^2, \beta^3$ must be as well. In particular, β has degree 4 over $\mathbb{Q}$.

Now, from $\alpha^4 - 1 = \alpha$, by squaring both sides, we get $\alpha^8 - 2\alpha^4 + 1 = \alpha^2$, that is, β is a root of the polynomial $h(x) = x^4 - 2x^2 - x + 1$. Since $h(x)$ is monic of degree 4, it is the minimal polynomial of β.

198 Since the polynomial $f(x) = x^4 - 6x^2 - 3$ gives a biquadratic equation, one can compute the roots of $f(x)$ by means of the quadratic formula and find that they are $\pm\sqrt{3 \pm 2\sqrt{3}}$. The splitting field of $f(x)$ over $\mathbb{Q}$ is therefore $\mathbb{K} = \mathbb{Q}(\sqrt{3 + 2\sqrt{3}}, \sqrt{3 - 2\sqrt{3}})$, and $[\mathbb{K} : \mathbb{Q}] = [\mathbb{K} : \mathbb{Q}(\sqrt{3 + 2\sqrt{3}})][\mathbb{Q}(\sqrt{3 + 2\sqrt{3}}) : \mathbb{Q}]$.

$$\mathbb{K} = \mathbb{Q}(\sqrt{3 + 2\sqrt{3}}, \sqrt{3 - 2\sqrt{3}})$$
$$|$$
$$\mathbb{Q}(\sqrt{3 + 2\sqrt{3}}) \subseteq \mathbb{R}$$
$$|$$
$$\mathbb{Q}$$

The polynomial f is irreducible in $\mathbb{Z}[x]$ by Eisenstein's criterion for the prime 3, so by Gauss's lemma it is also irreducible in $\mathbb{Q}[x]$. It is therefore the minimal polynomial of its roots; in particular, $[\mathbb{Q}(\sqrt{3 + 2\sqrt{3}}) : \mathbb{Q}] = 4$.

Moreover, remark that $3 - 2\sqrt{3} = (\sqrt{3 - 2\sqrt{3}})^2 \in \mathbb{Q}(\sqrt{3 - 2\sqrt{3}})$ and that $\sqrt{3 - 2\sqrt{3}} \notin \mathbb{Q}(\sqrt{3 + 2\sqrt{3}})$, because $\sqrt{3 - 2\sqrt{3}} \notin \mathbb{R}$ and $\mathbb{Q}(\sqrt{3 + 2\sqrt{3}}) \subseteq \mathbb{R}$. It follows that $[\mathbb{K} : \mathbb{Q}(\sqrt{3 + 2\sqrt{3}})] = 2$, so $[\mathbb{K} : \mathbb{Q}] = 8$.

In $\mathbb{F}_{13}[x]$ we have $f(x) = (x^2 - 8)(x^2 + 2)$ and the two factors are irreducible because 8 and -2 are not squares in $\mathbb{F}_{13}$. The splitting field of $f(x)$ over $\mathbb{F}_{13}$ is therefore $\mathbb{F}_{13^2}$ and we have $[\mathbb{F}_{13^2} : \mathbb{F}_{13}] = 2$.

199

(i) By taking the square of α we obtain that $\alpha^2 - 2 = \sqrt{7}$, hence $\alpha^4 - 4\alpha^2 - 3 = 0$, so α is a root of the polynomial $f(x) = x^4 - 4x^2 - 3$. If we show that $f(x)$ is irreducible in $\mathbb{Q}[x]$ we find that it is the minimal polynomial of α and therefore $[\mathbb{Q}(\alpha) : \mathbb{Q}] = 4$.

Since the polynomial $f(x)$ gives a biquadratic equation, it is easy to compute its roots, namely, $\pm\sqrt{2 \pm \sqrt{7}}$, which are not rational: if $\pm\sqrt{2 \pm \sqrt{7}}$ were rational, then by taking its square we would have that $\sqrt{7}$ is rational, but this is of course not the case. Moreover, $\pm\sqrt{2 + \sqrt{7}}$ are real whereas $\pm\sqrt{2 - \sqrt{7}}$ are non-real complex conjugates.

If $f(x)$ were the product of two irreducible polynomials of degree 2, one of the two factors would have to be $(x - \sqrt{2 - \sqrt{7}})(x + \sqrt{2 - \sqrt{7}})$, whose coefficients are not rationals (its constant coefficient, for example, is $-2 + \sqrt{7} \notin \mathbb{Q}$).

(ii) Let $\mathbb{K}$ be the splitting field of $f(x)$ over $\mathbb{Q}$.

$$\mathbb{K} = \mathbb{Q}(\sqrt{2 + \sqrt{7}}, \sqrt{2 - \sqrt{7}})$$
$$|$$
$$\mathbb{Q}(\sqrt{2 + \sqrt{7}}) \subseteq \mathbb{R}$$
$$4 \Big|$$
$$\mathbb{Q}$$

By our previous remarks, we have $\mathbb{K} = \mathbb{Q}(\sqrt{2 + \sqrt{7}}, \sqrt{2 - \sqrt{7}})$ so $[\mathbb{K} : \mathbb{Q}] = [\mathbb{K} : \mathbb{Q}(\alpha)][\mathbb{Q}(\alpha) : \mathbb{Q}]$. Note that $[\mathbb{K} : \mathbb{Q}(\alpha)] = 2$: we have $\mathbb{K} = \mathbb{Q}(\alpha)(\sqrt{2 - \sqrt{7}})$ and $(\sqrt{2 - \sqrt{7}})^2 = 2 - \sqrt{7} \in \mathbb{Q}(\alpha)$, and moreover since $\mathbb{K}$ is not real it cannot coincide with $\mathbb{Q}(\alpha)$, which is contained in $\mathbb{R}$. It follows that $[\mathbb{K} : \mathbb{Q}] = 8$.

200

(i) By taking the square of both sides of the equation defining α we get $\alpha^2 = 2 + i\sqrt{2}$, hence $\alpha^2 - 2 = i\sqrt{2}$. Squaring again, we have $\alpha^4 - 4\alpha^2 + 4 = -2$ and so α is a root of the polynomial $h(x) = x^4 - 4x^2 + 6$.

By Eisenstein's criterion for the prime 2, $h(x)$ is irreducible in $\mathbb{Z}[x]$ and thus, by Gauss's lemma, in $\mathbb{Q}[x]$. Since it is monic, $h(x)$ is the minimal polynomial of α over $\mathbb{Q}$.

From the expression above, we get that α^2 is a root of $p(x) = x^2 - 4x + 6$. It follows that the polynomial $q(x) = p(x - 1) = (x - 1)^2 - 4(x - 1) + 6 = x^2 - 6x + 11$ has $\alpha^2 + 1$ as a root. The polynomial $q(x)$ is monic and irreducible in $\mathbb{Q}[x]$ (its roots are not real, so they are not rational). It follows that $q(x)$ is the minimal polynomial of $\alpha^2 + 1$ over $\mathbb{Q}$.

(ii) From $h(\alpha) = 0$ we obtain $(\alpha^2 + 2\alpha)(\alpha^2 - 2\alpha) = \alpha^4 - 4\alpha^2 = -6$, hence $(\alpha^2 + 2\alpha)^{-1} = -(\alpha^2 - 2\alpha)/6$.

Consequently, the polynomial $f(x) = -(x^2 - 2x)/6$ satisfies the required condition.

201 The splitting field of a polynomial of the form $x^n - a$ over a field $\mathbb{K}$ of characteristic zero, or of (positive) characteristic coprime to n, is given by $\mathbb{F} = \mathbb{K}(\alpha, \zeta)$, where α is an nth root of a and ζ is a primitive nth root of 1.

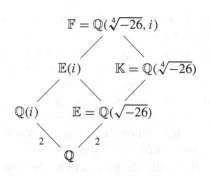

$$\mathbb{K} = \mathbb{Q}(\sqrt[4]{2}, i)$$
$$\mathbb{Q}(\sqrt[4]{2}) \subseteq \mathbb{R}$$
$$4$$
$$\mathbb{Q}$$

In our case, we can choose $\alpha = \sqrt[8]{4} = \sqrt[4]{2}$ and $\zeta = e^{2\pi i/8} = \frac{\sqrt{2}}{2}(1 + i)$. Since α is a root of the polynomial $x^4 - 2$, which is irreducible in $\mathbb{Q}[x]$ by Gauss's lemma and Eisenstein's criterion for the prime 2, we have $[\mathbb{Q}(\alpha) : \mathbb{Q}] = 4$. Remark that $\mathbb{Q}(\alpha) \subseteq \mathbb{R}$ and ζ is not real, so $\mathbb{Q}(\alpha) \neq \mathbb{F}$, that is, $[\mathbb{F} : \mathbb{Q}(\alpha)] > 1$. Moreover, $\sqrt{2} = \alpha^2 \in \mathbb{Q}(\alpha)$ so $\mathbb{F} \subseteq \mathbb{Q}(\alpha, i)$. The imaginary unit i is a root of the polynomial $x^2 + 1 \in \mathbb{Q}(\alpha)[x]$, so $[\mathbb{F} : \mathbb{Q}(\alpha)] \leq [\mathbb{Q}(\alpha, i) : \mathbb{Q}(\alpha)] \leq 2$.

It follows that $[\mathbb{F} : \mathbb{Q}] = [\mathbb{F} : \mathbb{Q}(\alpha)][\mathbb{Q}(\alpha) : \mathbb{Q}] = 2 \cdot 4 = 8$.

Over $\mathbb{K} = \mathbb{F}_3$ we can write the polynomial as $x^8 - 1$, so its splitting field is $\mathbb{F}_3(\zeta)$. A field $\mathbb{F}_{3^n}$ contains ζ if and only if its multiplicative group $\mathbb{F}_{3^n}^*$, which is cyclic and has order $3^n - 1$, contains a subgroup of order 8, that is, if and only if $8 \mid 3^n - 1$. The least n for which this holds is $n = 2$; therefore, the splitting field in question is $\mathbb{F}_9$ and has degree 2 over $\mathbb{F}_3$.

202 The polynomial $f(x) = x^4 + 26$ is irreducible in $\mathbb{Q}[x]$ by Gauss's lemma, because it is irreducible in $\mathbb{Z}[x]$ by Eisenstein's criterion for the prime 2. Therefore, given a complex root α of $f(x)$ and letting $\mathbb{K} = \mathbb{Q}(\alpha)$, we have $[\mathbb{K} : \mathbb{Q}] = 4$. The roots of $f(x)$ are $\pm\alpha, \pm i\alpha$, so the splitting field of $f(x)$ over $\mathbb{Q}$ is $\mathbb{F} = \mathbb{K}(i)$. Since i has degree 2 over $\mathbb{Q}$, we have $[\mathbb{F} : \mathbb{K}] \leq 2$.

$$\mathbb{F} = \mathbb{Q}(\sqrt[4]{-26}, i)$$
$$\mathbb{E}(i) \qquad \mathbb{K} = \mathbb{Q}(\sqrt[4]{-26})$$
$$\mathbb{Q}(i) \qquad \mathbb{E} = \mathbb{Q}(\sqrt{-26})$$
$$2 \qquad 2$$
$$\mathbb{Q}$$

Now consider $\beta = i\sqrt{26}$; let $\mathbb{E} = \mathbb{Q}(\beta)$ and set $\alpha = \pm\sqrt{\beta}$. Clearly, $[\mathbb{E} : \mathbb{Q}] = 2$ and $[\mathbb{K} : \mathbb{E}] = 2$. Remark that $\mathbb{E} \neq \mathbb{Q}(i) = \mathbb{Q}(\sqrt{-1})$. This is because two extensions of the same field obtained by adding the square roots of two elements coincide if and only if the product of the two elements is a square in the field, whereas $(-1)(-26) = 26$ is not the square of a rational number. Consequently, both $\mathbb{K}$ and $\mathbb{E}(i)$ are quadratic extensions of $\mathbb{E}$.

By the same argument, $\mathbb{K}$ and $\mathbb{E}(i)$ are distinct, since $(-1)(i\sqrt{26})$ is not a square in $\mathbb{E}$: a square root of $-i\sqrt{26}$ is a root of $f(x)$, so it has degree 4 over $\mathbb{Q}$ and therefore cannot belong to $\mathbb{E}$. So $i \notin \mathbb{K}$ and thus $[\mathbb{F} : \mathbb{Q}] = [\mathbb{F} : \mathbb{K}][\mathbb{K} : \mathbb{Q}] = 2 \cdot 4 = 8$.

As for the splitting field of $f(x)$ over a finite field, its degree is given by the least common multiple of the degrees of the irreducible factors of $f(x)$ over the field in question.

In $\mathbb{F}_5[x]$ we have $x^4 + 26 = x^4 - 4 = (x^2 - 2)(x^2 + 2)$, with the degree 2 factors being irreducible because they have no roots in $\mathbb{F}_5$. Therefore, the degree of the splitting field over $\mathbb{F}_5$ is 2.

In $\mathbb{F}_7[x]$ we have $x^4 + 26 = x^4 - 9 = (x^2 + 3)(x^2 - 3) = (x^2 - 4)(x^2 - 3) = (x + 2)(x - 2)(x^2 - 3)$, and the degree 2 factor is irreducible because it has no roots in $\mathbb{F}_7$. Therefore, the degree of the splitting field over $\mathbb{F}_7$ is 2.

203 We have $f(x) = x^6 - 12x^3 + 27 = (x^3 - 3)(x^3 - 9)$.

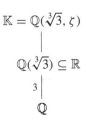

The polynomial $x^3 - 3$ is irreducible in $\mathbb{Z}[x]$ by Eisenstein's criterion for the prime 3; it is therefore irreducible in $\mathbb{Q}[x]$ bu Gauss's lemma. Denote by $\mathbb{K}$ the splitting field of $x^3 - 3$. We have $\mathbb{K} = \mathbb{Q}(\sqrt[3]{3}, \zeta)$, where ζ is a primitive third root of 1. The fact that $x^3 - 3$ is irreducible implies that $[\mathbb{Q}(\sqrt[3]{3}) : \mathbb{Q}] = 3$. Since we have $\mathbb{Q}(\sqrt[3]{3}) \subseteq \mathbb{R}$ and ζ is not real, the degree of $\mathbb{K}$ over $\mathbb{Q}$ must be strictly greater than 3. Finally, since this degree is at most $3! = 6$, we find it must be 6.

Similarly, the splitting field $\mathbb{F}$ of the polynomial $x^3 - 9$ is $\mathbb{Q}(\sqrt[3]{9}, \zeta)$. However, we have $\sqrt[3]{9} = \sqrt[3]{3}^2 \in \mathbb{K}$ and $\zeta \in \mathbb{K}$, so $\mathbb{F} \subseteq \mathbb{K}$. In conclusion, the splitting field of $f(x)$ over $\mathbb{Q}$ is equal to $\mathbb{K}$ and therefore has degree 6 over $\mathbb{Q}$.

In $\mathbb{F}_5[x]$ we have $f(x) = (x^3 - 3)(x^3 - 9) = (x - 2)(x^2 + 2x - 1)(x + 1)(x^2 - x + 1)$ because $2^3 \equiv 3$ and $(-1)^3 \equiv 9 \pmod{5}$. Moreover, $x^2 + 2x - 1$ and $x^2 - x + 1$ are irreducible because they have no roots in $\mathbb{F}_5$.

Therefore, the degree of the splitting field of $f(x)$ over $\mathbb{F}_5$ is 2, that is, the least common multiple of the degrees of the irreducible factors of $f(x)$.

204 By taking the square of $\alpha = \sqrt{2 + \sqrt{3}}$ we obtain $\alpha^2 = 2 + \sqrt{3}$, hence $\alpha^2 - 2 = \sqrt{3}$; squaring again yields $\alpha^4 - 4\alpha^2 + 4 = 3$. It follows that α is a root of the polynomial $f(x) = x^4 - 4x^2 + 1$.

Remark that applying the same procedure to $\beta = \sqrt{2 - \sqrt{3}}$ yields the same polynomial $f(x)$, so the roots of $f(x)$ are $\pm\alpha$ and $\pm\beta$.

The factorisation of $f(x)$ in $\mathbb{C}[x]$ is therefore $(x - \alpha)(x + \alpha)(x - \beta)(x + \beta)$. Since the squares of $\pm\alpha$ and $\pm\beta$ are clearly irrational, the numbers $\pm\alpha$ and $\pm\beta$ are themselves irrational; therefore, $f(x)$ has no factors of degree 1 in $\mathbb{Q}[x]$.

If $f(x)$ were the product of factors of degree 2, the factor having α as a root would have to be one of the following: $(x - \alpha)(x + \alpha)$, $(x - \alpha)(x - \beta)$, $(x - \alpha)(x + \beta)$.

The first of the polynomials above does not have rational coefficients because α^2 is irrational. The second does not have rational coefficients because $(\alpha + \beta)^2 = (\sqrt{2 + \sqrt{3}} + \sqrt{2 - \sqrt{3}})^2 = 6$, so $\alpha + \beta \notin \mathbb{Q}$. Finally, the third does not because $(\alpha - \beta)^2 = (\sqrt{2 + \sqrt{3}} - \sqrt{2 - \sqrt{3}})^2 = 2$.

The splitting field of $f(x)$ is $\mathbb{K} = \mathbb{Q}(\alpha, \beta)$, because clearly $-\alpha, -\beta \in \mathbb{K}$. Note, however, that $\alpha\beta = 1$, so $\beta = 1/\alpha \in \mathbb{Q}(\alpha)$. We therefore have $\mathbb{K} = \mathbb{Q}(\alpha)$ and, since the minimal polynomial of α over $\mathbb{Q}$ has degree 4, we have $[\mathbb{K} : \mathbb{Q}] = 4$.

205

(i) It is easy to check that the polynomial $x^3 - 7$ is irreducible in $\mathbb{Q}[x]$: it has degree 3 and, since ± 1 and ± 7 are not roots, it has no rational roots. In fact, its roots are $\sqrt[3]{7}, \sqrt[3]{7}\zeta, \sqrt[3]{7}\zeta^2$, where $\zeta = (-1 + \sqrt{-3})/2$ is a primitive third root of 1.

$\mathbb{K} = \mathbb{Q}(\sqrt[3]{7}, \zeta)$

$\mid$

$\mathbb{Q}(\sqrt[3]{7}) \subseteq \mathbb{R}$

$3 \mid$

$\mathbb{Q}$

Let $\mathbb{K}$ be the splitting field of $x^3 - 7$ over $\mathbb{Q}$; since the polynomial has degree 3 we have $[\mathbb{K} : \mathbb{Q}] \leq 3! = 6$, and since it is irreducible we have $[\mathbb{Q}(\sqrt[3]{7}) : \mathbb{Q}] = 3$. But $\mathbb{Q}(\sqrt[3]{7}) \subseteq \mathbb{K}$ and the inclusion is actually strict, because $\mathbb{Q}(\sqrt[3]{7}) \subseteq \mathbb{R}$ whereas $\mathbb{K}$ contains the two non-real roots of the polynomial. Therefore, we must have $[\mathbb{K} : \mathbb{Q}] = 6$. The polynomial $x^2 + 3$ is also irreducible in $\mathbb{Q}[x]$, and its splitting field is clearly $\mathbb{Q}(\sqrt{-3})$. Now, the fact that $\sqrt{-3} = \zeta - \zeta^2$ implies that $\sqrt{-3} \in \mathbb{K}$ and so $\mathbb{Q}(\sqrt{-3}) \subseteq \mathbb{K}$.

In conclusion, the splitting field of $f(x)$ coincides with $\mathbb{K}$ and has degree 6 over $\mathbb{Q}$.

(ii) We know that we can represent every element of $A = \mathbb{F}_5[x]/(f(x))$ as the class of a polynomial of degree at most 4, and the class of a polynomial $g(x)$ is a zero divisor in A if and only if $g(x)$ and $f(x)$ are not relatively prime. By checking for possible roots, we find that the factorisation of $f(x)$ in $\mathbb{F}_5[x]$ is $f(x) = (x - 3)(x^2 + 3x - 1)(x^2 + 3)$.

Remark that the classes that are multiples of some fixed $h(x)$ of degree d are of the form $h(x)k(x)$, where $k(x)$ is any polynomial of degree $4 - d$; there are therefore 5^{5-d} such classes. Thanks to this remark and the inclusion-exclusion principle, we can find the number of zero divisors by counting the multiples of $x - 3$, the multiples of $x^2 + 3x - 1$ and the multiples of $x^2 + 3$, subtracting the number of multiples of $(x - 3)(x^2 + 3x - 1)$, of $(x - 3)(x^2 + 3)$ and of $(x^2 + 3x - 1)(x^2 + 3)$, and finally summing the number of multiples of $(x - 3)(x^2 + 3x - 1)(x^2 + 3)$. We find that the number of zero divisors in A is

$$5^4 + 5^3 + 5^3 - (5^2 + 5^2 + 5) + 1 = 821.$$

206 First of all, we show that $f(x) = 2x^4 + 6x^2 - 5$ is irreducible in $\mathbb{Q}[x]$.

In $\mathbb{C}[x]$ we have $f(x) = 2(x^2 - \alpha)(x^2 - \beta) = 2(x - a)(x + a)(x - b)(x + b)$, where $\alpha = (-3 + \sqrt{19})/2$ and $\beta = (-3 - \sqrt{19})/2$ are the solutions of the equation $2y^2 + 6y - 5 = 0$, and $a^2 = \alpha$, $b^2 = \beta$. Since α and β are not rational numbers, neither are a and b, so $f(x)$ has no irreducible factors of degree 1 in $\mathbb{Q}[x]$.

Moreover, b is not real, so any polynomial with rational coefficients that has b as a root must also have among its roots the complex conjugate $\bar{b} = -b$ of b. It follows that the only potential factorisation of $f(x)$ as a product of two quadratic polynomials with rational coefficients is $2(x^2 - \alpha)(x^2 - \beta)$. But this is not a factorisation in $\mathbb{Q}[x]$, since α and β are not rational numbers.

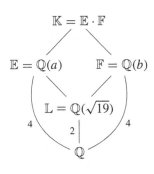

Set $\mathbb{L} = \mathbb{Q}(\alpha) = \mathbb{Q}(\beta) = \mathbb{Q}(\sqrt{19})$, $\mathbb{E} = \mathbb{Q}(a)$, $\mathbb{F} = \mathbb{Q}(b)$, $\mathbb{K} = \mathbb{EF} = \mathbb{Q}(a, b)$.

Clearly, $\mathbb{K}$ is the splitting field of $f(x)$. As $f(x)$ is irreducible we have $[\mathbb{E} : \mathbb{Q}] = [\mathbb{F} : \mathbb{Q}] = 4$, hence $[\mathbb{E} : \mathbb{L}] = [\mathbb{F} : \mathbb{L}] = 2$. Since $\mathbb{EF} = \mathbb{E}(b)$ and b has degree 2 over $\mathbb{L}$, the degree of b over $\mathbb{E}$ is at most 2; in particular, it is 1 if $b \in \mathbb{E}$, that is, if $\mathbb{E} = \mathbb{F}$, and 2 otherwise.

But $\mathbb{E} = \mathbb{L}(\sqrt{\alpha})$ and $\mathbb{F} = \mathbb{L}(\sqrt{\beta})$, so $\mathbb{E} = \mathbb{F}$ if and only if $\alpha\beta = -5/2$ is a square in $\mathbb{L}$. This is clearly not the case, because $\mathbb{L} \subseteq \mathbb{R}$ and squares in $\mathbb{R}$ are non-negative.

In conclusion, $\mathbb{E} \neq \mathbb{F}$ and $[\mathbb{EF} : \mathbb{E}] = 2$, so $[\mathbb{K} : \mathbb{Q}] = [\mathbb{EF} : \mathbb{E}][\mathbb{E} : \mathbb{Q}] = 2 \cdot 4 = 8$.

⟦One could alternatively show that $[\mathbb{E}(b) : \mathbb{E}] > 1$ by remarking that $E = \mathbb{Q}(a) = \mathbb{Q}(\sqrt{(-3 + \sqrt{19})/2}) \subseteq \mathbb{R}$, whereas $b = \sqrt{(-3 - \sqrt{19})/2} \notin \mathbb{R}$.⟧

Our initial computation of α and β shows that in $\mathbb{F}_{19}[x]$ we have $f(x) = 2(x^2 + 3 \cdot 2^{-1})^2 = 2(x^2 - 8)^2$. We can easily check that $x^2 - 8$ has no roots in $\mathbb{F}_{19}$, so $f(x)$ factors as a product of irreducible factors of degree 2 in $\mathbb{F}_{19}[x]$. Therefore, the degree of the splitting field of $f(x)$ over $\mathbb{F}_{19}$ is 2.

207 The polynomial $f(x)$ is irreducible in $\mathbb{Q}[x]$ by Eisenstein's criterion for the prime 3. We therefore have $[\mathbb{Q}(\alpha) : \mathbb{Q}] = 5$. Since $\alpha^7 \in \mathbb{Q}(\alpha)$, we have $\mathbb{Q} \subseteq \mathbb{Q}(\alpha^7) \subseteq \mathbb{Q}(\alpha)$ and so $[\mathbb{Q}(\alpha^7) : \mathbb{Q}] \mid 5$. If $[\mathbb{Q}(\alpha^7) : \mathbb{Q}]$ were 1, then α^7 would be a rational number. But the fact that $\alpha^5 + 3\alpha + 3 = 0$ implies that $\alpha^7 = -3\alpha^3 - 3\alpha^2$, and the latter is not a rational number because $1, \alpha^2, \alpha^3$ are linearly independent over $\mathbb{Q}$ (since $[\mathbb{Q}(\alpha) : \mathbb{Q}] = 5$). We thus also have $[\mathbb{Q}(\alpha^7) : \mathbb{Q}] = 5$.

Let us factor $f(x)$ in $\mathbb{F}_2[x]$. It is clear that $f(x)$ has no roots; on the other hand, it is divisible by the unique irreducible polynomial of degree 2, namely, $x^2 + x + 1$, hence $f(x) = (x^2 + x + 1)(x^3 + x^2 + 1)$. It follows that $[\mathbb{F}_2(\alpha) : \mathbb{F}_2]$ is either 2 or 3, depending on whether α is a root of the first or the second factor.

Remark that in the first case $\alpha \in \mathbb{F}_4^*$, which is a cyclic group with three elements; so $\alpha^7 = \alpha$ and $[\mathbb{F}_2(\alpha^7) : \mathbb{F}_2] = 2$. In the second case we have $\alpha \in \mathbb{F}_8^*$, which is a cyclic group with seven elements, so $\alpha^7 = 1$ and $[\mathbb{F}_2(\alpha^7) : \mathbb{F}_2] = 1$.

208 We know that an element $\overline{g(x)} \in \mathbb{K}[x]/(f(x))$ is a zero divisor if and only if $(g(x), f(x)) \neq 1$. In other words, an element $\overline{g(x)}$ is a zero divisor if and only if $g(x)$ is divisible by *at least one* of the irreducible factors of $f(x)$.

Moreover, we know that $\overline{g(x)}$ is nilpotent if and only if $g(x)$ is divisible by *all* irreducible factors of $f(x)$.

To summarise, if $f(x) = p(x)^k$ is a power of a single irreducible polynomial, then every zero divisor is represented by a polynomial $g(x)$ which is a multiple of $p(x)$ and therefore nilpotent. Conversely, if $f(x)$ is divisible by at least two distinct irreducible polynomials $p(x), q(x)$, then the element $\overline{p(x)}$ is a zero divisor and is not nilpotent.

209 The factorisation of $f(x)$ in $\mathbb{F}_5[x]$ is

$$x^3 - 2x + 1 = (x - 1)(x - 2)^2.$$

The zero divisors in $\mathbb{F}_5[x]/(x^3 - 2x + 1)$ are given by the union of classes of polynomials that are multiples of $x - 1$ and polynomials that are multiples of $x - 2$. The classes of multiples of $x - 1$ are represented by polynomials of the form $(x - 1)(ax + b)$ with $a, b \in \mathbb{F}_5$, so there are 25, as many as the ways to choose the pair (a, b). Similarly, there are 25 classes of multiples of $x - 2$. The intersection of classes of multiples of $x - 1$ and $x - 2$ is the set of classes of polynomials that are multiples of $(x - 1)(x - 2)$, that is, classes represented by $c(x - 1)(x - 2)$ with $c \in \mathbb{F}_5$: there are 5 of them. By the inclusion-exclusion principle, we have $25 + 25 - 5 = 45$ zero divisors.

The nilpotent elements are the classes of polynomials that are multiples of every irreducible factor of $f(x)$, that is, the five classes of multiples of $(x - 1)(x - 2)$. The answer is therefore $45 - 5 = 40$.

210

(i) Let α be the positive fourth root of a. In $\mathbb{C}[x]$ the polynomial $f(x) = x^4 - a$ factors as $f(x) = (x - \alpha)(x + \alpha)(x - i\alpha)(x + i\alpha)$. We know that $f(x)$ is reducible in $\mathbb{Z}[x]$. There are two possibilities: either $f(x)$ has a root in $\mathbb{Z}$ or $f(x)$ factors as the product of two irreducible quadratic polynomials in $\mathbb{Z}[x]$. In the first case, since $\pm i\alpha$ are not real numbers, the only possible roots are $\pm \alpha$ and in fact, if one of the two is an integer then the other must be as well. But if $\alpha = k \in \mathbb{N}$ then $a = \alpha^4 = k^4$, so we have $a = b^2$ for some $b = k^2 \in \mathbb{N}$.
In the second case, the degree 2 factors must be $x^2 - \alpha^2$ and $x^2 + \alpha^2$, because the factor that has a non-real root must also have its complex conjugate as a root. It follows that $\alpha^2 = b \in \mathbb{N}$ and again $a = \alpha^4 = b^2$.

(ii) Let α be the positive fourth root of $-a$. The polynomial $f(x)$ factors in $\mathbb{C}[x]$ as $f(x) = (x - \zeta\alpha)(x - \bar{\zeta}\alpha)(x - \zeta^3\alpha)(x - \bar{\zeta}^3\alpha)$, where $\zeta = (1 + i)/\sqrt{2}$ is a primitive eighth root of 1. In this case the polynomial has no real roots: the only potential factorisation with integer coefficients comes from pairing up complex conjugate roots to obtain the factors $(x - \zeta\alpha)(x - \bar{\zeta}\alpha) = x^2 - \sqrt{2}\alpha x + \alpha^2$ and $(x - \zeta^3\alpha)(x - \bar{\zeta}^3\alpha) = x^2 + \sqrt{2}\alpha x + \alpha^2$. Again $\alpha^2 = c$ must be a natural number and we must have $-a = \alpha^4 = c^2$. Moreover, the condition that $\sqrt{2}\alpha$ is an integer yields that $\alpha = d'/\sqrt{2}$ for some $d' \in \mathbb{N}$. By squaring both sides we obtain $c = d'^2/2$, so d' must be even, say $d' = 2d$ with $d \in \mathbb{N}$, and $c = 2d^2$.

211 First of all, remark that the polynomial $f(x) = x^4 + 5x^2 + 5$ is irreducible in $\mathbb{Q}[x]$ by Gauss's lemma and by Eisenstein's criterion for the prime 5. By the same argument, the polynomial $g(y) = y^2 + 5y + 5$ is irreducible. Its roots are $\alpha_1, \alpha_2 = (-5 \pm \sqrt{5})/2$, and its splitting field is $\mathbb{F} = \mathbb{Q}(\alpha_1, \alpha_2) = \mathbb{Q}(\sqrt{5})$, which has degree 2 over $\mathbb{Q}$.

The splitting field of $f(x)$ is thus $\mathbb{K} = \mathbb{F}(\sqrt{\alpha_1}, \sqrt{\alpha_2})$. Remark that $\alpha_1\alpha_2$, being the product of the two roots of $g(x)$, is equal to 5, which is a square in $\mathbb{F}$. So the two extensions $\mathbb{F}(\sqrt{\alpha_1})/\mathbb{F}$ and $\mathbb{F}(\sqrt{\alpha_2})/\mathbb{F}$ coincide and the degree of the splitting field $\mathbb{K}$ over $\mathbb{Q}$ is at most 4.

On the other hand, since $f(x)$ is irreducible in $\mathbb{Q}[x]$, the degree of its splitting field is a multiple of 4, which is the degree of the extension obtained by adding a single root of $f(x)$. Therefore, $[\mathbb{K} : \mathbb{Q}] = 4$.

In $\mathbb{F}_{11}[x]$ the polynomial $f(x)$ can be written as $x^4 - 6x^2 + 5 = (x^2 - 1)(x^2 - 5) = (x + 1)(x - 1)(x^2 - 5)$. Since $5 \equiv 4^2 \pmod{11}$, we have $x^2 - 5 = (x + 4)(x - 4)$, so $f(x) = (x + 1)(x - 1)(x + 4)(x - 4)$ and thus the degree of the splitting field over $\mathbb{F}_{11}$ is 1.

212

(i) Since α is a root of $f(x) = x^3 - x^2 - 2x - 1$, we have $\alpha^3 = \alpha^2 + 2\alpha + 1$. Multiplying both sides by α and substituting into the expression for β yields

$$\begin{aligned}
\beta &= \alpha^4 - 3\alpha^2 \\
&= \alpha^3 + 2\alpha^2 + \alpha - 3\alpha^2 \\
&= \alpha^3 - \alpha^2 + \alpha \\
&= \alpha^2 + 2\alpha + 1 - \alpha^2 + \alpha \\
&= 3\alpha + 1.
\end{aligned}$$

Now, remark that the polynomial $f(x)$ has no rational roots because $f(\pm 1) \neq 0$. So, since its degree is 3, $f(x)$ is irreducible in $\mathbb{Q}[x]$. It follows that $[\mathbb{Q}(\alpha) : \mathbb{Q}] = 3$.

Clearly, $\beta \in \mathbb{Q}(\alpha)$ and $\alpha = (\beta - 1)/3 \in \mathbb{Q}(\beta)$, so $\mathbb{Q}(\beta) = \mathbb{Q}(\alpha)$ and therefore the degree of the minimal polynomial of β over $\mathbb{Q}$ is 3.

Substituting $x = (y - 1)/3$ into the polynomial $f(x)$, we find that the polynomial

$$\left(\frac{y-1}{3}\right)^3 - \left(\frac{y-1}{3}\right)^2 - \frac{y-1}{3} - 1 = \frac{1}{27}y^3 - \frac{2}{9}y^2 - \frac{1}{3}y - \frac{13}{27}$$

has β as a root, so $y^3 - 6y^2 - 9y - 13$ is the minimal polynomial of β over $\mathbb{Q}$.

(ii) Using the minimal polynomial of β, we have that $\beta^3 - 6\beta^2 - 9\beta = 13$. By dividing both sides of the equality by 13, we obtain

$$\beta \frac{\beta^2 - 6\beta - 9}{13} = 1;$$

substituting $\beta = 3\alpha + 1$ yields

$$\frac{\beta^2 - 6\beta - 9}{13} = \frac{9\alpha^2 - 12\alpha - 14}{13}.$$

It follows that $g(x) = (9x^2 - 12x - 14)/13$ satisfies the requirements of the problem.

⟦The second part of the problem can also be solved as follows. Every element of $\mathbb{Q}(\alpha)$ can be written in the form $u_2\alpha^2 + u_1\alpha + u_0$ for some $u_0, u_1, u_2 \in \mathbb{Q}$. In order to find a polynomial $g(x)$ such that $\beta g(\alpha) = 1$, we can simply compute $h = (3\alpha + 1)(u_2\alpha^2 + u_1\alpha + u_0) - 1$ and solve the linear system in u_0, u_1, u_2 obtained by setting $h = 0$.⟧

213 The roots of $x^2 + 3$ are $\pm\sqrt{-3}$ and the roots of $x^3 - 5$ are $\sqrt[3]{5}, \sqrt[3]{5}\zeta, \sqrt[3]{5}\zeta^2$, where $\zeta = (-1 + \sqrt{-3})/2$. The splitting field in question is therefore $\mathbb{K} \doteq \mathbb{Q}(\sqrt{-3}, \sqrt[3]{5}, \zeta)$.

Remark that $\mathbb{K} = \mathbb{Q}(\sqrt{-3}, \sqrt[3]{5})$: clearly, we have $\mathbb{Q}(\sqrt{-3}, \sqrt[3]{5}) \subseteq \mathbb{K}$, and moreover $\zeta = (-1 + \sqrt{-3})/2 \in \mathbb{Q}(\sqrt{-3})$ so we also have $\sqrt[3]{5}\zeta, \sqrt[3]{5}\zeta^2 \in \mathbb{Q}(\sqrt{-3}, \sqrt[3]{5})$.

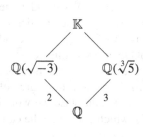

The field $\mathbb{K}$ contains the two subextensions $\mathbb{Q}(\sqrt{-3})$ and $\mathbb{Q}(\sqrt[3]{5})$, whose degrees over $\mathbb{Q}$ are 2 and 3, respectively. Indeed, the respective minimal polynomials of their generators are $x^2 + 3$ e $x^3 - 5$, and both polynomials are irreducible in $\mathbb{Q}[x]$ because they have no rational roots and their degrees are at most 3. This implies that $2 \mid [\mathbb{K} : \mathbb{Q}]$ and $3 \mid [\mathbb{K} : \mathbb{Q}]$, so $6 \mid [\mathbb{K} : \mathbb{Q}]$. On the other hand, $[\mathbb{K} : \mathbb{Q}] = [\mathbb{K} : \mathbb{Q}(\sqrt[3]{5})][\mathbb{Q}(\sqrt[3]{5}) : \mathbb{Q}] \le 2 \cdot 3 = 6$, and therefore $[\mathbb{K} : \mathbb{Q}] = 6$.

214

(i) For $p = 3$ we have $x^{15} - 1 = (x^5 - 1)^3$, so factoring $f(x)$ is the same as factoring $x^5 - 1$. Since 5 and 3 are relatively prime, the splitting field of $x^5 - 1$ over $\mathbb{F}_3$ is $\mathbb{F}_{3^d}$, where d is the multiplicative order of 3 modulo 5. We have $d = 4$ and $(x - 1)^3(x^4 + x^3 + x^2 + x + 1)^3$ is the factorisation of $f(x)$ as a product of irreducible factors.

For $p = 5$ the argument is similar: $f(x) = (x^3 - 1)^5$ and the splitting field is $\mathbb{F}_{5^2}$, because 5 has order 2 in $\mathbb{F}_3^*$. The factorisation if $f(x)$ as a product of irreducible factors is given by $f(x) = (x - 1)^5(x^2 + x + 1)^5$.

(ii) If p is neither 3 nor 5 then 15 and p are relatively prime. We can find the splitting field of $f(x)$ as before: it is $\mathbb{F}_{p^d}$, where d is the multiplicative order of p modulo 15. Since $(\mathbb{Z}/15\mathbb{Z})^* \simeq (\mathbb{Z}/3\mathbb{Z})^* \times (\mathbb{Z}/5\mathbb{Z})^* \simeq \mathbb{Z}/2\mathbb{Z} \times \mathbb{Z}/4\mathbb{Z}$, the order of p in $(\mathbb{Z}/15\mathbb{Z})^*$ is a divisor of 4.

(iii) For $p = 31$ we have $d = 1$ because $31 \equiv 1 \pmod{15}$, for $p = 11$ we have $d = 2$ because $11^2 = 121 \equiv 1 \pmod{15}$, and for $p = 2$ we have $d = 4$ because the multiplicative order of 2 modulo 15 is 4.

215 First, consider the case of the field $\mathbb{F}_7$. One can check that $x^2 + 2$ has no roots in $\mathbb{F}_7$ and is therefore irreducible; moreover, $x^4 - 2 = (x^2 - 3)(x^2 + 3)$. So, independently of whether or not the two polynomials $x^2 - 3$ and $x^2 + 3$ are irreducible, the least common multiple of the degrees of the irreducible factors of $(x^2 + 2)(x^4 - 2)$ is 2. The splitting field of the polynomial over $\mathbb{F}_7$ has thus degree 2.

The roots in $\mathbb{C}$ of the polynomial $(x^2 + 2)(x^4 - 2)$ are $\pm i\sqrt{2}$, $\pm\sqrt[4]{2}$ and $\pm i\sqrt[4]{2}$, so its splitting field over $\mathbb{Q}$ is $\mathbb{K} = \mathbb{Q}(i\sqrt{2}, \sqrt[4]{2}, i\sqrt[4]{2}) = \mathbb{Q}(\sqrt[4]{2}, i)$.

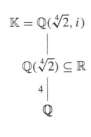

In order to compute the degree $[\mathbb{K} : \mathbb{Q}]$, consider the tower of extensions $\mathbb{Q} \subseteq \mathbb{Q}(\sqrt[4]{2}) \subseteq \mathbb{Q}(\sqrt[4]{2})(i) = \mathbb{K}$. The first extension has degree 4 because the minimal polynomial of $\sqrt[4]{2}$ over $\mathbb{Q}$ is $x^4 - 2$: it is an irreducible polynomial in $\mathbb{Z}[x]$ by Eisenstein's criterion for the prime 2 and therefore is also irreducible in $\mathbb{Q}[x]$ by Gauss's lemma. The second extension has degree 2 because i is a root of $x^2 + 1$ and the extension is nontrivial, since $\mathbb{Q}(\sqrt[4]{2}) \subseteq \mathbb{R}$ and $i \notin \mathbb{R}$. We therefore have $[\mathbb{K} : \mathbb{Q}] = 8$.

216

(i)

$$F = \mathbb{Q}(\sqrt{5}, \sqrt{-5})$$

$$\mathbb{Q}(\sqrt{5}) \subseteq \mathbb{R}$$

Let $F = \mathbb{Q}(\sqrt{5}, \sqrt{-5}) = \mathbb{Q}(\alpha, \beta)$. Consider the extensions $\mathbb{Q} \subseteq \mathbb{Q}(\sqrt{5}) \subseteq F$. The first has degree 2 over $\mathbb{Q}$ and is a subextension of $\mathbb{R}/\mathbb{Q}$; the second has degree at most 2: it is obtained by adding $\sqrt{-5}$ to $\mathbb{Q}(\sqrt{5})$, where $\sqrt{-5}$ has degree 2 over $\mathbb{Q}$ and therefore degree at most 2 over $\mathbb{Q}(\sqrt{5})$. However, the degree of the second extension cannot be 1, otherwise we would have $\mathbb{Q}(\sqrt{5}) = \mathbb{Q}(\sqrt{5}, \sqrt{-5})$, which cannot be the case because the field $\mathbb{Q}(\sqrt{5}, \sqrt{-5})$ is not real.

We have thus proven that F has degree 4 over $\mathbb{Q}$. Clearly, $\alpha + \beta \in F$. We want to show that, in fact, $\mathbb{Q}(\alpha + \beta) = F$, and thus that the minimal polynomial of $\alpha + \beta$ over $\mathbb{Q}$ has degree 4.

Remark that $i = (\alpha+\beta)^2/10 \in \mathbb{Q}(\alpha+\beta)$ and so the element $-\alpha+\beta = i(\alpha+\beta)$ also belongs to $\mathbb{Q}(\alpha + \beta)$. But then it is clear that $\alpha, \beta \in \mathbb{Q}(\alpha + \beta)$, that is, $\mathbb{Q}(\alpha + \beta) = \mathbb{Q}(\alpha, \beta) = F$.

(ii) The degree of $\mathbb{F}_p(\alpha)/\mathbb{F}_p$ is at most 2, as is the degree of $\mathbb{F}_p(\beta)/\mathbb{F}_p$, so both $\mathbb{F}_p(\alpha)$ and $\mathbb{F}_p(\beta)$ are contained in $\mathbb{F}_{p^2}$ because any fixed algebraic closure of $\mathbb{F}_p$ contains a unique extension of $\mathbb{F}_p$ of degree 2. So $\alpha + \beta \in \mathbb{F}_{p^2}$ and thus the minimal polynomial of $\alpha + \beta$ has degree at most 2 over $\mathbb{F}_p$.

For example, if $p = 5$ then $\alpha = \beta = 0$ and $\alpha + \beta$ obviously has degree 1 over $\mathbb{F}_5$. If $p = 3$ then $\alpha = \pm\sqrt{2} \in \mathbb{F}_{3^2} \setminus \mathbb{F}_3$, whereas $\beta = \pm 1$, so $\alpha + \beta \in \mathbb{F}_{3^2} \setminus \mathbb{F}_3$ and $\alpha + \beta$ has degree 2 over $\mathbb{F}_3$.

(iii) Since 2011 is congruent to 3 modulo 4, we have that -1 is not a square in $\mathbb{F}_{2011}$. So exactly one among 5 and -5 is a square in $\mathbb{F}_{2011}$: one of the two extensions $\mathbb{F}_{2011}(\alpha)$ and $\mathbb{F}_{2011}(\beta)$ has degree 1 and the other has degree 2. In any case, the degree of the minimal polynomial of $\alpha + \beta$ is 2.

217 The polynomial $f(x)$ is irreducible over $\mathbb{Q}$ by Eisenstein's criterion, so it is the minimal polynomial of each of its roots. It follows that $[\mathbb{Q}(\alpha) : \mathbb{Q}] = 4$. The roots of $f(x)$ in an algebraic closure of $\mathbb{Q}$ are $\pm\sqrt[4]{3}$, $\pm i\sqrt[4]{3}$, so the splitting field of $f(x)$ over $\mathbb{Q}$ is $F = \mathbb{Q}(\sqrt[4]{3}, i\sqrt[4]{3}) = \mathbb{Q}(\sqrt[4]{3}, i)$.

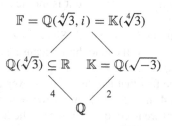

$$\mathbb{F} = \mathbb{Q}(\sqrt[4]{3}, i) = \mathbb{K}(\sqrt[4]{3})$$

Using the fact that the degree is multiplicative in towers, we have $[\mathbb{F} : \mathbb{Q}] = [\mathbb{Q}(\sqrt[4]{3})(i) : \mathbb{Q}(\sqrt[4]{3})][\mathbb{Q}(\sqrt[4]{3}) : \mathbb{Q}] = 2 \cdot 4 = 8$. Indeed, we already found the degree of $\mathbb{Q}(\sqrt[4]{3})/\mathbb{Q}$, and the other degree is 2: it is at most 2 because i is a root of $x^2 + 1 \in \mathbb{Q}(\sqrt[4]{3})[x]$, and it is more than 1 because $i \notin \mathbb{Q}(\sqrt[4]{3}) \subseteq \mathbb{R}$.

Now, let $\mathbb{K} = \mathbb{Q}(\sqrt{-3}) = \mathbb{Q}(i\sqrt{3})$. Since $i\sqrt{3}$ is not real we have that $[\mathbb{Q}(\sqrt[4]{3}, i\sqrt{3}) : \mathbb{Q}] = 8$, so $[\mathbb{K}(\sqrt[4]{3}) : \mathbb{K}] = [\mathbb{Q}(\sqrt[4]{3}, i\sqrt{3}) : \mathbb{Q}]/[\mathbb{K} : \mathbb{Q}] = 8/2 = 4$. It follows that the polynomial $f(x)$ is also irreducible in $\mathbb{K}[x]$, and thus every root α of $f(x)$ has degree 4 over $\mathbb{K}$. As before, the splitting field of $f(x)$ over $\mathbb{K}$ is $\mathbb{K}(\sqrt[4]{3}, i)$. On the other hand, $\mathbb{K}(\sqrt[4]{3}) = \mathbb{K}(\sqrt[4]{3}, i) = \mathbb{Q}(i\sqrt{3}, \sqrt[4]{3}, i) = \mathbb{Q}(\sqrt[4]{3}, i) = \mathbb{F}$ because $i\sqrt{3} = i(\sqrt[4]{3})^2 \in \mathbb{F}$, so the degree we are computing is $[\mathbb{F} : \mathbb{K}] = [\mathbb{K}(\sqrt[4]{3}) : \mathbb{K}] = 4$.

218

(i) Remember that the splitting field of a polynomial over a finite field $\mathbb{F}_{p^n}$ is $\mathbb{F}_{p^{nd}}$, where d is the least common multiple of the degrees of the irreducible factors of $f(x)$ in $\mathbb{F}_{p^n}[x]$.

In $\mathbb{F}_2[x]$ we have $f(x) = x(x^4 + x + 1)$; moreover, the polynomial $x^4 + x + 1$ is irreducible because it has no roots and it is not the square of the unique degree 2 irreducible polynomial in $\mathbb{F}_2[x]$, namely, $x^2 + x + 1$. The splitting field of $f(x)$ over $\mathbb{F}_2$ is therefore $\mathbb{F}_{2^4}$.

In $\mathbb{F}_3[x]$ we have $f(x) = (x^2 + 1)(x^3 - x + 1)$ and each of the two factors is irreducible because it has no roots. The splitting field of $f(x)$ over $\mathbb{F}_3$ is therefore $\mathbb{F}_{3^6}$.

(ii) We need to check whether or not $x^2 + 1$ and $x^3 - x + 1$ are irreducible in $\mathbb{F}_{3^k}[x]$. The roots of $x^2 + 1$ generate $\mathbb{F}_{3^2}$, so the polynomial factors in $\mathbb{F}_{3^k}[x]$ if and only if $\mathbb{F}_{3^2} \subseteq \mathbb{F}_{3^k}$, that is, if and only if $2 \mid k$. Similarly, $x^3 - x + 1$ factors in $\mathbb{F}_{3^k}[x]$ if and only if $3 \mid k$.

In conclusion, if $k \equiv 0 \pmod 6$ then $f(x)$ factors as a product of degree 1 polynomials in $\mathbb{F}_{3^k}[x]$. If $k \equiv 2, 4 \pmod 6$ then $f(x)$ has two roots in $\mathbb{F}_{3^k}$ and an irreducible factor of degree 3. If $k \equiv 3 \pmod 6$ then $f(x)$ has three roots and an irreducible factor of degree 2. Finally, if $k \equiv 1, 5 \pmod 6$ then $f(x)$ has an irreducible factor of degree 2 and one of degree 3 in $\mathbb{F}_{3^k}[x]$ (as in $\mathbb{F}_3[x]$).

219 Reducing the polynomial $f(x) = x^4 + 2x^3 + 2x^2 + x + 3$ modulo 2 yields $x^4 + x + 1$, which is irreducible because it has no roots in $\mathbb{F}_2$ and is not a square of $x^2 + x + 1$, which is the only irreducible quadratic polynomial in $\mathbb{F}_2[x]$. Therefore, $f(x)$ is irreducible in $\mathbb{Z}[x]$ and hence, by Gauss's lemma, in $\mathbb{Q}[x]$.

This implies that $f(x)$ is the minimal polynomial of α over $\mathbb{Q}$, so $[\mathbb{Q}(\alpha) : \mathbb{Q}] = 4$. Since $\mathbb{Q}(\alpha + 1) = \mathbb{Q}(\alpha)$, the minimal polynomial of $\alpha + 1$ over $\mathbb{Q}$ also has degree 4. Clearly, the polynomial $f(x - 1) = x^4 - 2x^3 + 2x^2 - x + 3$ has $\alpha + 1$ as a root, so it must be a multiple of the minimal polynomial of $\alpha + 1$; since it is monic and has degree 4, it *is* the polynomial we are looking for.

We now set out to find the minimal polynomial of $\alpha^2 + \alpha$ over $\mathbb{Q}$. Since $\mathbb{Q}(\alpha^2 + \alpha) \subseteq \mathbb{Q}(\alpha)$, the degree of $\alpha^2 + \alpha$ over $\mathbb{Q}$ is a divisor of 4. It cannot be 1 because that would imply $\alpha^2 + \alpha = q \in \mathbb{Q}$ and thus $x^2 + x - q$ would be a degree 2 polynomial of $\mathbb{Q}[x]$ that has α as a root, which contradicts the fact that the degree of α over $\mathbb{Q}$ is 4. The degree of $\alpha^2 + \alpha$ is therefore either 2 or 4: it is 2 if and only if $(\alpha^2 + \alpha)^2$, $(\alpha^2 + \alpha)$ and 1 are linearly dependent over $\mathbb{Q}$. What we need to determine is whether there are solutions $a, b \in \mathbb{Q}$ of

$$(\alpha^2 + \alpha)^2 + a(\alpha^2 + \alpha) + b = 0.$$

Simple calculations yield the equation $\alpha^4 + 2\alpha^3 + (a+1)\alpha^2 + a\alpha + b = 0$ and, by replacing α^4 with the expression obtained from $f(\alpha) = 0$, we get

$$(a-1)\alpha^2 + (a-1)\alpha + b - 3 = 0$$

whose solution is $a = 1$, $b = 3$. This shows that the polynomial $h(x) = x^2 + x + 3$ has $\alpha^2 + \alpha$ as a root and, since it has the least possible degree, it is the desired minimal polynomial.

220

(i) Let us show that $f(x)$ is irreducible in $\mathbb{Q}[x]$. First of all, it has no rational roots, as we can see by checking that $f(1)$ and $f(-1)$ are both nonzero. By Gauss's lemma, it is enough to prove that $f(x)$ is irreducible in $\mathbb{Z}[x]$. Now, if in $\mathbb{Z}[x]$ the polynomial $f(x)$ were a product of two quadratic factors, which we may assume to be monic, we would have $f(x) = (x^2 + ax + b)(x^2 + cx + d)$ for some $a, b, c, d \in \mathbb{Z}$. By equating the terms of degree three of the left and right hand side, we get $a + c = 0$, while for the constant coefficients we have $bd = 1$, hence $c = -a$ and $d = b = \pm 1$. So we would have

$$f(x) = (x^2 + ax + b)(x^2 - ax + b) = (x^2 + b)^2 - (ax)^2 = x^4 + (2b - a^2)x^2 + 1.$$

Finally the terms of degree two give the equality $2b - a^2 = 3$, that is, $2b - 3 = a^2$. But if $b \in \{1, -1\}$ then $2b - 3 \in \{-1, -5\}$ so it can't be the square of an integer. It follows that $f(x)$ is irreducible, and since it is also monic it is the minimal polynomial of α over $\mathbb{Q}$. Therefore, $[\mathbb{Q}(\alpha) : \mathbb{Q}] = 4$.

(ii) Let β_1, β_2 be the roots of the quadratic polynomial $y^2 + 3y + 1$. The field $\mathbb{E} = \mathbb{Q}(\beta_1) = \mathbb{Q}(\beta_2)$ has degree 2 over $\mathbb{Q}$: we have already checked that this polynomial has no rational roots. If we let $\mathbb{K}$ be the splitting field of $f(x)$, we have $\mathbb{K} = \mathbb{E}(\sqrt{\beta_1}, \sqrt{\beta_2})$. But the fields $\mathbb{E}(\sqrt{\beta_1})$ and $\mathbb{E}(\sqrt{\beta_2})$ coincide (because $\beta_1\beta_2 = 1$ is a square in $\mathbb{E}$), so $\mathbb{K} = \mathbb{E}(\sqrt{\beta_1})$. Without loss of generality, we may assume that $\mathbb{K} = \mathbb{Q}(\alpha)$ and obtain $[\mathbb{K} : \mathbb{Q}] = 4$.

(iii) The polynomial $f(x-1) = (x-1)^4 + 3(x-1)^2 + 1 = x^4 - 4x^3 + 9x^2 - 10x + 5$ vanishes at $\alpha + 1$. Its reciprocal polynomial, that is, $g(x) = 5x^4 - 10x^3 + 9x^2 -$

$4x + 1$, vanishes at $1/(\alpha + 1)$. Moreover, it is clear that $\mathbb{Q}(\alpha) = \mathbb{Q}(1/(\alpha + 1))$, so the minimal polynomial of $1/(\alpha + 1)$ over $\mathbb{Q}$ has degree 4. It follows that the minimal polynomial we want is $g(x)/5$.

221 The polynomial $f(x) = x^4 - 4x^2 + 2$ is irreducible in $\mathbb{Q}[x]$ by Gauss's lemma and by Eisenstein's criterion for the prime 2, so each of its roots has degree 4 over $\mathbb{Q}$.

The complex solutions of the equation $y^2 - 4y + 2 = 0$ are $2 \pm \sqrt{2}$, so the complex roots of $f(x)$ are $\pm\alpha, \pm\beta$, where $\alpha^2 = 2 + \sqrt{2}$ and $\beta^2 = 2 - \sqrt{2}$. Clearly, both $\mathbb{Q}(\alpha)$ and $\mathbb{Q}(\beta)$ are degree 2 extensions of $\mathbb{Q}(\sqrt{2})$. Moreover, they coincide because $(2 + \sqrt{2})(2 - \sqrt{2}) = 2$ is a square in $\mathbb{Q}(\sqrt{2})$. So the splitting field of $f(x)$ over $\mathbb{Q}$ is $\mathbb{K} = \mathbb{Q}(\pm\alpha, \pm\beta)$ and we have $[\mathbb{K} : \mathbb{Q}] = 4$.

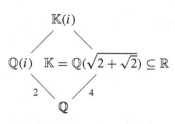

Remark that the roots of $f(x)$ are all real because they are the square roots of positive numbers, so $i \notin \mathbb{K}$. It follows that $[\mathbb{K}(i) : \mathbb{K}] = 2$ and, since the degree is multiplicative in towers, $[\mathbb{K}(i) : \mathbb{Q}] = 8$. On the other hand, $\mathbb{K}(i) = \mathbb{Q}(i, \alpha, \beta)$ is the splitting field of $f(x)$ over $\mathbb{Q}(i)$, so $[\mathbb{K}(i) : \mathbb{Q}(i)] = [\mathbb{K}(i) : \mathbb{Q}]/[\mathbb{Q}(i) : \mathbb{Q}] = 8/2 = 4$.

In $\mathbb{F}_7$ we have $(\pm 3)^2 = 2$, so the roots of $y^2 - 4y + 2 = 0$ are $2 \pm 3 = -1, 5$. It follows that $y^2 - 4y + 2 = (y + 1)(y - 5)$ and $x^4 - 4x^2 + 2 = (x^2 + 1)(x^2 - 5)$. One can check that $x^2 + 1$ and $x^2 - 5$ have no roots in $\mathbb{F}_7$, so they are irreducible in $\mathbb{F}_7[x]$. Therefore, the degree of the splitting field of $f(x)$ over $\mathbb{F}_7$ is 2.

222

(i) We show that $\mathbb{E}\mathbb{F} \subseteq \mathbb{K}$. The roots of $x^8 - 1$ and of $x^3 - 1$ are also roots of $x^{24} - 1$, so the generators of $\mathbb{E}$ and $\mathbb{F}$ belong to $\mathbb{K}$.
Let $\zeta_8 = e^{2\pi i/8}$ and $\zeta_3 = e^{2\pi i/3}$ be a primitive eighth and third root of unity in $\mathbb{C}$, respectively. The product $\zeta_8 \cdot \zeta_3$ is a primitive 24th root of unity, so we also have $\mathbb{K} \subseteq \mathbb{E}\mathbb{F}$.

(ii)

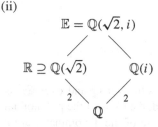

The eighth roots of 1 are $\pm 1, \pm i, (\pm 1 \pm i)/\sqrt{2}$, so $\mathbb{E}$ is contained in $\mathbb{Q}(\sqrt{2}, i)$. On the other hand, $\mathbb{E}$ must contain the root i and also $\sqrt{2} = (1 + i)\sqrt{2}/(1 + i)$, so $\mathbb{E}$ coincides with $\mathbb{Q}(\sqrt{2}, i)$. Note that $\mathbb{Q}(\sqrt{2})$ and $\mathbb{Q}(i)$ are different extensions of $\mathbb{Q}$ of degree 2 (one is a subextension of the reals and the other is not) and therefore we have $[\mathbb{E} : \mathbb{Q}] = 4$.

The third roots of 1 are $(-1 \pm \sqrt{-3})/2$, so $\mathbb{F} = \mathbb{Q}(\sqrt{-3})$ and $[\mathbb{F} : \mathbb{Q}] = 2$. It follows that $\mathbb{K} = \mathbb{Q}(\sqrt{2}, i, \sqrt{-3})$.
Remark that both $\sqrt{2}$ and $\sqrt{3} = i \cdot \sqrt{-3}$ belong to $\mathbb{K}$ and therefore we have $\mathbb{K} = \mathbb{T}(i)$, where $\mathbb{T} = \mathbb{Q}(\sqrt{2}, \sqrt{3})$ is a subfield of $\mathbb{R}$. Since $2 \cdot 3 = 6$ is not a square in $\mathbb{Q}$, we have $[\mathbb{T} : \mathbb{Q}] = 4$, while $\mathbb{K}$ is an extension of degree 2 of $\mathbb{T}$, so $[\mathbb{K} : \mathbb{Q}] = 8$.

Finally, $\mathbb{T} \subseteq \mathbb{K} \cap \mathbb{R} \subseteq \mathbb{K}$ and, since $[\mathbb{T} : \mathbb{Q}] = 4$ and $[\mathbb{K} : \mathbb{Q}] = 8$, we have $\mathbb{K} \cap \mathbb{R} = \mathbb{T}$ or $\mathbb{K} \cap \mathbb{R} = \mathbb{K}$. Since $\mathbb{K} \cap \mathbb{R} \subseteq \mathbb{R}$ whereas $\mathbb{K} \not\subseteq \mathbb{R}$, we must have $\mathbb{K} \cap \mathbb{R} = \mathbb{T}$.

We can find a basis of $\mathbb{K} \cap \mathbb{R}$ by taking the products of the elements of a basis of $\mathbb{Q}(\sqrt{2})$ and a basis of $\mathbb{Q}(\sqrt{3})$; for instance, one is given by $1, \sqrt{2}, \sqrt{3}, \sqrt{6}$.

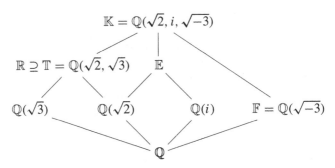

223

(i) The roots of the polynomial $x^2 + x + 1$ are ζ, ζ^2, where $\zeta \in \mathbb{C}$ is a primitive third root of 1. In order to have the required divisibility, ζ and ζ^2 must also be roots of $f(x) = x^{2n} + x^n + 1$. In fact, it is enough to check that one of them is a root, because the two are complex conjugates; if one is a root then the other one must be as well. Let us evaluate $f(x)$ at ζ. We have

$$f(\zeta) = \begin{cases} 1 + 1 + 1 = 3 & \text{if } n \equiv 0 \ (\text{mod } 3); \\ \zeta^2 + \zeta + 1 = 0 & \text{if } n \equiv 1 \ (\text{mod } 3); \\ \zeta + \zeta^2 + 1 = 0 & \text{if } n \equiv 2 \ (\text{mod } 3). \end{cases}$$

So the values we want are the natural numbers $n \not\equiv 0 \ (\text{mod } 3)$.

(ii) The equality

$$\frac{x^{12} - 1}{x^4 - 1} = x^8 + x^4 + 1$$

implies that the splitting field of $x^8 + x^4 + 1$, both over $\mathbb{Q}$ and over $\mathbb{F}_7$, is generated by a primitive 12th root of unity. Indeed, the roots of the polynomial $x^{12} - 1$ form a cyclic group of order 12, and those of the denominator are a subgroup of order 4 of the cyclic group in question. On the other hand, adding to the base field any 12th root of unity also adds its powers, and so all other 12th roots of unity.

In order to find the degree of the splitting field over $\mathbb{Q}$, let us compute the complex roots of the polynomial; they are

$$\frac{\pm 1 \pm \sqrt{3}i}{2}, \frac{\pm\sqrt{3}\pm i}{2}$$

so the splitting field is $\mathbb{Q}(\sqrt{3}, i)$. Indeed, the splitting field must be contained in $\mathbb{Q}(\sqrt{3}, i)$ given the form of the roots; moreover, the opposite inclusion follows from the fact that summing the two complex conjugate roots $(\sqrt{3} \pm i)/2$ yields $\sqrt{3}$, and taking their difference shows that i is in the splitting field as well. The degree of the splitting field of $x^8 + x^4 + 1$ over $\mathbb{F}_7$ is the least positive integer k for which $\mathbb{F}_{7^k}$ contains the 12th roots of 1; in other words, it is the least positive integer k for which $12 \mid 7^k - 1$, that is, $k = 2$.

224 We first show that $f(x) = x^4 - x^3 + x^2 - x + 1$ is irreducible in $\mathbb{Q}[x]$. We can check that $f(x)$ has no rational roots: it is enough to see that $f(\pm 1) \neq 0$. Moreover, by Gauss's lemma it is enough to check that $f(x)$ is irreducible in $\mathbb{Z}[x]$. Suppose by contradiction that

$$f(x) = (x^2 + ax + b)(x^2 + cx + d),$$

that is, that $f(x)$ factors as the product of two quadratic polynomials in $\mathbb{Z}[x]$. By computing the coefficients of the right hand side, we get

$$\begin{cases} a + c = -1 \\ b + d + ac = 1 \\ ad + bc = -1 \\ bd = 1. \end{cases}$$

The last equation implies that $b = d = \pm 1$. Substituting into the third equation, we find that we must have $b = d = 1$, otherwise the first equation has no solutions. On the other hand, the solutions of the first two equations $a + c = -1$ and $ac = -1$ are solutions of $t^2 + t - 1 = 0$, but it is clear that the latter polynomial in t has no integer roots.

⟦One could also show that $f(x)$ is irreducible in $\mathbb{Z}[x]$ by considering its coefficients modulo 2. The reduction yields the polynomial $x^4 + x^3 + x^2 + x + 1$ which clearly has no roots in $\mathbb{F}_2$ and is not the square of $x^2 + x + 1$, which is the only irreducible polynomial of degree 2 in $\mathbb{F}_2[x]$.⟧

The degree $[\mathbb{Q}(\alpha) : \mathbb{Q}]$ is thus 4, and the degree $d = [\mathbb{Q}(\alpha + c\alpha^{-1}) : \mathbb{Q}]$ is a divisor of 4 because $\alpha + c\alpha^{-1} \in \mathbb{Q}(\alpha)$. We can immediately remark that $d \neq 1$, because if we had $d = 1$ we would have $\alpha + c\alpha^{-1} = q \in \mathbb{Q}$, that is, $\alpha^2 - q\alpha + c = 0$; the element α would then satisfy an equation of degree 2 with rational coefficients, which would contradict the fact that $[\mathbb{Q}(\alpha) : \mathbb{Q}] = 4$. So we have $d = 2$ or $d = 4$.

We have $d = 2$ if and only if the elements $(\alpha + c\alpha^{-1})^2, \alpha + c\alpha^{-1}, 1$ are linearly dependent over $\mathbb{Q}$. Multiplying by α^2, the condition becomes that $(\alpha^2 + c)^2, \alpha^3 + c\alpha, \alpha^2$ are linearly dependent over $\mathbb{Q}$.

Let us write all the terms as linear combinations of 1, α, α^2, α^3. The vectors $\alpha^3 + c\alpha$, α^2 are already expressed in terms of this basis, and we have $(\alpha^2 + c)^2 = \alpha^4 + 2c\alpha^2 + c^2 = \alpha^3 + (2c-1)\alpha^2 + \alpha + (c^2 - 1)$. By looking at the coefficients of α^3 we find that a linear combination $r(\alpha^3 + (2c-1)\alpha^2 + \alpha + (c^2 - 1)) + s(\alpha^3 + c\alpha) + t\alpha^2$ can be 0 only if $r = -s$. If $r = -s = 0$ we also get $t = 0$, so the coefficients are all zero. If $r = -s \neq 0$ then setting the coefficient of α to zero yields $c = 1$, hence $t = -r$, which is a solution.

In conclusion, $d = 2$ if $c = 1$ and $d = 4$ if $c \neq 1$.

225

(i) Since $f(x)$ has degree 3, it is irreducible in $\mathbb{Q}[x]$ if and only if it has no rational roots; moreover, the only potential roots are ± 1. Since neither 1 nor -1 is a root, $f(x)$ is irreducible in $\mathbb{Q}[x]$. Let α be a real root of $f(x)$; we have $[\mathbb{Q}(\alpha) : \mathbb{Q}] = 3$. Now, since $f(x)$ is irreducible and has degree 3, the degree of its splitting field $\mathbb{K}$ over $\mathbb{Q}$ is a divisor $3! = 6$. But we have chosen $\alpha \in \mathbb{R}$ and, since the derivative $f'(x) = 3x^2 + 3$ is always positive, there is a unique real root of $f(x)$, so $\mathbb{K}$ is not contained in $\mathbb{R}$, that is, $[\mathbb{K} : \mathbb{Q}] = 6$.

(ii) By the derivative criterion, $f(x) \in \mathbb{F}_p[x]$ has multiple roots in $\mathbb{F}_p$ if and only if $(f(x), f'(x)) \neq 1$. For $p = 3$, the polynomial $f(x) = x^3 + 1 = (x+1)^3$ has the triple root 1. For $p \neq 3$, since $f'(x) = 3x^2 + 3 = 3(x^2 + 1)$ and 3 is invertible modulo p, we can take the greatest common divisor of $f(x)$ and $x^2 + 1$. The first division of Euclid's algorithm yields $x^3 + 3x + 1 = x(x^2 + 1) + (2x + 1)$. Remark that for $p = 2$ the polynomial $f(x)$ is irreducible, because its degree is 3 and it has no roots. We can therefore assume that $p \neq 2$ and multiply the polynomial $x^2 + 1$ by the invertible constant 4, thus obtaining $4x^2 + 4 = (2x - 1)(2x + 1) + 5$ from the next Euclidean division. It follows that the greatest common divisor of $f(x)$ and $f'(x)$ is not 1 if and only if $p = 3$ or $p = 5$. Indeed, for $p = 5$ we have $f(x) = (x - 1)(x - 2)^2$.

In conclusion, $f(x)$ has a multiple root in $\mathbb{F}_p$ if and only if $p = 3$ or $p = 5$.

226

(i) The polynomial $f(x) = x^9 - 1$ factors in $\mathbb{Z}[x]$ as $f(x) = (x - 1)(x^2 + x + 1)(x^6 + x^3 + 1)$; we therefore need to show that $h(x) = x^6 + x^3 + 1$ is irreducible in $\mathbb{F}_{11}[x]$.

By the theorem about cyclotomic extensions of finite fields, the splitting field of $f(x)$ over $\mathbb{F}_{11}$ is $\mathbb{F}_{11^d}$, where d is the multiplicative order of 11 modulo 9; we find that $d = 6$. It follows that, denoting by C the set of roots of $f(x)$ in some fixed algebraic closure of $\mathbb{F}_{11}$, we have $\mathbb{F}_{11}(C) = \mathbb{F}_{11^6}$. Moreover, we know that C is a finite multiplicative group, so it is cyclic; letting $C = \langle \alpha \rangle$, we have $\mathbb{F}_{11^6} = \mathbb{F}_{11}(\alpha)$. This implies that the minimal polynomial of α, which is a divisor of $f(x)$, has degree 6; it is therefore $h(x)$, which is thus irreducible.

(ii) By our arguments above, $x^2 + x + 1$ has no roots in $\mathbb{F}_{11}$, so the factors in $f(x) = (x - 1)(x^2 + x + 1)(x^6 + x^3 + 1)$ are irreducible in $\mathbb{Z}[x]$ because they are irreducible modulo 11. Moreover, by Gauss's lemma they are also irreducible in $\mathbb{Q}[x]$. Let $\eta \in \mathbb{C}$ be a primitive ninth root of 1; the roots of $f(x)$

are $1, \eta, \eta^2, \ldots, \eta^8$. The splitting field of $f(x)$ over $\mathbb{Q}$ is therefore $\mathbb{Q}(\eta)$, and its degree is the degree of the minimal polynomial $\mu(x)$ of η over $\mathbb{Q}$. We know that $\mu(x) \mid f(x)$ and we have $\eta^3 \neq 1$, so $\mu(x) \mid f(x)/(x^3 - 1) = x^6 + x^3 + 1$. Since the latter polynomial is irreducible, we have $\mu(x) = x^6 + x^3 + 1$ so the degree of the splitting field of $f(x)$ over $\mathbb{Q}$ is 6.

Letting $\mathbb{K} = \mathbb{Q}(\zeta)$, the splitting field of $f(x)$ over $\mathbb{K}$ is $\mathbb{K}(\eta) = \mathbb{Q}(\eta)$. Since the minimal polynomial of ζ over $\mathbb{Q}$ is $x^2 + x + 1$ we have $[\mathbb{K} : \mathbb{Q}] = 2$, and thus $[\mathbb{Q}(\eta) : K] = 3$.

227

(i) We show that the polynomial $f(x)$ is irreducible modulo 2. It is obvious that 0 and 1 are not roots of the polynomial in $\mathbb{F}_2$, so $f(x)$ has no factors of degree 1. If it were reducible it would be the product of two irreducible factors of degree 2, but the only irreducible polynomial of degree 2 in $\mathbb{F}_2[x]$ is $x^2 + x + 1$, and we have $(x^2 + x + 1)^2 = x^4 + x^2 + 1 \neq f(x)$, so $f(x)$ is irreducible.

Since it is irreducible modulo 2, the polynomial $f(x)$ is also irreducible in $\mathbb{Z}[x]$ and, by Gauss's lemma, in $\mathbb{Q}[x]$. This implies that $f(x)$ is the minimal polynomial of α, so $[\mathbb{Q}(\alpha) : \mathbb{Q}] = 4$; since $\mathbb{Q}(1/(\alpha + 1)) = \mathbb{Q}(\alpha)$, the minimal polynomial of $1/(\alpha + 1)$ also has degree 4. The minimal polynomial of $\alpha + 1$ is $f(x - 1) = (x - 1)^4 + (x - 1) + 1 = x^4 - 4x^3 + 6x^2 - 3x + 1$, so the minimal polynomial of $1/(\alpha + 1)$ is its reciprocal polynomial $x^4 - 3x^3 + 6x^2 - 4x + 1$. Remark that $\mathbb{Q}(\alpha) = \mathbb{Q}(\alpha^2)$: clearly, we have $\mathbb{Q}(\alpha^2) \subseteq \mathbb{Q}(\alpha)$, and the equality $f(\alpha) = 0$ yields $\alpha = -\alpha^4 - 1 \in \mathbb{Q}(\alpha^2)$, which implies the opposite inclusion. Consequently, the minimal polynomial of α^2 over $\mathbb{Q}$ has degree 4. There are many ways to compute this minimal polynomial. We can, for example, remark that $\alpha^4 + 1 = -\alpha$, hence by squaring both sides $\alpha^8 + 2\alpha^4 + 1 = \alpha^2$, so the polynomial $x^4 + 2x^2 - x + 1$ vanishes at α^2, and since it has degree 4 it is the minimal polynomial of α^2.

⟦Here is a different way to compute the minimal polynomial of α^2: we know that it is of the form $\mu(x) = x^4 + ax^3 + bx^2 + cx + d$ and we must find $a, b, c, d \in \mathbb{Q}$ such that $\mu(\alpha) = \alpha^8 + a\alpha^6 + b\alpha^4 + c\alpha^2 + d = 0$. The expression given by the minimal polynomial of α yields that $\alpha^4 = -\alpha - 1$, $\alpha^6 = -\alpha^3 - \alpha^2$ and $\alpha^8 = \alpha^2 + 2\alpha + 1$. We therefore have $\mu(\alpha^2) = -a\alpha^3 + (1 - a + c)\alpha^2 + (2 - b)\alpha + 1 - b + d = 0$. Since the vectors $1, \alpha, \alpha^2, \alpha^3$ over $\mathbb{Q}$ are linearly independent, the equation is satisfied if and only if $a = 0, b = 2, c = -1$ e $d = 1$.⟧

(ii) The splitting field of $f(x)$ over $\mathbb{F}_5$ is $\mathbb{F}_{5^d}$, where d is the least common multiple of the degrees of the irreducible factors of $f(x)$ in $\mathbb{F}_5[x]$. Evaluating the polynomial at $0, \pm 1, \pm 2$ yields that the only root in $\mathbb{F}_5$ is -2. By Ruffini's theorem, $x + 2 \mid f(x)$ and one immediately gets $f(x) = (x + 2)(x^3 + 3x^2 - x + 3)$. The only possible roots of the factor of degree 3 are those of $f(x)$, so the only potential root -2; but we have $(-2)^3 + 3(-2) - (-2) + 3 \neq 0$, so $x^3 + 3x^2 - x + 3$ has no roots in $\mathbb{F}_5$ and therefore is irreducible. The splitting field of $f(x)$ over $\mathbb{F}_5$ is thus $\mathbb{F}_{5^3}$.

228

(i) Let α be a root of $f(x)$ in an algebraic closure of $\mathbb{F}_p$; we have $\alpha^4 = \pm a \in \mathbb{F}_p^*$, so $\mathrm{ord}(\alpha) \mid 4(p-1) \mid p^2-1$, where the last divisibility follows by the assumption that $p \equiv 3 \pmod 4$. We therefore have that every root α of $f(x)$ belongs to $\mathbb{F}_{p^2}$. All that is left to show is that $f(x)$ does not have all of its roots in $\mathbb{F}_p$. Indeed, $x^4 - a$ has a root in $\mathbb{F}_p$ if and only if $a = b^4$ for some $b \in \mathbb{F}_p$, in which case $-a$ is not a fourth power in $\mathbb{F}_p$: it is not even a square, because -1 is not a square in $\mathbb{F}_p$ for $p \equiv 3 \pmod 4$. A similar argument can be applied after exchanging a and $-a$, so the splitting field of $f(x)$ is $\mathbb{F}_{p^2}$.

(ii) If $a = 1$, then $f(x) = x^8 - 1$ and its splitting field over $\mathbb{F}_p$ is $\mathbb{F}_{p^k}$, where k is the order of p in $(\mathbb{Z}/8\mathbb{Z})^*$. It follows that, for $p \equiv 1 \pmod 8$, for example for $p = 17$, we have $k = 1$, and for $p \equiv 5 \pmod 8$, for example $p = 5$, we have $k = 2$.

We need to show that one can find a and p for which the splitting field has degree 4. Consider $a = 2$ and $p = 5$; we have $f(x) = (x^4 - 2)(x^4 + 2)$. Since neither 2 nor -2 are squares modulo 5, they are not fourth powers, so $f(x)$ has no roots in $\mathbb{F}_5$. We need to exclude the possibility that both polynomials $x^4 - 2$ and $x^4 + 2$ factor as the product of two irreducible polynomials of degree 2. In fact, neither polynomial does, as we can show by a direct computation.

Suppose that $x^4 \pm 2 = (x^2 + ax + b)(x^2 + cx + d) = x^4 + (a+c)x^3 + (b + ac + d)x^2 + (ad + bc)x + bd$ con $a, b, c, d \in \mathbb{F}_5$. Equating coefficients on the two sides yields

$$
\begin{cases}
a + c = 0 \\
b + ac + d = 0 \\
ad + bc = 0 \\
bd = \pm 2
\end{cases}
$$

and some calculations show that neither system has any solutions in $\mathbb{F}_5$.

⟦Another way to show that $x^4 - 2$ and $x^4 + 2$ are irreducible is to remark that, given a root $\alpha \in \mathbb{F}_{5^k}$ of $f(x)$, we have $\alpha^4 = \pm 2$, hence $\mathrm{ord}(\alpha^4) = 4$ and so $\mathrm{ord}(\alpha) = 4r$. The formula $\mathrm{ord}(\alpha^4) = \mathrm{ord}(\alpha)/(4, \mathrm{ord}(\alpha))$ yields $r = 4$, that is, $\mathrm{ord}(\alpha) = 16$. It follows that $16 \mid 5^k - 1$ and so $k = 4$.⟧

229

(i) Set $\Delta = a - 4b^2$, $\alpha = (-a + \sqrt{\Delta})/2$ and $\beta = (-a - \sqrt{\Delta})/2$. We have $\mathbb{F}_p(\sqrt{\Delta}) \subseteq \mathbb{F}_{p^2}$, so $f(x) = (x^3 - \alpha)(x^3 - \beta)$ in $\mathbb{F}_{p^2}[x]$. Remark that a polynomial of the form $x^3 - \gamma$ in $\mathbb{F}_{p^2}[x]$ is either irreducible or the product of three factors of degree 1. Indeed, if $p = 3$ then $x^3 - \gamma = (x - \gamma^3)^3$. If $p > 3$ then $3 \mid p^2 - 1$, so the homomorphism $\mathbb{F}_{p^2}^* \ni z \longmapsto z^3 \in \mathbb{F}_{p^2}^*$, is a 3-to-1 map, that is, cubes have three distinct third roots in $\mathbb{F}_{p^2}$. We can therefore conclude that the degree of the splitting field of $f(x)$ over $\mathbb{F}_{p^2}$ is 1 if both α and β are cubes, and 3 otherwise.

(ii) Let $\mathbb{F}_{p^k}$ be the splitting field of $f(x)$ over $\mathbb{F}_p$. Because of the previous argument, the splitting field of $f(x)$ over $\mathbb{F}_{p^2}$ is contained in $\mathbb{F}_{p^6}$, so the one over $\mathbb{F}_p$ is as well. The inclusion $\mathbb{F}_{p^k} \subseteq \mathbb{F}_{p^6}$ implies $k \mid 6$, and in particular $k \neq 4, 5$.

(iii) As before, let $\mathbb{F}_{p^k}$ be the splitting field of $f(x)$ over $\mathbb{F}_p$. We have seen that $\mathbb{F}_p(\sqrt{\Delta}) \subseteq \mathbb{F}_{p^k}$, so if we had $k = 3$ we would have $\sqrt{\Delta} \in \mathbb{F}_p$, and $f(x) = (x^3 - \alpha)(x^3 - \beta)$ in $\mathbb{F}_p[x]$. For $p \equiv 2 \pmod 3$, the map $z \longmapsto z^3$ is an isomorphism in $\mathbb{F}_p^*$, so both $x^3 - \alpha$ and $x^3 - \beta$ factor as a product of a linear factor and an irreducible factor of degree 2, so in this case the splitting field cannot have degree 3.

230 Remember that, by the theorem about cyclotomic extensions of finite fields, if $(n, p) = 1$ then the degree of the splitting field of the polynomial $x^n - 1$ over $\mathbb{F}_p$ coincides with the multiplicative order of p modulo n. It follows that if $p \neq 2, 3, 5$ then the degree of the splitting field of $f(x) = (x^{15} - 1)(x^{12} - 1)$ over $\mathbb{F}_p$ is the least common multiple of the multiplicative order of p modulo 15 and the multiplicative order of p modulo 12, that is, the smallest positive solution of the system

$$\begin{cases} p^x \equiv 1 \pmod{15} \\ p^x \equiv 1 \pmod{12}. \end{cases}$$

By the Chinese remainder theorem, the system is equivalent to

$$\begin{cases} p^x \equiv 1 \pmod 3 \\ p^x \equiv 1 \pmod 4 \\ p^x \equiv 1 \pmod 5. \end{cases}$$

One can immediately check that $x = 4$ is a solution of the system, so the minimal solution is a divisor of 4. We now show that all divisors of 4 are possible degrees.

For $p = 7$, by the arguments above, one can immediately see that the degree of the splitting field is 4.

The degree of the splitting field over $\mathbb{F}_p$ is 2 if $p \equiv -1 \pmod 5$: indeed, $p^2 \equiv 1 \pmod 3$ and $p^2 \equiv 1 \pmod 4$ for all primes larger than 3, for example for $p = 19$.

The degree of the splitting field is 1 if and only if $3 \mid p - 1$, $4 \mid p - 1$ and $5 \mid p - 1$ that is, if and only if $60 \mid p - 1$. Since 61 is prime, the splitting field of $f(x)$ over $\mathbb{F}_{61}$ has degree 1.

The last case to discuss is $p = 2$. In $\mathbb{F}_2[x]$ we have $f(x) = (x^{15} - 1)(x^3 - 1)^4$ and, since $3 \mid 15$, we have $x^3 - 1 \mid x^{15} - 1$, so the degree of the splitting field of $f(x)$ is the order of 2 in $(\mathbb{Z}/15\mathbb{Z})^*$, which one can easily show to be 4.

⟦In order to cover every prime, we could also discuss 3 and 5. For $p = 3$ we have $f(x) = (x^5 - 1)^3(x^4 - 1)^3$ and, since the multiplicative order of 3 modulo 5 is 4 and its multiplicative order modulo 4 is 2, the degree of the splitting field is 4. Finally, for $p = 5$ we have $f(x) = (x^3 - 1)^5(x^{12} - 1)$ and a similar argument yields that the degree of the splitting field is 2.⟧

231　The prime factorisation of 1635 is $3 \cdot 5 \cdot 109$, so $\mathbb{Z}/1635\mathbb{Z}$ is not a field and the equation can have more than four solutions. By the Chinese remainder theorem, $\mathbb{Z}/1635\mathbb{Z} \simeq \mathbb{Z}/3\mathbb{Z} \times \mathbb{Z}/5\mathbb{Z} \times \mathbb{Z}/109\mathbb{Z}$.

Let us now look for a factorisation of $f(x) = 2x^4 - 41x^3 + 201x^2 - 71x - 91$ in $\mathbb{Z}[x]$, which will induce factorisations in $\mathbb{Z}/3\mathbb{Z}[x], \mathbb{Z}/5\mathbb{Z}[x], \mathbb{Z}/109\mathbb{Z}[x]$ by passing to the quotient.

Since $91 = 13 \cdot 7$, any rational root of f is of the form a/b, where a is a divisor of 91 and b is a divisor of 2. It is easy to check that $f(1) = f(7) = f(13) = 0$ and so we have that $(x - 1)(x - 7)(x - 13)$ divides $f(x)$. By performing the division we obtain that $f(x) = (x-1)(x-7)(x-13)(2x+1)$ in $\mathbb{Z}[x]$, and so $-1/2$ is also a root. This factorisation yields: $f(x) = -(x-1)^4$ in $\mathbb{Z}/3\mathbb{Z}[x]$, $f(x) = (x-1)(x-2)^2(x-3)$ in $\mathbb{Z}/5\mathbb{Z}[5]$ and $(x-1)(x-7)(x-13)(2x+1)$ in $\mathbb{Z}/109\mathbb{Z}[x]$. Since $\mathbb{Z}/3\mathbb{Z}, \mathbb{Z}/5\mathbb{Z}$ and $\mathbb{Z}/109\mathbb{Z}$ are fields, these factorisations immediately give us the respective sets of roots: we have $x = 1$, of multiplicity 4, in $\mathbb{Z}/3\mathbb{Z}[x]$; $x = 1, -2, 2$, where 2 has multiplicity 2, in $\mathbb{Z}/5\mathbb{Z}[x]$, and finally $x = -1/2, 1, 7, 13$ in $\mathbb{Z}/109\mathbb{Z}[x]$ (and it is easy to check that these roots are distinct in $\mathbb{Z}/109\mathbb{Z}$). We therefore have $1 \cdot 3 \cdot 4 = 12$ distinct solutions in $\mathbb{Z}/3\mathbb{Z} \times \mathbb{Z}/5\mathbb{Z} \times \mathbb{Z}/109\mathbb{Z}$ and thus in $\mathbb{Z}/1635\mathbb{Z}$.

Since the factorisation in $\mathbb{Z}[x]$ remains valid in the quotient ring $\mathbb{Z}/1635\mathbb{Z}[x]$, three roots of $f(x)$ are immediately found and are $x \equiv 1, 7, 13 \pmod{1635}$. Moreover, remark that $(2, 1635) = 1$, so the element 2 is invertible in $\mathbb{Z}/1635\mathbb{Z}$ and we have the solution $x = -1/2 \equiv 817 \in \mathbb{Z}/1635\mathbb{Z}$, which is clearly different from the previous ones.

We need to construct two more solutions and we may do it using the Chinese remainder theorem, for example from the triples of solutions $(1, 1, 7)$ and $(1, 2, 13)$ in $\mathbb{Z}/3\mathbb{Z} \times \mathbb{Z}/5\mathbb{Z} \times \mathbb{Z}/109\mathbb{Z}$. Simple calculations yield that 661 and 667 are the corresponding classes modulo 1635, and are thus roots of $f(x)$.

⟦For the sake of completeness, by remarking that $-1/2 \equiv 54 \pmod{109}$, we can give the full correspondence between solution triples in $\mathbb{Z}/3\mathbb{Z} \times \mathbb{Z}/5\mathbb{Z} \times \mathbb{Z}/109\mathbb{Z}$ and solutions in $\mathbb{Z}/1635\mathbb{Z}$:

$$
\begin{array}{llll}
(1, 1, 1) &\longleftrightarrow 1 & (1, 1, 7) &\longleftrightarrow 661 \\
(1, 1, 13) &\longleftrightarrow 1321 & (1, 1, 54) &\longleftrightarrow 1471 \\
(1, 2, 1) &\longleftrightarrow 982 & (1, 2, 7) &\longleftrightarrow 7 \\
(1, 2, 13) &\longleftrightarrow 667 & (1, 2, 54) &\longleftrightarrow 817 \\
(1, 3, 1) &\longleftrightarrow 328 & (1, 3, 7) &\longleftrightarrow 988 \\
(1, 3, 13) &\longleftrightarrow 13 & (1, 3, 54) &\longleftrightarrow 163.
\end{array}
$$

⟧

232

(i) First of all, we show that $f(x)$ is irreducible in $\mathbb{Q}[x]$. Since $\deg(f) = 3$, it is enough to show that $f(x)$ has no rational roots. Any rational root must have a numerator that divides the constant coefficient, which is -1, and a denominator that divides the leading coefficient, which is 1; it is therefore enough to check that ± 1 are not roots. But $f(1) = f(-1) = -1$, so the polynomial has no rational roots and is therefore irreducible.

It is now clear that the field $\mathbb{Q}(\alpha)$ has degree 3 over $\mathbb{Q}$ and thus each of its elements can be expressed as a linear combination of 1, α and α^2 with rational coefficients. We write $1/(\alpha + 2) = a\alpha^2 + b\alpha + c$, where a, b, c are rational numbers to be determined. We get

$$1 = (a\alpha^2 + b\alpha + c)(\alpha + 2) = a\alpha^3 + b\alpha^2 + c\alpha + 2a\alpha^2 + 2b\alpha + 2c$$
$$= (b + 2a)\alpha^2 + (a + c + 2b)\alpha + (a + 2c)$$

where we used $\alpha^3 = \alpha + 1$ to obtain the last equality. Equating coefficients of the basis vectors on the two sides yields the system

$$\begin{cases} b + 2a = 0 \\ a + c + 2b = 0 \\ a + 2c = 1 \end{cases}$$

whose unique solution is $a = 1/7$, $b = -2/7$, $c = 3/7$. In conclusion, $1/(\alpha + 2) = (\alpha^2 - 2\alpha + 3)/7$.

⟦Alternatively, we could divide the polynomial $x^3 - x - 1$ by $x + 2$ and obtain that $x^3 - x - 1 = (x^2 - 2x + 3)(x + 2) - 7$. Replacing x with α, we have $(\alpha^2 - 2\alpha + 3)(\alpha + 2) - 7 = 0$, so dividing by $7(\alpha + 2)$ yields the desired expression.⟧

(ii) We have the obvious inclusions $\mathbb{Q}(\alpha^2) \subseteq \mathbb{Q}(\alpha)$ and $\mathbb{Q}(\alpha^3) \subseteq \mathbb{Q}(\alpha)$, so the degrees to be computed are at most 3, and moreover they are divisors of 3 (because degrees are multiplicative in towers), so each is either 1 or 3.

If we had $[\mathbb{Q}(\alpha^2) : \mathbb{Q}] = 1$ then we would have $\alpha^2 = c \in \mathbb{Q}$ and so α would be a root of the degree 2 polynomial $x^2 - c \in \mathbb{Q}[x]$, which contradicts the fact that the minimal polynomial of α is $x^3 - x - 1$ and has degree 3. Therefore, $[\mathbb{Q}(\alpha^2) : \mathbb{Q}] = 3$.

From the equation $\alpha^3 = \alpha + 1$ we get that $\alpha = \alpha^3 - 1 \in \mathbb{Q}(\alpha^3)$, so $\mathbb{Q}(\alpha) = \mathbb{Q}(\alpha^3)$ and $[\mathbb{Q}(\alpha^3) : \mathbb{Q}] = 3$.

233

(i) Reducing the polynomial $f(x)$ modulo 2 yields $x^4 + x + 1$, which clearly has no roots and is not the square of the unique irreducible polynomial of degree 2 in $\mathbb{F}_2[x]$, namely, $x^2 + x + 1$. The polynomial $f(x)$ is thus irreducible modulo 2 and so irreducible in $\mathbb{Z}[x]$. By Gauss's lemma, it is also irreducible in $\mathbb{Q}[x]$.

(ii) Since it is irreducible, $f(x)$ is the minimal polynomial of α over $\mathbb{Q}$, so $[\mathbb{Q}(\alpha) : \mathbb{Q}] = 4$. It is clear that $2\alpha - 3 \in \mathbb{Q}(\alpha)$; on the other hand, $\alpha = (2\alpha - 3)/2 + 3/2 \in \mathbb{Q}(2\alpha - 3)$, hence $\mathbb{Q}(2\alpha - 3) = \mathbb{Q}(\alpha)$ and $[\mathbb{Q}(2\alpha - 3) : \mathbb{Q}] = 4$. The minimal polynomial of $2\alpha - 3$ over $\mathbb{Q}$ has thus degree 4, and is the unique monic polynomial of degree 4 in $\mathbb{Q}[x]$ that vanishes at $2\alpha - 3$. The polynomial

$$g(x) = f\left(\frac{x + 3}{2}\right) = \frac{1}{16}(x^4 + 12x^3 + 54x^2 + 84x - 71)$$

has degree 4, and clearly $g(2\alpha - 3) = 0$. The minimal polynomial of $2\alpha - 3$
over $\mathbb{Q}$ is thus $16g(x) = x^4 + 12x^3 + 54x^2 + 84x - 71$.

(iii) Clearly, $\alpha^2 \in \mathbb{Q}(\alpha)$. On the other hand, the equality $\alpha^4 - 3\alpha - 5 = 0$ implies
that $\alpha = (\alpha^4 - 5)/3 = ((\alpha^2)^2 - 5)/3 \in \mathbb{Q}(\alpha^2)$.

So we have $\mathbb{Q}(\alpha^2) = \mathbb{Q}(\alpha)$ and thus $[\mathbb{Q}(\alpha^2) : \mathbb{Q}] = 4$, and the minimal
polynomial of α^2 is the unique monic polynomial of degree 4 that vanishes at
α^2. Squaring both sides of $\alpha^4 - 5 = 3\alpha$, we immediately obtain that $x^4 -
10x^2 - 9x + 25$ has α^2 as a root, so it is the minimal polynomial in question.

234 We have $5 = 2x^2 + 17 - 2(x^2 + 6)$ and so $5 \in I = (2x^2 + 17, x^2 + 6)$; in
fact, we have $I = (5, x^2 + 1)$.

The map $\mathbb{F}_5 \ni \bar{a} \longmapsto \bar{a} + I \in A = \mathbb{Z}[x]/I$ is well defined, because the ideal
generated by 5 in $\mathbb{Z}[x]$ is contained in I; moreover, it is a ring homomorphism. In
order to show it is injective, remark that $I \cap \mathbb{Z}$ is an ideal of $\mathbb{Z}$ that contains $5\mathbb{Z}$; it is
therefore either $5\mathbb{Z}$ or $\mathbb{Z}$. But if we had $I \cap \mathbb{Z} = \mathbb{Z}$ then we would have $1 \in I$, that
is, $\bar{1} \in (x^2 + 1)$ in $\mathbb{F}_5[x]$, which is not the case. Therefore, if $\bar{a} + I = \bar{a}' + I$ then
we have $a - a' \in I \cap \mathbb{Z} = 5\mathbb{Z}$, so $\bar{a} = \bar{a}'$.

Having shown that $\mathbb{F}_5 \subseteq A$, we conclude that A is a vector space over $\mathbb{F}_5$.

We now show that $1, x$ form a basis of A as a vector space over $\mathbb{F}_5$. It is clear that
they generate A, because every element of A can be expressed in the form $\bar{a}x + \bar{b}$
for some $a, b \in \mathbb{Z}$. In order to show that they are linearly independent, suppose that
$\bar{a}x + \bar{b} = 0$ in $A = \mathbb{Z}[x]/(5, x^2 + 1)$. We then have $ax + b \in (5, x^2 + 1)$ in $\mathbb{Z}[x]$
and so $\bar{a}x + \bar{b} \in (x^2 + 1)$ in $\mathbb{F}_5[x]$, which implies $\bar{a} = \bar{b} = 0$ in $\mathbb{F}_5$.

Since it has a basis consisting of two elements, A is isomorphic to $\mathbb{F}_5^2$, as are all
vector spaces of dimension 2 over $\mathbb{F}_5$.

235

(i)   Since degrees are multiplicative in towers, we have $[\mathbb{K} :
$\mathbb{Q}] = [\mathbb{K} : \mathbb{Q}(\sqrt[3]{2})][\mathbb{Q}(\sqrt[3]{2}) : \mathbb{Q}]$. Now, $[\mathbb{Q}(\sqrt[3]{2}) : \mathbb{Q}] =$

$\mathbb{Q}(\sqrt[3]{2}, i)$ 3; this is because the polynomial $x^3 - 2$ is irreducible in
$\mathbb{Z}[x]$ by Eisenstein's criterion, hence by Gauss's lemma it
| is irreducible in $\mathbb{Q}[x]$, and it has $\sqrt[3]{2}$ as a root, so it is the
$\mathbb{Q}(\sqrt[3]{2})$ minimal polynomial of $\sqrt[3]{2}$ over $\mathbb{Q}$. In order to compute
$[\mathbb{K} : \mathbb{Q}(\sqrt[3]{2})]$, we shall find the minimal polynomial of i
3 | over $\mathbb{Q}(\sqrt[3]{2})$. Since we know how to factor $x^2 + 1$ in $\mathbb{C}[x]$,
                                                                        we can immediately see that it is irreducible in $\mathbb{R}[x]$ and
$\mathbb{Q}$ therefore in $\mathbb{Q}(\sqrt[3]{2})[x] \subseteq \mathbb{R}[x]$.

Since the irreducible polynomial $x^2 + 1 \in \mathbb{Q}(\sqrt[3]{2})[x]$ does have i as a root, it
is its minimal polynomial over $\mathbb{Q}(\sqrt[3]{2})$, and we have $[\mathbb{K} : \mathbb{Q}(\sqrt[3]{2})] = 2$.

In conclusion, $[\mathbb{K} : \mathbb{Q}] = 2 \cdot 3 = 6$.

(ii) We show that $\mathbb{K} = \mathbb{Q}(\sqrt[3]{2}, i) = \mathbb{Q}(\sqrt[3]{2} + i)$. Since $\sqrt[3]{2} + i \in \mathbb{Q}(\sqrt[3]{2}, i)$, it is
clear that $\mathbb{Q}(\sqrt[3]{2} + i) \subseteq \mathbb{Q}(\sqrt[3]{2}, i)$.

Let $u = \sqrt[3]{2} + i$. By taking the cube of both sides of the equality $u - i = \sqrt[3]{2}$ we get

$$i = \frac{u^3 - 3u - 2}{3u^2 - 1} \in \mathbb{Q}(u) = \mathbb{Q}(\sqrt[3]{2} + i).$$

Moreover, $\sqrt[3]{2} = u - i \in \mathbb{Q}(\sqrt[3]{2} + i)$ and thus $\mathbb{Q}(\sqrt[3]{2}, i) \subseteq \mathbb{Q}(\sqrt[3]{2} + i)$.

(iii) We need to find the minimal polynomial of $\sqrt[3]{2} + i$ over $\mathbb{Q}$. By the results shown above, we know that its degree must be 6. It is thus enough to find a monic polynomial of degree 6 with rational coefficients that vanishes at $\sqrt[3]{2} + i$.
As before, let $u = \sqrt[3]{2} + i$. Squaring both sides of the equality $i = u - \sqrt[3]{2}$, isolating $\sqrt[3]{2}$ and taking cubes, we immediately get that $u^6 + 3u^4 - 4u^3 + 3u^2 + 12u + 5 = 0$, so the polynomial $x^6 + 3x^4 - 4x^3 + 3x^2 + 12x + 5 \in \mathbb{Q}(x)$ is monic, has degree 6 and has $\sqrt[3]{2} + i$ as a root. It is therefore the minimal polynomial of $\sqrt[3]{2} + i$ over $\mathbb{Q}$.

236

(i)

$$\mathbb{Q}(\sqrt{3}, \sqrt{5})$$
$$|$$
$$\mathbb{Q}(\sqrt{3})$$
$$2 \,|$$
$$\mathbb{Q}$$

Consider the extensions $\mathbb{Q} \subseteq \mathbb{Q}(\sqrt{3}) \subseteq \mathbb{Q}(\sqrt{3}, \sqrt{5})$. Since $x^2 - 3$ and $x^2 - 5$ are the minimal polynomials of $\sqrt{3}$ and $\sqrt{5}$ over $\mathbb{Q}$, the first extension has degree 2 and the second has degree at most 2. If we had $\sqrt{5} \in \mathbb{Q}(\sqrt{3})$, then we would have $\sqrt{5} = a + b\sqrt{3}$ for some nonzero rationals a and b. But squaring the expression gives $5 = a^2 + 2ab\sqrt{3} + 3b^2$, that is, $\sqrt{3} = (5 - a^2 - 3b^2)/2ab \in \mathbb{Q}$, which is a contradiction.

⟦Another way to prove that $\sqrt{5} \notin \mathbb{Q}(\sqrt{3})$ is the following. If we had $\sqrt{5} \in \mathbb{Q}(\sqrt{3})$ then the two quadratic extensions $\mathbb{Q}(\sqrt{3})$ and $\mathbb{Q}(\sqrt{5})$ would coincide; but this is not the case, because $3 \cdot 5 = 15$ is not a square in $\mathbb{Q}$.⟧

This implies that the second extension has degree 2 and thus

$$[\mathbb{Q}(\sqrt{3}, \sqrt{5}) : \mathbb{Q}] = [\mathbb{Q}(\sqrt{3}, \sqrt{5}) : \mathbb{Q}(\sqrt{3})][\mathbb{Q}(\sqrt{3}) : \mathbb{Q}] = 4.$$

Now consider the extensions $\mathbb{Q} \subseteq \mathbb{Q}(\sqrt{3} - \sqrt{5}) \subseteq \mathbb{Q}(\sqrt{3}, \sqrt{5})$. Since $\sqrt{3} - \sqrt{5}$ is irrational (its square is irrational) the first extension has degree at least 2 and thus can have degree 2 or 4, since its degree must be a divisor of $[\mathbb{Q}(\sqrt{3}, \sqrt{5}) : \mathbb{Q}] = 4$.
Suppose it has degree 2, that is, there exists a polynomial $f(x) = x^2 + ax + b \in \mathbb{Q}[x]$ such that $f(\sqrt{3} - \sqrt{5}) = 0$. This would imply $(\sqrt{3} - \sqrt{5})^2 + a(\sqrt{3} - \sqrt{5}) + b = 0$, hence $3 + 5 - 2\sqrt{15} + a\sqrt{3} - a\sqrt{5} + b = 0$, that is, $8 + b + a\sqrt{3} = (a + 2\sqrt{3})\sqrt{5}$. Squaring both sides yields $64 + b^2 + 3a^2 + 16b + 16a\sqrt{3} + 2ab\sqrt{3} = 5a^2 + 60 + 20a\sqrt{3}$, hence $b^2 + 16b - 2a^2 + 4 + (2ab - 4a)\sqrt{3} = 0$. Since $\sqrt{3}$ is irrational, this implies $2ab - 4a = 0$, that is, $a = 0$ or $b = 2$. In the first case we get $b^2 + 16b + 4 = 0$ and in the second $a^2 = 20$, but neither equation has any rational solutions. In conclusion, such a polynomial does not exist, hence $[\mathbb{Q}(\sqrt{3} - \sqrt{5}) : \mathbb{Q}] = 4$.

(ii) Set $\alpha = \sqrt{3} - \sqrt{5}$. Taking the square yields $\alpha^2 = 3 + 5 - 2\sqrt{15}$, that is, $\alpha^2 - 8 = -2\sqrt{15}$. Squaring again, we have $\alpha^4 - 16\alpha^2 + 4 = 0$, so $\sqrt{3} - \sqrt{5}$ is a root of the polynomial $x^4 - 16x^2 + 4 \in \mathbb{Q}[x]$. Since $[\mathbb{Q}(\sqrt{3} - \sqrt{5}) : \mathbb{Q}] = 4$, this polynomial is irreducible and is thus the minimal polynomial of $\sqrt{3} - \sqrt{5}$ over $\mathbb{Q}$.

Consider now $\sqrt{\sqrt{3} - \sqrt{5}} - 1$. Setting $f(x) = x^8 - 16x^4 + 4 \in \mathbb{Q}[x]$, it is easy to see that $f(\sqrt{\sqrt{3} - \sqrt{5}}) = 0$, so the polynomial $g(x) = f(x + 1) = (x + 1)^8 - 16(x + 1)^4 + 4 \in \mathbb{Q}[x]$ is monic and has $\sqrt{\sqrt{3} - \sqrt{5}} - 1$ as a root. In order to show that $g(x)$ is the minimal polynomial of $\sqrt{\sqrt{3} - \sqrt{5}} - 1$ over $\mathbb{Q}$, we need to show that $g(x)$ is irreducible, or equivalently that the extension $\mathbb{Q} \subseteq \mathbb{Q}(\sqrt{\sqrt{3} - \sqrt{5}} - 1) = \mathbb{Q}(\sqrt{\sqrt{3} - \sqrt{5}})$ has degree 8.

$$\begin{array}{c} \mathbb{Q}(\beta) \\ | \\ \mathbb{Q}(\sqrt{3} - \sqrt{5}) \subseteq \mathbb{R} \\ 4\ | \\ \mathbb{Q} \end{array}$$

Set $\beta = \sqrt{\sqrt{3} - \sqrt{5}}$. Since we know that $[\mathbb{Q}(\beta) : \mathbb{Q}] = [\mathbb{Q}(\beta) : \mathbb{Q}(\sqrt{3} - \sqrt{5})] \cdot [\mathbb{Q}(\sqrt{3} - \sqrt{5}) : \mathbb{Q}]$ and that $[\mathbb{Q}(\sqrt{3} - \sqrt{5}) : \mathbb{Q}] = 4$, we have that $[\mathbb{Q}(\beta) : \mathbb{Q}] = 8$ is equivalent to $\mathbb{Q}(\sqrt{3} - \sqrt{5}) \neq \mathbb{Q}(\beta)$. But this immediately follows from the fact that $\mathbb{Q}(\sqrt{3} - \sqrt{5})$ is a subextension of $\mathbb{R}$, whereas $\mathbb{Q}(\beta)$ is not, since $\beta = \sqrt{\sqrt{3} - \sqrt{5}}$ is the square root of a negative number.

237 Set $f(x) = x^7 + x^6 + x^5 + x^4 + x^3 + x^2 + x + 1$; notice that $(x - 1)f(x) = x^8 - 1$. It is thus clear that in $\mathbb{C}[x]$ the roots of the polynomial $f(x)$ are all the 8th roots of unity except for 1, that is,

$$f(x) = (x - \frac{1 + i}{\sqrt{2}})(x - i)(x - \frac{-1 + i}{\sqrt{2}})(x + 1)(x - \frac{-1 - i}{\sqrt{2}})(x + i)(x - \frac{1 - i}{\sqrt{2}}).$$

In $\mathbb{Z}[x]$ we can factor $x^8 - 1$ as $(x - 1)(x + 1)(x^2 + 1)(x^4 + 1)$, as one immediately finds by repeatedly factoring differences of squares. This factorisation also holds in $\mathbb{F}_5[x]$ and $\mathbb{F}_{13}[x]$, because both rings are quotients of $\mathbb{Z}[x]$.

Now, remark that $x^2 + 1$ is irreducible in $\mathbb{Z}[x]$ because it has no real roots and is monic. The polynomial $(x + 1)^4 + 1 = x^4 + 4x^3 + 6x^2 + 4x + 2$ is irreducible by Eisenstein's criterion for the prime 2, so $x^4 + 1$ is also irreducible. The factorisation of $f(x)$ in irreducible factors in $\mathbb{Z}[x]$ is therefore

$$f(x) = (x + 1)(x^2 + 1)(x^4 + 1).$$

In $\mathbb{F}_5$ we have $\overline{2}^2 = -\overline{1}$, whereas $\pm \overline{2}$ are not squares. So $x^2 + 1 = (x - 2)(x + 2)$ and $x^4 + 1 = (x^2 - 2)(x^2 + 2)$, and the two quadratic factors are irreducible because they have no roots. So $f(x)$ factors in $\mathbb{F}_5[x]$ as

$$f(x) = (x + 1)(x - 2)(x + 2)(x^2 - 2)(x^2 + 2).$$

In $\mathbb{F}_{17}$ we have $\overline{4}^2 = -\overline{1}$, so $x^2+1 = (x-4)(x+4)$ and $x^4+1 = (x^2-4)(x^2+4) = (x-2)(x+2)(x-8)(x+8)$. The polynomial is therefore a product of linear factors:

$$f(x) = (x+1)(x-4)(x+4)(x-2)(x+2)(x-8)(x+8).$$

238

(i) Remark that 1 is a root of $f(x)$ over $\mathbb{F}_7$, so we can divide $f(x)$ by $x-1$ and obtain $f(x) = (x-1)(x^3-x^2-3x+3)$. Since 1 is also a root of x^3-x^2-3x+3, we divide once again to get $f(x) = (x-1)^2(x^2-3)$.

  In order to check whether x^2-3 is irreducible, we compute all squares in $\mathbb{F}_7^*$: we have $(\pm 1)^2 = 1$, $(\pm 2)^2 = -3$, $(\pm 3)^2 = 2$, so 3 is not a square and therefore x^2-3 is irreducible. The factorisation of $f(x)$ in irreducible factors is thus $(x-1)^2(x^2-3)$.

(ii) Since $\mathbb{F}_7[x]/(f(x))$ is a finite ring, its elements are either invertible or zero divisors; we can therefore just count zero divisors. We know that each zero divisor is represented by a polynomial of degree less than 4 that is not coprime to $f(x)$. The set of zero divisors is thus in bijection with the set of polynomials of degree at most 3 that are multiples of $x-1$ or of x^2-3. By the inclusion-exclusion principle,

$$|\{\text{zero divisors}\}| = |\{\text{multiples of } x-1\}| + |\{\text{multiples of } x^2-3\}|$$
$$- |\{\text{multiples of } (x-1)(x^2-3)\}|.$$

The multiples of $x-1$ are represented by polynomials of the form $(a_2x^2 + a_1x + a_0)(x-1)$, with no conditions on $a_0, a_1, a_2 \in \mathbb{F}_7$, so there are $7^3 = 343$ of them. Similarly, there are $7^2 = 49$ multiples of x^2-3 and 7 multiples of $(x-1)(x^2-3)$.

Therefore, there are $7^3 + 7^2 - 7 = 385$ zero divisors in $\mathbb{F}_7[x]/(f(x))|$, and the number of invertible elements is

$$|\mathbb{F}_7[x]/(f(x))| - 385 = 7^4 - 385 = 2016.$$

239 Let us find the minimal polynomial of α over $\mathbb{Q}$. By repeatedly isolating square roots and squaring we get

$$\alpha = 2 + \sqrt{5 + \sqrt{-5}}$$
$$(\alpha - 2)^2 = 5 + \sqrt{-5}$$
$$((\alpha - 2)^2 - 5)^2 = -5$$
$$\alpha^4 - 8\alpha^3 + 14\alpha + 8\alpha + 6 = 0,$$

so α is a root of the polynomial $f(x) = x^4 - 8x^3 + 14x^2 + 8x + 6$. Now, $f(x) \in \mathbb{Z}[x]$ is irreducible in $\mathbb{Z}[x]$ by Eisenstein's criterion for $p = 2$. By Gauss's lemma, $f(x)$ is also irreducible in $\mathbb{Q}[x]$ and, being monic, it is the minimal polynomial of α over $\mathbb{Q}$. It follows that $[\mathbb{Q}(\alpha) : \mathbb{Q}] = \deg(f) = 4$.

Consider the tower of extensions $\mathbb{Q} \subseteq \mathbb{Q}(\alpha^2) \subseteq \mathbb{Q}(\alpha)$. We know that $[\mathbb{Q}(\alpha) : \mathbb{Q}] = 4$ and that $[\mathbb{Q}(\alpha) : \mathbb{Q}(\alpha^2)] \leq 2$, because α is a root of the polynomial $x^2 - \alpha^2$, which has degree 2 and whose coefficients are in $\mathbb{Q}(\alpha^2)$. The fact that the degree is multiplicative in towers implies that $[\mathbb{Q}(\alpha^2) : \mathbb{Q}]$ is either 2 or 4. If we had $[\mathbb{Q}(\alpha^2) : \mathbb{Q}] = 2$ then the minimal polynomial of α^2 over $\mathbb{Q}$ would be of the form $x^2 + ax + b$ for some a and b in $\mathbb{Q}$, so α would be a root of the polynomial $g(x) = x^4 + ax^2 + b \in \mathbb{Q}[x]$.

But then $g(x)$ would be a multiple of $f(x)$. In order for this to be the case, since f and g are monic polynomial of the same degree, they would need to coincide: but this is not possible because g gives a biquadratic equation, whereas f does not. In conclusion, $[\mathbb{Q}(\alpha^2) : \mathbb{Q}] = 4$.

240

(i) In $\mathbb{F}_2[x]$ we have

$$f(x) = x^4 + 3x^3 + x + 1$$
$$= x^4 + x^3 + x + 1$$
$$= (x + 1)^2 (x^2 + x + 1)$$

and the latter is a product of irreducible factors because $x^2 + x + 1$ has degree 2 and no roots in $\mathbb{F}_2$.

Similarly, in $\mathbb{F}_3[x]$ we have $f(x) = (x - 1)(x^3 + x^2 + x - 1)$ and $x^3 + x^2 + x - 1$ is an irreducible polynomial in $\mathbb{F}_3[x]$, because it has degree 3 and no roots in $\mathbb{F}_3$.

It follows that the splitting field of $f(x)$ over $\mathbb{F}_2$ is $\mathbb{F}_{2^2}$ and the splitting field of $f(x)$ over $\mathbb{F}_3$ is $\mathbb{F}_{3^3}$.

The splitting field of $f(x)$ over $\mathbb{F}_{2^k}$ is, by definition, the smallest extension of $\mathbb{F}_{2^k}$ that contains the roots of $f(x)$. Now, an extension of $\mathbb{F}_{2^k}$ contains the roots of $f(x)$ if and only if it contains the extension that they generate over $\mathbb{F}_2$, so what we need to find is the degree of the smallest extension of $\mathbb{F}_{2^k}$ that contains $\mathbb{F}_{2^2}$. From the theory concerning inclusions of finite fields, we know that the degree of the splitting field in question is $2/(k, 2)$, that is, 1 if k is even and 2 if k is odd.

Similarly, the splitting field of $f(x)$ over $\mathbb{F}_{3^k}$ is the smallest extension of $\mathbb{F}_{3^k}$ that contains $\mathbb{F}_{3^3}$, which has thus degree $3/(k, 3)$.

(ii) We have $[\mathbb{Q}(\alpha) : \mathbb{Q}] = 4$ because $f(x)$ is an irreducible polynomial over $\mathbb{Q}$. Indeed, $f(x)$ has no roots in $\mathbb{Q}$, because the only candidates are ± 1 and one can check that $f(\pm 1) \neq 0$. So if $f(x)$ were reducible over $\mathbb{Q}$ it would be the product of two irreducible factors of degree 2, but this contradicts the factorisation found over $\mathbb{F}_3$.

⟦Alternatively, we can remark that if $f(x)$ factored over $\mathbb{Q}$ as the product of two irreducible polynomials of degree 2, we would have

$$x^4 + 3x^3 + x + 1 = (x^2 + ax \pm 1)(x^2 + bx \pm 1)$$

for some $a, b \in \mathbb{Z}$. Looking at the coefficients of x and x^3, we would have $a + b = \pm 1$ and $a + b = 3$, which are clearly incompatible. ⟧

241 First of all, remark that the polynomial $f(x) = x^4 - 2$ is irreducible over $\mathbb{Q}$ by Eisenstein's criterion for the prime 2 and Gauss's lemma. Its complex roots are the numbers $i^k \sqrt[4]{2}$ for $k = 0, 1, 2, 3$. Setting $\alpha = \sqrt[4]{2}$, we have that the splitting field of $f(x)$ over $\mathbb{Q}$ is given by $\mathbb{Q}(i, \alpha)$.

$\mathbb{Q}(\alpha, i)$

$\quad|$

$\mathbb{Q}(\alpha)$

$4 \:|$

$\mathbb{Q}$

Consider the tower of extensions $\mathbb{Q} \subseteq \mathbb{Q}(\alpha) \subseteq \mathbb{Q}(\alpha, i)$. The first extension has degree 4 because $f(x)$ is irreducible. The second extension has degree 2: its degree is at most 2 because the polynomial $x^2 + 1$ has coefficients in $\mathbb{Q}(\alpha)$ and has i as a root; moreover, its degree is not 1 because $\mathbb{Q}(\alpha)$ is contained in $\mathbb{R}$, so it cannot coincide with $\mathbb{Q}(\alpha, i)$. In particular, the degree of the splitting field of $f(x)$ over $\mathbb{Q}$ is 8.

In $\mathbb{F}_3[x]$ we have

$$f(x) = x^4 - 2 = x^4 + 4 = (x^4 + 4x^2 + 4) - (2x)^2 = (x^2 - 2x + 2)(x^2 + 2x + 2)$$

and the polynomials $x^2 \pm 2x + 2$ are irreducible over $\mathbb{F}_3$ because their discriminant is -1, and therefore not a square in $\mathbb{F}_3$. In particular, the degree of the splitting field of $f(x)$ over $\mathbb{F}_3$ is 2.

In $\mathbb{F}_{17}[x]$ we have

$$f(x) = x^4 - 2 = x^4 - 36 = (x^2 - 6)(x^2 + 6).$$

The polynomials $x^2 \pm 6$ are irreducible over $\mathbb{F}_{17}$ because 6 and -6 are not squares in $\mathbb{F}_{17}$, as one can check by listing all squares of $\mathbb{F}_{17}$. In particular, the degree of the splitting field of $f(x)$ over $\mathbb{F}_{17}$ is 2.

⟦The factorisation $a^4 + 4b^4 = (a^2 + 2b^2 + 2ab)(a^2 + 2b^2 - 2ab)$, which we used above with $b = 1$ and over $\mathbb{F}_3$, is called Sophie Germain's identity, in honour of the famous French mathematician.

We could have alternatively computed the degree over $\mathbb{F}_{17}$ by remarking that, since 2 has order 8 in $(\mathbb{Z}/17\mathbb{Z})^*$, any root α of $f(x)$ in the algebraic closure of $\mathbb{F}_{17}$ is such that $(\alpha^4)^8 = 2^8 = 1$, so we have $\alpha^{32} = 1$ and $\alpha^{16} \neq 1$. What we have to compute is therefore the degree of the field generated by the 32nd roots of unity over $\mathbb{F}_{17}$; this field is $\mathbb{F}_{17^d}$, where d is the order of 17 in $(\mathbb{Z}/32\mathbb{Z})^*$. Since $17 \not\equiv 1 \pmod{32}$ and $17^2 \equiv 1 \pmod{32}$, we have $d = 2$. ⟧

Index

© Springer Nature Switzerland AG 2020
R. Chirivì et al., *Selected Exercises in Algebra*, UNITEXT 119,
https://doi.org/10.1007/978-3-030-36156-3

Printed in the United States
By Bookmasters